A First Look at Coding in Chemistry
Solving Problems Using MATLAB

A First Look at Coding in Chemistry
Solving Problems Using MATLAB

By

Tamas Bansagi

University of Southampton, UK
Email: t.bansagi@soton.ac.uk

Paperback ISBN: 978-1-83767-733-7
EPUB ISBN: 978-1-83767-734-4
PDF ISBN: 978-1-83767-735-1

A catalogue record for this book is available from the British Library

© Tamas Bansagi 2025

All rights reserved

Apart from fair dealing for the purposes of research for non-commercial purposes or for private study, criticism or review, as permitted under the Copyright, Designs and Patents Act 1988 and the Copyright and Related Rights Regulations 2003, this publication may not be reproduced, stored or transmitted, in any form or by any means, without the prior permission in writing of the Royal Society of Chemistry or the copyright owner, or in the case of reproduction in accordance with the terms of licences issued by the Copyright Licensing Agency in the UK, or in accordance with the terms of the licences issued by the appropriate Reproduction Rights Organization outside the UK. Enquiries concerning reproduction outside the terms stated here should be sent to the Royal Society of Chemistry at the address printed on this page.

Whilst this material has been produced with all due care, The Royal Society of Chemistry cannot be held responsible or liable for its accuracy and completeness, nor for any consequences arising from any errors or the use of the information contained in this publication. The publication of advertisements does not constitute any endorsement by The Royal Society of Chemistry or Authors of any products advertised. The views and opinions advanced by contributors do not necessarily reflect those of The Royal Society of Chemistry which shall not be liable for any resulting loss or damage arising as a result of reliance upon this material.

The Royal Society of Chemistry is a charity, registered in England and Wales, Number 207890, and a company incorporated in England by Royal Charter (Registered No. RC000524), registered office: Burlington House, Piccadilly, London W1J 0BA, UK, Telephone: +44 (0)20 7437 8656.

For further information see our website at www.rsc.org

For general enquiries, please contact books@rsc.org

For EU product safety enquiries, please email books@rsc.org or contact Royal Society of Chemistry Worldwide (Germany) GmbH, Römischer Hof, Unter den Linden 10, 10117 Berlin.

Printed in the United Kingdom by CPI Group (UK) Ltd, Croydon, CR0 4YY, UK

For dad

Preface

The aim of this book is to help chemistry students and chemistry professionals develop basic coding skills. Computer jargon is kept to a minimum while focusing on solving chemistry problems that university students usually encounter early on in a chemistry degree course. Its chemistry-first, scenario-based approach combined with a workbook-style delivery is intended to enhance engagement with the material and make programming feel more natural to use in chemistry. Along this journey, additional broadly-applicable, high-value skills are also going to be developed; and while learning to code, some topics closely related to both chemistry and coding will be visited, such as uncertainties in measurements, data analysis and a few topics in maths necessary to move the discussion forward. A basic understanding of calculus is assumed.

The programming language chosen for this introductory text is MATLAB; however, much of what is covered is not specific to MATLAB. What will be learnt is easily transferrable to other programming languages; so, by the end, the reader will be able to pick up other languages relatively quickly.

Acknowledging that coding can feel intimidating at first for many just starting on their programming journey, we will begin with step-by-step instructions, and only gradually increase the pace after the first few chapters. For those who have already done some coding and/or wish to further delve into it, the Appendices at the end of chapters and the book are there to provide extra opportunities.

As the title suggests, this book merely touches the surface of coding in chemistry (and the capabilities of MATLAB) and is intended to assist the reader in building a robust general understanding of the basics, which could then serve as a stepping stone towards specializing in an area of interest. The reading list at the end provides additional resources for further exploration.

I would like to give a special thanks to Dr Sarah Horswell for providing feedback on an early version of the manuscript. I would also like to thank Dr Hossein Z. Jooya for checking the code.

Tamas Bansagi
School of Chemistry and Chemical Engineering
University of Southampton

A First Look at Coding in Chemistry: Solving Problems Using MATLAB
By Tamas Bansagi
© Tamas Bansagi 2025
Published by the Royal Society of Chemistry, www.rsc.org

Introduction

With the rapid advances in lab automation, a growing number of chemists have begun spending less time carrying out experiments while becoming more involved in designing workflows for increasingly automated lab processes and analysing the growing amount of data these produce. In the meantime, computational chemistry has been steadily evolving in sophistication, speed and accessibility to everyone in chemistry. Making better predictions quicker with less effort is now enabling more and more chemists to add computational tools to their research. These trends increase the demand for chemistry graduates who not only have strong IT skills but also some experience in coding.

Coding is much more than using a computer programming language to give instructions to a computer for performing a task. It is a way of thinking, seeing problems and conceiving solutions. Coding means dissecting problems and devising algorithms in code to solve them. It promotes abstraction and viewing problems from different angles. These days, we are surrounded by flat screens, even in chemistry laboratories, displaying colourful graphic user interfaces (GUI) where seemingly everything can be solved in a few taps. Despite what this visually rich experience may suggest, these applications are usually designed to perform a relatively narrow set of tasks, unwittingly constraining the mind of the user rather than freeing it. Coding, often portrayed as typing into a featureless command-line interface, works the other way around. It enables the flexible exploration of strategies which develops algorithmic thinking and problem decomposition skills, as well as fosters creativity and encourages learning more about how our digital world works. Chemists are well placed to learn coding to meet the demand for extended digital problem-solving skills in the chemistry sector and beyond.

Coding becoming increasingly desired in chemistry is part of a broader trend seen recently across many areas of science and industry, likely resulting from the advent of modern programming languages and integrated development environments (IDE) which significantly lowered the entry barrier into coding, making it more widely accessible to people with non-specialist backgrounds. This shift had the same effect on computational chemistry, and the field, once dominated by the language FORTRAN

A First Look at Coding in Chemistry: Solving Problems Using MATLAB
By Tamas Bansagi
© Tamas Bansagi 2025
Published by the Royal Society of Chemistry, www.rsc.org

(and to a lesser extent C) only a small group of chemists had the dedication to master, is now for all chemists to explore. These developments are recognised by the chemistry education community and introductory coding courses are now included in the core first-year chemistry curricula, instead of being offered as an elective in later years, at a growing number of universities worldwide. Early exposure to coding also enables extending the computational chemistry syllabus, elevating its prominence to the level traditional core chemistry subjects have long enjoyed.

In chemistry, coding is used for many different purposes. Data acquisition and processing; finding trends, patterns and clusters in data; fitting curves to series of data points; enhancing and analysing images; controlling equipment; solving equations; calculating the most advantageous combination of conditions to satisfy simultaneous criteria; approximating how concentrations change in a reaction over time and space, fluid motion, the distribution of electrons in molecules, molecular trajectories and collisions in the presence of intermolecular forces, the structures and energies of transition states and so on to name some examples.

Contents

1	**MATLAB Basics**	**1**
	1.1 MATLAB Desktop	2
	1.2 Watch What You Type	3
	1.3 Assigning Variables	4
	1.4 Mathematical Functions	6
	1.5 Arrays	7
2	**Scripts and their Applications to Chemistry**	**14**
	2.1 Commands *versus* Scripts	14
	2.2 Advantages of Using Scripts	17
	2.3 Scripts as User-defined MATLAB Functions	20
	2.4 Appendix: Figure Formatting Using Arrays	27
3	**Presenting Data in Chemistry: Plots and Charts**	**29**
	3.1 Scatter and Line Plots	30
	3.2 Log Plots	35
	3.3 Combining Graphs and Plots	36
	3.4 Data Uncertainty in Plots	45
	3.5 Plots with Two y-Axes	47
	3.6 Histograms	49
	3.7 Appendix: Plot Stacking	51

A First Look at Coding in Chemistry: Solving Problems Using MATLAB
By Tamas Bansagi
© Tamas Bansagi 2025
Published by the Royal Society of Chemistry, www.rsc.org

4 Curve Fitting in Chemistry — 53

- 4.1 Fitting Curves to Data: *Lines* — 54
- 4.2 Fitting Curves to Data: *Polynomials* — 61
- 4.3 Fitting Curves to Data: *Splines* — 62
- 4.4 Fitting Curves to Data: *Custom Expressions* — 64
- 4.5 Quoting Fitting Parameters — 67
- 4.6 Appendix: Displaying Data Points Neglected from Fitting — 73

5 Measurements and Their Uncertainties — 74

- 5.1 Normal Distribution — 75
- 5.2 Population Mean and Standard Deviation — 78
- 5.3 Sample Mean and Standard Deviation — 79
- 5.4 Standard Error — 81
- 5.5 Small Sample Sizes and Student's t-distribution — 83
- 5.6 Size of Error Bars — 85
- 5.7 Comparing Values: Are They Different or Not? — 86
- 5.8 Identifying Outliers — 89

6 Propagation of Uncertainties (Errors) — 94

- 6.1 Addition and Subtraction — 97
- 6.2 Multiplication and Division — 98
- 6.3 Constants with Negligible Uncertainties — 98
- 6.4 Propagating Uncertainties in MATLAB: *Addition and Subtraction* — 100
- 6.5 Propagating Uncertainties in MATLAB: *Multiplication and Division* — 102
- 6.6 Propagating Uncertainties in MATLAB: *Functions* — 104
- 6.7 Appendix: Method of Least Squares for Linear Fitting — 104

7 Vectors and Their Uses in Chemistry — 112

- 7.1 Vector Algebra — 113
- 7.2 Drawing Vectors in MATLAB — 117
- 7.3 Vector Operations in MATLAB: *Addition and Subtraction* — 120
- 7.4 Vector Operations in MATLAB: Dot (Scalar) Product and Cross Product — 121
- 7.5 Vector Operations in MATLAB: *Calculating Volumes* — 124
- 7.6 Appendix: Vector Fields (Force Fields) — 130

8 Matrices in Chemistry — 132

- 8.1 Matrix Algebra — 135
- 8.2 Matrices in MATLAB — 137
- 8.3 Matrix Algebra in MATLAB: *Addition* and *Subtraction* — 137
- 8.4 Matrix Algebra in MATLAB: *Multiplication* — 138
- 8.5 Matrix Algebra in MATLAB: *Matrix Inversion* — 143
- 8.6 Applications of Matrices — 144
- 8.7 Molecular Symmetry and Transformation Matrices — 150

9 One-line Functions and Kinetic Modelling — 165

- 9.1 Kinetic Modelling — 169
- 9.2 Case Study: Nitration of Aniline — 177
- 9.3 Case Study: Self-replicating Peptides — 178
- 9.4 Appendix: Euler's Method (I) — 181

10 Self-controlling Code — 187

- 10.1 Loops — 189
- 10.2 Conditional Statements — 197
- 10.3 Case Study: Monitoring Polymerisation — 200
- 10.4 Appendix: Euler's Method (II) — 210

11 Maths for Chemistry with MATLAB I — 212

- 11.1 Introduction to Symbolic Maths — 212
- 11.2 Symbolic Maths: *Unit Algebra* — 218
- 11.3 Appendix: Intermolecular Forces between Two Dipoles — 225

12 Maths for Chemistry with MATLAB II — 227

- 12.1 Symbolic Maths: *Differentiation* — 229
- 12.2 Symbolic Maths: *Integration* — 233
- 12.3 Symbolic Maths: Solving Equations — 237
- 12.4 Case Study: Propagation of Uncertainties — 242
- 12.5 Appendix A: Exact Solutions, Approximations, and Simplifications — 246
- 12.6 Appendix B: Bisection Method — 259

Appendix 263
Predicting Titration Curves 263
Kinetic Modelling in Elucidating Reaction Mechanisms 268

Reading List 279

Subject Index 280

1 MATLAB Basics

MATLAB is a proprietary programming language and programming environment. Its development started in the 1970s and over the years it has become a powerful platform for technical computing widely-used across science and engineering. The main appeal of MATLAB to students studying chemistry and chemistry professionals comes from MATLAB being a high-level programming language, the intuitive user interface of the MATLAB programming environment, its ability to connect to, and control instruments, as well as the comprehensive supporting documentation and on-line training courses provided by MathWorks the company behind MATLAB.

Most scientists and engineers heavily rely on computers. They use software designed for specific tasks, or for problems that cannot be conveniently handled by these software applications they use general-purpose programming languages and environments. MATLAB belongs to the latter family of software, enabling users to interactively execute commands or create processes by combining commands for automating tasks. It is a high-level, interpreted programming language, which means that coding in MATLAB does not require detailed knowledge of how computers work, and that despite their simplicity and easy readability, MATLAB commands can perform complex tasks directly upon execution without having to translate them first to machine language instructions. The MATLAB programming environment has a rich, graphic, user-friendly interface which not only allows users to execute commands (in the *Command Window*) but also to explore and interact with inputs and outputs.

For example, the MATLAB programming environment allows users to read in raw data from text files, spreadsheets and measuring instruments connected to the computer. The data are displayed (in *Workspace*) for convenient pre-screening and amendment by users, before they can pre-process (scale, reduce the noise in the data and/or remove undesirable trends, *etc.* all interactively in the *Live Editor*). Then, a user-defined complex statistical analysis involving high-level maths can be performed by executing a sequence of MATLAB commands created and saved previously by the user or others. The output of this analysis can be explored and post-processed just like the input data prior to saving or being fed back to the instruments directly through

---

A First Look at Coding in Chemistry: Solving Problems Using MATLAB
By Tamas Bansagi
© Tamas Bansagi 2025
Published by the Royal Society of Chemistry, www.rsc.org

MATLAB to change their settings and/or control them. These wide-ranging functionalities are all integrated into the MATLAB programming environment, which makes it an immensely powerful platform for scientists and engineers.

Mathworks provides an unparalleled amount of supporting documentation including a very large amount of worked examples which are invaluable to people learning to code, as are the many on-line training courses, videos and technical support available on their website `mathworks.com`. With this vast amount of information and help provided in one place, MATLAB users very rarely need to resort to trying to find answers to their questions on internet discussion forums.

The desktop version of MATLAB can be downloaded from the Mathworks website and is commonly available pre-installed on centrally-managed university computers. Toolboxes designed to handle specific types of problems (such as Image Processing, Statistics and Machine Learning, Bioinformatics, Instrument Control and so on) can be downloaded and installed together with MATLAB or separately added later. MATLAB Online offers easy access from web browsers to the MATLAB programming environment, cloud storage with synchronisation, on-line code sharing and publishing.

A free alternative to MATLAB is GNU Octave which is mostly drop-in compatible with MATLAB. With the occasional tweak, your MATLAB code will run in Octave, and the other way around. Octave can be downloaded from `octave.org`, and just like MATLAB, it can also be extended by adding Packages (although Octave Packages are usually more limited than comparable MATLAB Toolboxes and cover fewer areas). Octave is also available through web browsers at `octave-online.net` with no registration required, which makes it ideal for light coding. In some cases, comments on the differences between MATLAB and Octave are added as footnotes.

1.1 MATLAB Desktop

When MATLAB is started on a computer or *via* MATLAB Online in a web browser, the Desktop appears (Figure 1.1). It comprises of three Panels

- *Current folder* (to directly access your files – left)
- *Command window* (for entering commands – centre)
- *Workspace* (to interactively explore the data you generate or import from files – right)

Above the Panels is the *Toolstrip* where desktop functions and menus are organised. It has three primary tabs, HOME, PLOTS and APPS. The tabs are composed of sections, each containing a specific group of features. We are going to discuss only some of the sections in detail, as most of them are beyond the scope of an introductory text.

A wide range of tutorials are available for users new to MATLAB on the MathWorks website, which you can access by clicking on "Getting Started" under the *Command Window* header.

MATLAB Basics 3

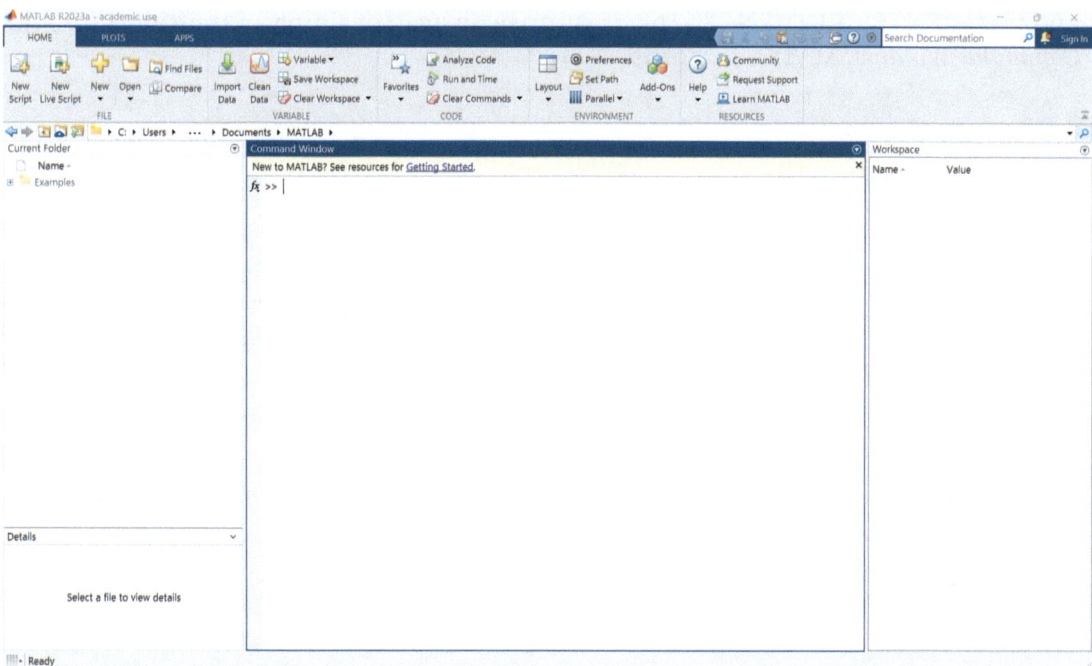

Figure 1.1 MATLAB default Desktop layout, everything in one place.

1.2 Watch What You Type

We start our journey of coding by exploring the *Command Window*. You can think of it as a conduit between you and MATLAB, a communication channel through which you can give text commands to MATLAB, and MATLAB can respond *via* text output. As long as MATLAB is able to interpret what you have typed into the *Command Window*, you will receive an answer written into the *Command Window* or some other types of response, for example, a diagram appearing on your screen, or saving your data on the computer. Otherwise, you will get an error message which will concisely tell you what MATLAB thinks went wrong. It is often the case when you first start coding that you mistype something or enter things in the wrong order and MATLAB throws an error that may make little sense to you. When this happens, carefully read the error message (instead of getting annoyed with MATLAB for not understanding what you meant) and inspect what you have entered into the *Command Window* in light of what the error message says. Then, try fixing the mistake you made; typing or pasting the exact same command again thinking it might work this time will not change the outcome.

As a first exercise, we type something into the *Command Window* after the prompt (">>") that looks perfectly correct (to us):

```
>> 5 - 2
```

and press "ENTER/RETURN/↵". To our surprise, instead of simply outputting 3 into the *Command Window*, MATLAB returns

```
5 – 2
↑
Error: Invalid text character. Check for unsupported symbol,
invisible character, or pasting of non-ASCII characters.
```

Looking at which character the arrow is pointing at and reading the error message makes us suspect that we must have used the so-called "en-dash" character (–) where MATLAB was expecting a minus sign (−). With that in mind, we re-type the expression, but this time with a minus sign

```
>> 5 - 2
```

which appears in the *Command Window* just like what we had before. Now, however, pressing "ENTER/RETURN/↵" leads to the expected output

```
ans =
     3
```

This peculiar behaviour can be understood when we realise that programming environments use monospaced (fixed-width) fonts which increase the readability of code and tabulated data by making characters perfectly line up vertically.

The same problem often arises when copying text including long lines of code displayed in multiple lines and single quotation marks (') from a pdf document or an e-book into MATLAB. Single quotation marks are sometimes misinterpreted by pdf or e-book viewers as apostrophes or ellipses, and when these are pasted into MATLAB it throws an invalid text character (or syntax) error. Next time you paste a command into MATLAB from a pdf file or e-book, remember to check, and re-type if necessary, any characters MATLAB flags up as invalid. Even when the error is triggered by something other than a wrong character, still apply the principle outlined above: carefully read the error message, and try finding and fixing the mistake you made. With practice and patience, you will gradually become better at debugging your code.

1.3 Assigning Variables

Now, we continue with creating variables. After the prompt, type the name of the variable followed by the '=' character and the value you want to assign to that variable. Please note that variable names cannot start with numbers and contain white spaces. For example, type

```
>> b = 3
```

then press "ENTER/RETURN/↵". MATLAB will output the result into the *Command Window* as

```
b =
     3
```

while creating the new variable in the memory of the computer which will be displayed in *Workspace*. Double clicking on b in *Workspace* will open up an interactive spreadsheet to show the content of b, where its value can be changed (right-click the variable and select "Edit Value"). The spreadsheet can be closed by clicking "x" in the upper right corner.

A new value can be assigned to an existing variable by the same method used to create it.

```
>> b = 5
```

will overwrite the previous value of b and generate the output

```
b =
     5
```

To delete a variable (*via* the *Command Window*), use the clear command followed by the name of the variable. For example,

```
>> clear b
```

will remove b from the memory of the computer and consequently from *Workspace*. Note that clear all and clearvars will delete all your variables. You can also delete variables interactively in *Workspace*.

We shall now re-create b

```
>> b = 3
b =
     3
```

and define another variable (k),

```
>> k = 0.4
k =
     0.4000
```

then multiply the two variables as

```
>> b*k
ans =
     1.2000
```

We can combine operations with creating new variables. Defining a variable to hold the result of the above multiplication can be done as

```
>> p = b*k
p =
     1.2000
```

You can combine as many operations as you need. For example, we can multiply p by k and square the result while creating a new variable (q) for the end result:

```
>> q = (b*k)^2
q =
    1.4400
```

1.4 Mathematical Functions

Calling mathematical functions in MATLAB works like this:

```
>> w = sin(q/3.14)
w =
    0.4427
```

Trigonometric functions - `sin()`, `cos()`, `tan()`... - take values in radians; whereas `sind()`, `cosd()`, `tand()`... take values in degrees; hence

```
>> sin(3.14)
ans =
    0.001592652916487
>> sind(180)
ans =
    0
```

Some important constants, such as π are available as built-in constants. `pi` returns the 16-digit (so-called double precision) numeric value 3.141592653589793. Using `pi` in place of 3.14 above, MATLAB returns a value much closer to the expected zero.

```
>> sin(pi)
ans =
    1.2246e-16
```

which means $1.2246 \times 10^{-16} = 0.00000000000000012246 \approx 0$.

Taking exponentials is done by using the `exp()` function; whereas, for taking the natural logarithm of a number, call the `log()` function. If you require base 10 logarithm, for example when calculating pH, use the `log10()` function.

```
>> g = exp(1)
g =
    2.7183
>> log(g)
ans =
    1
```

```
>> log10(g)
ans =
    0.4343
```

Notice the difference between the outputs of `log(g)` and `log10(g)`.

> **Exercise 1.1**
>
> For any given time (t) during nucleophile substitution of a tertiary haloalkane (R_3CX), its concentration can be predicted by the equation $[R_3CX]=[R_3CX]_0 e^{-kt}$, where $[R_3CX]_0$ is the initial concentration of the haloalkane and k is the rate constant of the reaction. What is the concentration of the haloalkane 10 minutes after the start of the reaction, if the initial concentration of R_3CX is 0.500 mol dm^{-3} and the rate constant is $k=0.24$ min^{-1}.
>
> **Answer:**
>
> ```
> >> c_R3CX0=0.500; k=0.24; t=10;
> >> c_R3CX=c_R3CX0*exp(-k*t)
> c_R3CX =
>     0.045359
> ```
>
> We estimate that 10 minutes after the start of the reaction the concentration of R_3CX will be 0.0454 mol dm^{-3} (we rounded the result to the same number of significant figures as the initial concentration).

1.5 Arrays

Arrays of numbers are created by using square brackets. Separate the rows of the array by semicolons as follows:

```
>> M=[1  2  3; 4  5  6; 7  8  9]
M =
     1   2   3
     4   5   6
     7   8   9
```

If you would like to supress the output to the *Command Window* (while a variable or array is still created by MATLAB and appear in *Workspace*), put a semicolon at the end of the command:

```
>> N=[3  2  1; 6  5  4; 9  8  7];
```

which generates no output into the *Command Window*, yet array N is still created, as seen in the *Workspace*.

MATLAB has clever ways of generating sequences of numbers saving us from having to type in every single member of the sequence. With the help of the *colon operator*, (:), creating rows of numbers is easy, so when we need to enter a series of pH values, for example, 1 through 14 we can simply execute

```
>> pH=1:14
pH =
     1   2   3   4   5   6   7   8   9   10   11   12   13   14
```

If you only need every other pH value in this sequence, run

```
>> pH=1:2:14
pH =
     1   3   5   7   9   11   13
```

Non-integer steps can be also prescribed.

```
>> pH=1:3.5:14
pH =
    1.0000    4.5000    8.0000   11.5000
```

Negative increments generate descending sequences. For instance, a pH sequence starting on 14 and decreasing stepwise by 3.5 is created as

```
pH_rev=14:-3.5:1
pH_rev =
   14.0000   10.5000    7.0000    3.5000
```

Exercise 1.2

In the alpha form of polonium (α-Po), atoms are arranged in a simple cubic crystal structure with edge length 335 picometers (335×10^{-12} m). Create an array that holds the x coordinates of the first five Po atoms starting from the left edge (where $x=0$) in any rows parallel to the x axis.

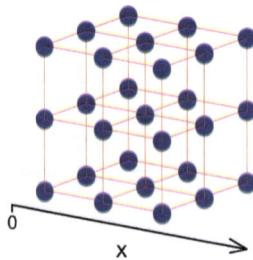

Answer:

```
>> Po_x_crd=(0:4)*335
Po_x_crd =
     0    335    670    1005    1340
```

First, 0:4 creates the sequence 0 1 2 3 4, then each member is multiplied by 335 pm. Another way of achieving the same output:

```
>> Po_x_crd=0:335:4*335
Po_x_crd=
         0    335    670   1005   1340
```

Looking at arrays pH and pH_rev, we realise that we need to be careful when specifying the last member of a series, because the series will terminate before the desired endpoint, if the final step exceeds the defined range. To avoid this, we must specify the endpoints or the increment carefully. Alternatively, we can use the linspace() function which generates sequences that contain both endpoints. For example, if we need five equally-spaced time points for the first 13 minutes of a reaction for taking samples, we can just give MATLAB the two endpoints of the time interval (0,13) and the required number of members (5) in the sequence as

```
>> time=linspace(0,13,5)
time=
    0.0000    3.2500    6.5000    9.7500   13.0000
```

Exercise 1.3

Estimate the concentration of R_3CX for the reaction discussed in Exercise 1.1 at seven equally-spaced instances in time between 1 and 16 minutes after the start of the reaction.

Answer:

```
>> c_R3CX0=0.500; k=0.24;
>> t=linspace(1,16,7)
t=
    1.0000    3.5000    6.0000    8.5000   11.0000   13.5000   16.0000
>> c_R3CX=c_R3CX0*exp(-k*t)
c_R3CX=
    0.3933    0.2159    0.1185    0.0650    0.0357    0.0196    0.0107
```

It is necessary sometimes to space the members in a sequence logarithmically, which happens in chemistry as chemists often prefer logarithmic scales. For example, when we are interested in the equilibrium concentrations of some ligands binding to a metal ion or protein at a series of equally spaced pH values. To prepare these mixtures, we would first need to have the corresponding H^+ concentrations which will be spaced logarithmically. The logspace() function is ideally suited for this task: we only need to pass the exponents of the endpoints (a and b for endpoints 10^a and 10^b) to MATLAB (low followed by high, $a<b$) along with the desired number of members in the sequence. For instance, a 6-member logarithmic pressure sequence between 1 Pa to 0.1 MPa is generated by

```
>> P=logspace(0,5,6)
P =
     1    10   100   1000   10000   100000
```

where the input arguments 0, 5, 6 for `logspace()` represent $10^0 = 1$, $10^5 = 100\,000$ and the number of pressure values within the required range including the two endpoints.

Exercise 1.4

We are investigating the binding of a drug molecule to a protein at five logarithmically-spaced H^+ concentrations between pH 2 and 13. Generate the required $[H^+]$ sequence recalling that $pH = -\log_{10}[H^+]$ which rearranges to $[H^+] = 10^{-pH}$.

Answer:

$[H^+]$ at pH 13 is smaller than $[H^+]$ at pH 2 because 1×10^{-13} mol dm$^{-3} < 1 \times 10^{-2}$ mol dm^{-3}, therefore

```
>> H=logspace(-13,-2,5)
H =
    1.0000e-13   5.6234e-11   3.1623e-08   1.7783e-05   1.0000e-02
```

Uniform arrays can be created by using the `zeros()` function. On its own, it only creates arrays of zeros, to which the desired uniform value can be added as

```
>> z1=zeros(4,3)
z1 =
     0   0   0
     0   0   0
     0   0   0
     0   0   0
z2=zeros(4,3)+pi
z2 =
    3.1416   3.1416   3.1416
    3.1416   3.1416   3.1416
    3.1416   3.1416   3.1416
    3.1416   3.1416   3.1416
```

For accessing an element of an array, pass the row and column numbers of the element to MATLAB as:

```
>> n=N(2,3)
n =
     4
```

which for a one-dimensional array simplifies to providing only one number in parenthesis:

```
>> h=H(3)
h =
    3.1623e-08
```

Accessing multiple elements is done similarly but with the help of the colon operator.

```
>> m=M(2:3,1:2)
m =
    4   5
    7   8
```

Not only numbers, but also characters can be assigned to MATLAB arrays. These so-called *character arrays* can be defined by enclosing some text in single quotes. (If you are copying code from a pdf file or e-book, please check that you have pasted single quotes into MATLAB as opposed to other similar-looking characters such as apostrophes, primes or acute accents. Retype if necessary.) We now define two character arrays text1 and text2

```
>> text1='Hello,'
text1 =
    'Hello,'
>> text2=' World!'
text2 =
    'World!'
```

and combine them by simply using square brackets as:

```
>> text3=[text1,text2]
text3 =
    'Hello, World!'
```

Outputting a character array (and other types of arrays) into the *Command Window* without having the name of the array written out as well can be achieved by using the disp() function.

```
>> disp(text3)
Hello, World!
```

Traditionally, programs generating the output "Hello, World!" are often the first students learn to write in a particular language. Congratulations on passing this milestone.

Exercise 1.5

Condense the displaying of "Hello, World!" into a single command without predefining character arrays. Clear the *Command Window* by first issuing clc.

> **Answer:**
>
> Nest the character array assignment directly inside `disp()` as
>
> ```
> >> clc; disp('Hello, World!')
> Hello, World!
> ```
>
> Do not forget the semicolon after `clc` if you are putting the two commands in one line. The more experienced in coding you become the more concise codes you will be able to create.

This simple way of joining character arrays is advantageous, for example, when trying to generate paths for files to be loaded into MATLAB. Imagine that you have carried out three small series of similar experiments to produce a library of compounds. You organised the experimental conditions, your observations and the spectroscopic data collected into three folders called "My_Exp_A", "My_Exp_B" and "My_Exp_C" on the D: drive of your computer. Each contains three subfolders: "Experiment_1_3", "Experiment_4_6" and "Experiment_7_9" storing data on individual experiments. Each subfolder contains nine data files: *text files* (ending with extension ".txt") with details of the conditions for each experiment along with your observations, *data files* (ending with extension ".dat") containing spectroscopic data acquired on each product, and *image files* (ending with extension ".jpg") taken of the crystals of the compounds produced after work-up (Figure 1.2).

While analysing your data and writing your report, you will need to load various files into MATLAB. Instead of having all the folders open and continuously switching between them to select the file you need, you decide to generate the path and filename and pass them to MATLAB to load in the file. You define the main components as character arrays: the roots of the folder and file names, and the file extensions.

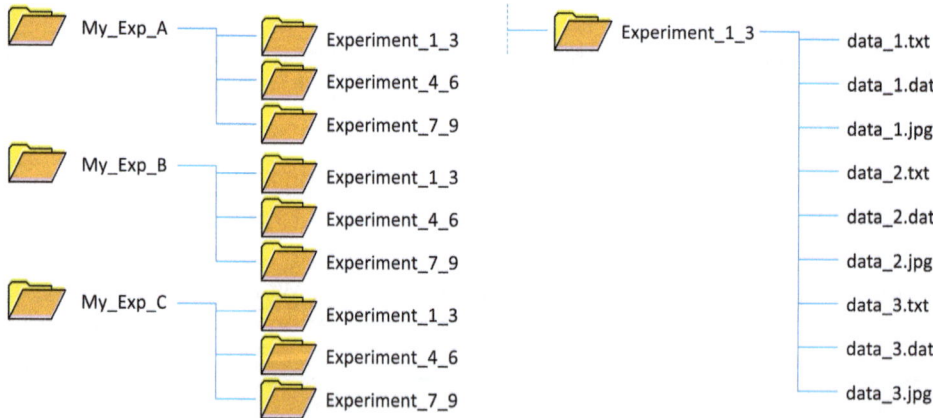

Figure 1.2 Example file system. With the path to a file specified, MATLAB can load in the file directly without the user having to navigate to the folder in the bar above the *Command Window*

```
>> dr='D:';
>> dir='My_Exp_';
>> s_dir='Experiment_';
>> exp='data_';
>> ext1='.txt'; ext2='.dat'; ext3='.jpg';
>> bs='\';
```

Now, whenever you need to load a file into MATLAB you can just edit the relevant sections in quotes in the command below and generate the path to the file merged with the file name in one step, before passing array PathAndFileName to MATLAB which it will use to locate the file on your computer and load it in.

```
>> PathAndFileName=[dr,bs,dir,'A',bs,s_dir,'7_9',bs,exp,'2',ext1]
PathAndFileName =
                'D:\My_Exp_A\Experiment_7_9\data_2.txt'
```

MATLAB is not only able to fetch files from your computer but from the internet as well, the latter requiring specifying the type of protocol needed for the data transfer. For example, an image of the molecular structure of methane can be downloaded from its Wikipedia page *via* the "https" protocol using the imread() function and subsequently displayed with imshow(). The commands below (entered as single lines in the *Command Window* after the prompt)

```
>> methane_image=
imread('https://upload.wikimedia.org/wikipedia/commons/1/1f/Methane-
CRC-MW-3D-balls.png');
>> imshow(methane_image)
```

should locate the image using the URL and download it into the array methane_image, followed by displaying the array in a new window.[†] Generating URLs to access files on the internet is a common use of character arrays.

[†] The URL must be valid. If the image is moved, MATLAB will not find the file and throw an error.

2 Scripts and their Applications to Chemistry

Now being familiar with the basics of MATLAB: the Desktop layout, how to use commands for defining variables and arrays, and with carrying out simple calculations in the *Command Window*, you probably wonder if there is a way to save sequences of commands so that we (or others) can reuse them later instead of having to type them into the *Command Window* every single time we need to do something we have already done in the past.

2.1 Commands *versus* Scripts

We have seen how commands work and how practical they can be in creating smooth workflows when communicating complex tasks to a computer. The advantages of using commands can only be fully exploited if we save the commands we use to solve a problem, so that next time the same problem arises we just have to load in the commands we used last time. Saving commands not only enables us to use them again in the future, but it also makes sharing commands with others easy.

A collection of commands saved on the hard drive of a computer is called a *script*. Scripts need to be loaded into the memory of the computer through the programming environment they were written in so that they can perform their intended sets of tasks. The names of script files always end with the extension assigned to the programming environment they were created in. The extension of MATLAB script files is ".m". (Please note that Octave also uses the ".m" ending.)

To begin learning about MATLAB scripts, we will be looking at displaying the outcome of an environmental chemistry project, where dissolved oxygen levels in water were measured at a few locations along the east shore of Lake Windermere in the Lake District National Park. The geographic coordinates of each spot where a water sample was collected from for analysis were recorded and now we would like to display the measured dissolved oxygen levels on a map. (We will use for this task commands

---

A First Look at Coding in Chemistry: Solving Problems Using MATLAB
By Tamas Bansagi
© Tamas Bansagi 2025
Published by the Royal Society of Chemistry, www.rsc.org

available in MATLAB's Mapping Toolbox,[†] so before we can continue, please make sure you have it downloaded and installed. Otherwise, use MATLAB Online.) We load the latitude and longitude of each sample collection location as a row into the array we call GeoCoord and the corresponding dissolved oxygen levels in another array called DO. The numbers in DO need to be converted into their text character equivalents before they can be passed to the wmmarker() function which displays markers on a map.

```
>> GeoCoord=[54.420660  -2.963455; 54.399078  -2.946233; 54.384354
-2.922945; 54.362652,-2.926088; 54.310322  -2.948620]; % latitudes and
longitudes
>> DO=[5.2  4.8  4.7  4.2  3.8]; % Dissolved Oxygen levels in mg/l (numbers)
>> DOS = string(DO); % convert numbers into characters to be displayed as text
>> wmmarker(GeoCoord(:,1),GeoCoord(:,2),'FeatureName','Dissolved Oxygen
(mg/l)','Description',DOS)
```

After executing these four commands one after the other, MATLAB opens an interactive map and marks the sample collection locations (Figure 2.1). By clicking on a pin representing a sample collection point, the Dissolved Oxygen level in that sample is revealed in a bubble, as seen in Figure 2.1.

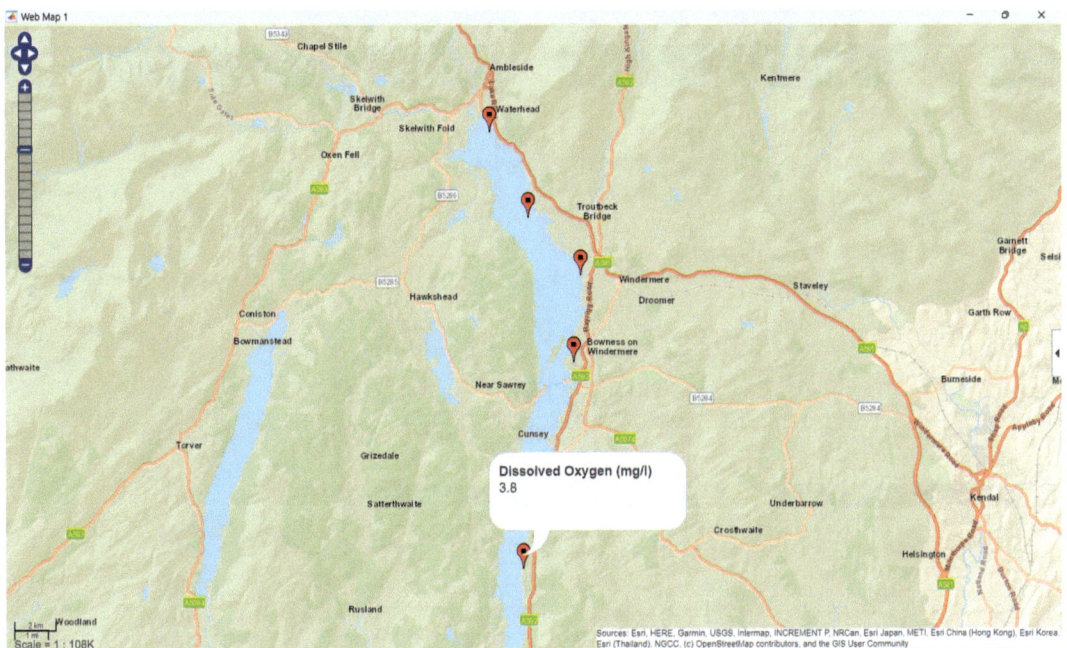

Figure 2.1 Pins on a MATLAB-generated interactive map representing environmental sample locations.

[†] As of mid-2024, Octave's Mapping Package is under development. Please carry on reading and from Exercise 3.3 onwards resume writing and testing code.

This seems like an effective and convenient way of presenting environmental chemistry data; however, we would prefer not to type in the commands every single time we need to make a map like the one seen in Figure 2.1. Saving these commands into a script file would make it quick and easy to generate maps for displaying environmental data in the future. For that, first go to the HOME tab and click on "New Script" (□) and create a blank script (shown in Figure 2.2 on the left). Alternatively, when a script is already open and the EDITOR tab is added to the Toolstrip, within the FILE section of the EDITOR tab click on "New" (□ 1), then select "Script" from the drop-down list (□ 2) (shown in Figure 2.2 on the right).

After the blank script has opened, copy the four commands from the *Command Window* and paste them into the new script window. Then, click "Save" in the *Toolstrip* (□ 1), which opens a window for entering the filename of the script (□ 2). Please note that MATLAB script file names cannot start with numbers, nor can they contain white spaces. (Do not use the filename "matlab" or the names of built-in MATLAB functions.) Click OK (□ 3) when you, and MATLAB, are happy with the file name you entered (Figure 2.3).

With our commands now saved, we will not need to remember them when we are trying to create an interactive map displaying sample collection points with analyte concentrations overlaid in bubbles. Next time we just need to open "dissolved_oxygen_disp_on_map.m", replace the latitudes, longitudes and the dissolved oxygen levels, then press "Run" in the *Toolstrip* (□ 4) to execute the edited commands to create an interactive map showing our most recent environmental data.

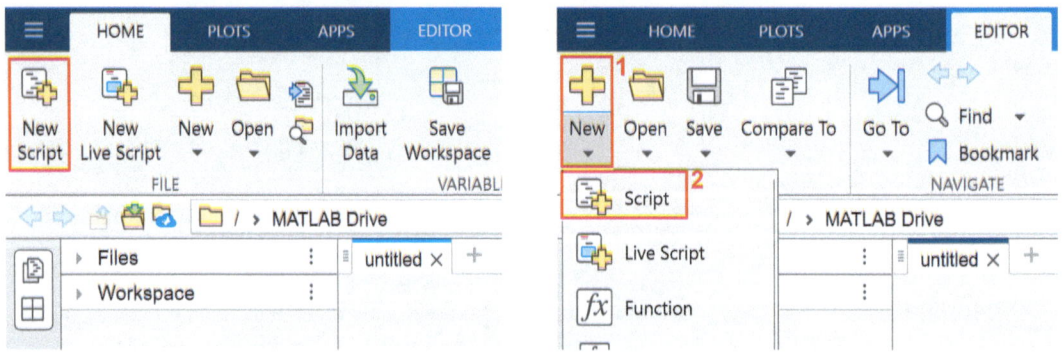

Figure 2.2 Ways of opening a new MATLAB script (in MATLAB Online).

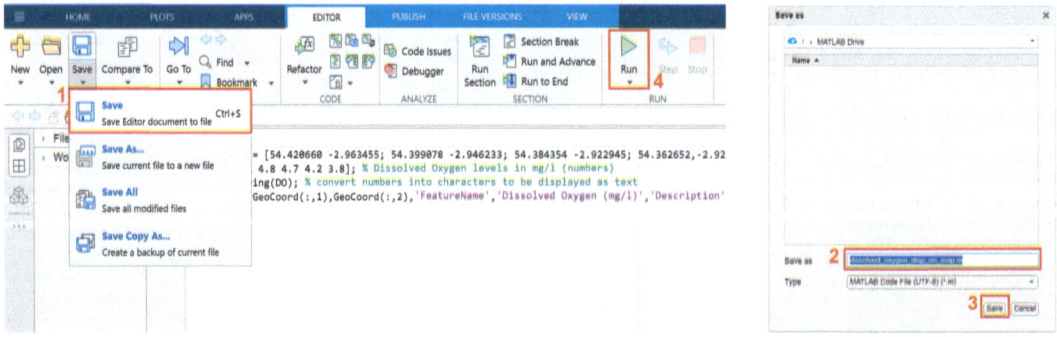

Figure 2.3 Naming a MATLAB script (in MATLAB Online). Some restrictions apply.

2.2 Advantages of Using Scripts

Another advantage of using scripts is that we can **add comments** which can be very helpful in the future when we are trying to understand our own (or somebody else's) approach to producing a particular output. To add an in-line comment to your code, click on the location, press the "%" key on your keyboard and type your comment. The characters after % will be written in green and will not be interpreted by MATLAB.[‡] (Right now, you may not necessarily see the usefulness of comments, but as the number and complexity of your scripts increase, the time will inevitably come when you have difficulty understanding your own scripts written in the past, and then you will be glad that you did add comments.)

Using scripts also enables us to break up the list of commands into **wrapped sections**[§] within a script which we can execute as a block of commands separately from the rest of the script. This feature is advantageous when we wish to keep a collection of script sections in one place, each handling a separate aspect of a complex task, or just generating variations on a theme.

For example, if we want differently-styled outputs without having to save multiple near-identical versions of the "dissolved_oxygen_disp_on_map.m" script, we can wrap each slightly different version of the original script by simply typing "%%" in between them, followed by adding some useful comment in the same line to help distinguish the sections. As a result, MATLAB will draw horizontal lines between the sections to visually separate them while highlighting the active block being edited or used. Here we create two sets of commands generating slightly different versions of the diagram displaying dissolved oxygen levels. Copying the script below

```
%% Default Map
GeoCoord=[54.420660  -2.963455; 54.399078  -2.946233; 54.384354
-2.922945; 54.362652,-2.926088; 54.310322  -2.948620]; % latitudes
and longitudes
DO=[5.2  4.8  4.7  4.2  3.8]; % Dissolved Oxygen levels in mg/l (numbers)
DOS = string(DO); % convert numbers into characters to be displayed as text
wmmarker(GeoCoord(:,1),GeoCoord(:,2),'FeatureName','Dissolved Oxygen
(mg/l)','Description',DOS) % create map

%% Non-default, Satellite Map (pins: size 2, blue)
GeoCoord=[54.420660  -2.963455; 54.399078  -2.946233; 54.384354
-2.922945; 54.362652,-2.926088; 54.310322  -2.948620]; % latitudes
and longitudes
DO=[5.2  4.8  4.7  4.2  3.8]; % Dissolved Oxygen levels in mg/l
(numbers)
DOS = string(DO); % convert numbers into characters to be displayed
as text
```

[‡] In Octave, "#" can also be used.
[§] As of mid-2024, wrapped sections are not available in Octave.

```
webmap('World Imagery'); % load non-default map
wmmarker(GeoCoord(:,1),GeoCoord(:,2),'FeatureName','Dissolved Oxygen
(mg/l)','Description',DOS,'IconScale',2,'Color','blue')
% create map
```

and pasting it over the "dissolved_oxygen_disp_on_map.m" script already open in MATLAB will result in a layout similar to the layout shown in Figure 2.4.

The first wrapped set of commands is the original script, whereas the second wrapped block of script (highlighted) is a slightly altered copy of the original. It places large blue location pins over a satellite image of the sample collection area, but otherwise it is the same as the wrapped section above it. To execute a wrapped section separately from the rest of the script, click somewhere within the block (it becomes highlighted) and then

- either press "Run" in the EDITOR Tab or
- hold down the "CTRL" key on your keyboard and press "ENTER/RETURN/↵".

(Having only two similar scripts normally would not warrant for section wrapping, but you can appreciate how useful wrapping can be if you worked for a water company obligated to regularly produce reports to several stakeholders and government agencies all requiring similar diagrams in slightly different formats.)

Wrapping scripts is also very advantageous when working with long scripts. They might be composed of hundreds or thousands of lines, and wrapping could provide a convenient way of compartmentalising giant scripts into manageable chunks, each performing a distinct sub-task within the overall grand task, making it much clearer which block of the script does what. This is often the case in science and engineering,

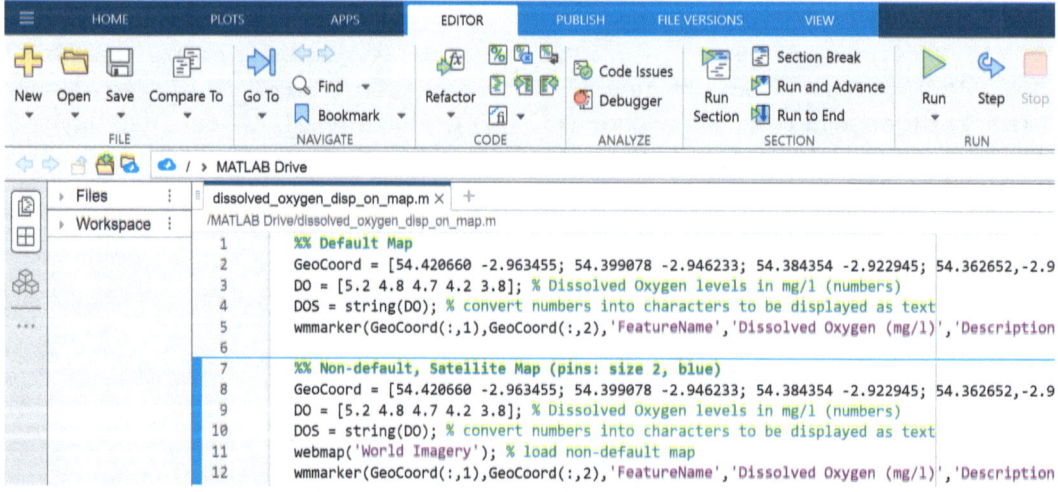

Figure 2.4 Wrapped code sections in MATLAB Online. To execute a section, first select it by clicking anywhere in it, then press Ctrl+Enter.

Scripts and their Applications to Chemistry

where we first have a section that deals with loading data into MATLAB, followed by some complex, multi-step analysis performed on the data, and at the end a section displaying and saving the results of the analysis. It is good practice to wrap each section (with a comment in each section heading) to increase the readability of the script. We can choose to run the entire script or execute wrapped sections individually one after the other, which is an efficient way of inspecting, testing and debugging each separately.

Exercise 2.1

Change the "dissolved_oxygen_disp_on_map.m" script to present dissolved oxygen levels (you make up) in samples collected from the African Great Lakes: Victoria, Kyoga, Albert, Edward, Kivu, Tanganyika and Rukwa

Use size 1.5 green pins. Geographic coordinates can be found on websites, for example, www.latlong.net.

Answer:

```
GeoCoord=[−1.124653  32.833748; 1.470976  32.989828; 1.671058
30.894620; −0.323752  29.602853; −1.887933  29.153133; −6.156939
29.513538; −3.59094  35.10346]; % Geocoordinates of the 7 sample
collection locations
DO=[5.2  4.8  4.7  4.2  3.8  2.5  6.2]; % 7 dissolved oxygen
concentrations
DOS=string(DO);
wmmarker(GeoCoord(:,1),GeoCoord(:,2),'FeatureName','Dissolved
Oxygen (mg/l)','Description',DOS,'IconScale',1.5,'Color','green')
```

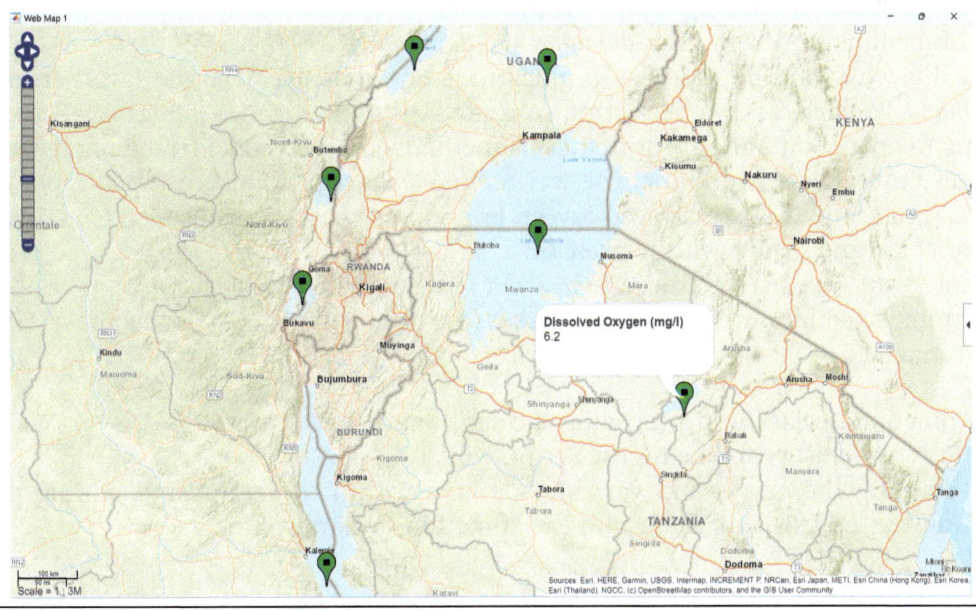

2.3 Scripts as User-defined MATLAB Functions

For anything other than quick tasks, scripts are more practical to use than typing into the *Command Window*. However, they still need to be adapted as tasks change even if the purpose of the script remains exactly the same. In our scenario, each time new dissolved oxygen levels are measured, we need to open the `dissolved_oxygen_disp_on_map` script and overwrite the dissolved oxygen levels to the latest values before executing the script. Similarly, when the number of sample collection points changes, or when they are moved, our script would again need to be updated prior to running it. This inflexibility stems from scripts being self-contained and thus not being able to take any input. Unlike built-in MATLAB functions, our `dissolved_oxygen_disp_on_map` script can only use variables and arrays defined within its scope. For example, when we need the sine of $\alpha = 3\pi/2$, we assign $3\pi/2$ to variable `alpha` and call function `sin()` with `alpha` as an input argument:

```
>> alpha = 3*pi/2;
>> sin(alpha)
ans =
    -1
```

yet when we have a new set of dissolved oxygen levels to display on a map, we have to open `dissolved_oxygen_disp_on_map`, edit `GeoCoord` and `DO`, before saving and running the script by clicking "Run" on the EDITOR Tab. Now, type "dissolved_oxygen_disp_on_map" into the *Command Window* and press "ENTER/RETURN/↵" on your keyboard. The result is going to be the same as when you clicked "Run" on the EDITOR Tab. (You might have noticed that each time you ran a script the name of the script appeared in the *Command Window*. This is because MATLAB handles scripts executed by clicking "Run" on the EDITOR Tab as no-input, primitive user-defined functions run from the *Command Window*.)

Ideally, we would like to pass geolocations and dissolved oxygen levels from the *Command Window*, or another script, to `dissolved_oxygen_disp_on_map` without having to open, edit and save it first. This method would be considerably more flexible and result in a faster workflow. The way to enable a script to take external input arguments (geolocations and dissolved oxygen levels in our case) – as opposed to containing all variables and arrays within itself (like `dissolved_oxygen_disp_on_map` in its current form) – we need to *convert the script into a user-defined function*.

Turning a script into a function takes the following steps. Place the script in between a

- "function declaration" preceding the script and
- an "end" closure following the script, whilst
- removing the arrays and variables defined within the script and adding placeholders representing them into the "function declaration"

as shown below.

```
function dissolved_oxygen_disp_on_map(GeoCoord,DO)
DOS = string(DO); % convert numbers into characters to be displayed
as text
wmmarker(GeoCoord(:,1),GeoCoord(:,2),'FeatureName','Dissolved
Oxygen (mg/l)','Description',DOS)
end
```

Notice that we no longer have `GeoCoord` and `DO` as part of the block of script between the "function declaration" and "end". Instead, we declare these as required input arguments at the end of the "function declaration" line in parenthesis. After saving the above script as a user-defined function, whenever we execute `dissolved_oxygen_disp_on_map()` from the *Command Window*, or call it from another script, we must supply its prescribed two input arguments enclosed in parenthesis separated by a comma: an array storing geolocations, followed by an array containing dissolved oxygen levels. (If you save your function into a specific folder, other than the default, you will need to add the path to that folder, *via* "Set Path" in the HOME Tab, to the list of search paths MATLAB uses to locate functions when they are called.)

Next time you are in Patagonia on an expedition to monitor dissolved oxygen levels in Lago Argentino and Lago Viedma, you can use your used-defined `dissolved_oxygen_disp_on_map()` function to display your findings. If your mobile environmental sample analyser automatically saves all data in a single array called `env_data`, in which each row contains the latitude, longitude and measured dissolved oxygen level for a sample, you would only need to run

```
>> dissolved_oxygen_disp_on_map(env_data(:,1:2),env_data(:,3))
```

from the *Command Window* whenever you have a set of data to present.

```
>> env_data=[-1.124653  32.833748  5.2; 1.470976v32.989828  4.8; 1.671058
30.894620  4.7; -0.323752  29.602853  4.2; -1.887933  29.153133  3.8;
-6.156939  29.513538  2.5; -3.59094  35.10346  6.2]; % Geocoordinates for
sample locations and corresponding dissolved oxygen levels from Exercise 2.1
>> dissolved_oxygen_disp_on_map(env_data(:,1:2),env_data(:,3))
```

should produce the map seen in Exercise 2.1.

Turning scripts into user-defined functions also enables users to compartmentalise large and complex scripts by breaking them up into smaller user-defined functions, each given a narrowly-defined scope. This, in turn, will also improve the readability of your code. More importantly, using user-defined functions largely simplifies code maintenance and debugging because each user-defined function can be edited without having to update all scripts it is called from.

Exercise 2.2

Broaden the scope of the `dissolved_oxygen_disp_on_map()` function to also include the (i) name of analyte displayed, (ii) colour and (iii) size of location pins, as well as the (iv) map type used. Save the modified script under the name "analyte_disp_on_map".

To monitor fertilisers washed into natural water reserves from surrounding areas, display nitrate concentrations in millimoles per dm^3 (*i.e.*, mM) measured in samples collected at locations given in Exercise 2.1 by calling `analyte_disp_on_map()` from the *Command Window*. Mark sample locations with yellow pins of size 1.5 on a map called "World Imagery" (For more map types, visit www.mathworks.com/help/map/ref/webmap.html)

Answer:

```
function analyte_disp_on_map(GPS_Conc, Analyte, PinSize, PinColor, MapType)
ConcText = string(GPS_Conc(:,3));
webmap(MapType);   % Map types: www.mathworks.com/help/map/ref/webmap.html
wmmarker(GPS_Conc(:,1), GPS_Conc(:,2), 'FeatureName', Analyte, 'Description', ConcText, 'IconScale', PinSize, 'Color', PinColor)
end
```

```
>> analyte_disp_on_map(env_data,'nitrate (mM)',1.5,'yellow','World Imagery')
```

will generate the interactive map shown below.

In Chapter 10, we will return to this exercise and learn how to change the colour of pins based on the local analyte concentration to increase the amount of visual information provided.

Exercise 2.3

Write a user-defined function that converts energy (E) given in Joules into units used in spectroscopy: wavenumber ($\tilde{\nu}$) in cm^{-1}, wavelength (λ) in nm and frequency (ν) in Hz (s^{-1}).

$$E = h\nu = h\frac{c}{\lambda} = hc\tilde{\nu}$$

$$\tilde{\nu} = \frac{1}{\lambda} = \frac{\nu}{c}$$

Answer:

```
function energy_spec_units(E_in_J)
h = 6.6261e-34; % Plack's constant (Js)
c = 2.9979e10; % Speed of light in vacuum (cm/s)
nm_in_cm = 1e-2/1e-9; % nanometres in a centimetre
Wavenumber = E_in_J/h/c;
Frequency = E_in_J/h;
Wavelength = 1/Wavenumber*nm_in_cm;
% Display results:
disp(['E=',num2str(E_in_J),' J'])
disp(['Wavenumber=',num2str(Wavenumber),' cm^{-1}'])
disp(['Frequency=',num2str(Frequency),' Hz'])
disp(['Wavelength=',num2str(Wavelength),' nm'])
end
```

To test the function use 2.1799×10^{-18} J, which is the value of 1 Rydberg, a unit of energy equivalent to the ionisation energy of the hydrogen atom.

```
>> energy_spec_units(2.1799e-18)
E=2.1799e-18 J
Wavenumber=109739.1124 cm^{-1}
Frequency=3289868851964202 Hz
Wavelength=91.1252 nm
```

So far, the output of our user-defined functions were interactive maps and text written in the *Command Window*. We can also create functions that return numbers or arrays, which can be used in subsequent calculations, like the output of mathematical functions, for example `sin()`.

Exercise 2.4

Write a MATLAB function that calculates the standard reaction enthalpy of a reaction ($\Delta_r H^\circ_{298}$) from the standard molar formation enthalpies ($\Delta_f H^\circ_{298}$) of the reactants and products, and outputs the numeric result. The function should have two input arguments, both arrays containing stoichiometric coefficients and standard formation enthalpies, one for the reactants and the other for the products. Use the combustion of table sugar for testing your function.

Answer:

The standard reaction enthalpy of a reaction is the difference between the combined standard formation enthalpies of the products and the reactants. It is calculated according to the formula

$$\Delta_r H^\circ_{298} = \sum_{i=1}^{n} \nu_i \Delta_f H^\circ_{298}(\text{Product}_i) - \sum_{j=1}^{m} \nu_j \Delta_f H^\circ_{298}(\text{Reactant}_j)$$

where ν_i and ν_j are the stoichiometric coefficients of Product$_i$ and Reactant$_j$, respectively.

In the function declaration, we create a variable (H) for outputting the result of the calculations. Our function will accept two arrays, R and P, which hold the stoichiometric coefficients and standard formation enthalpies in this order for the reactants and the products, respectively, with each row corresponding to a different compound. The total standard enthalpies of the reactants and the products are computed by first multiplying the stoichiometric coefficient and the standard formation enthalpy together row after row – using element-wise multiplication (.*) – and then summing the results. The multiplications are nested in the sum() functions for brevity. Finally, the standard reaction enthalpy is computed, by subtracting the total standard enthalpy of the products from that of the reactants, and loaded into variable H.

```
function H = DrH_st(R, P)
% ΔrH° = ΣνΔfH° (products) - ΣνΔfH° (reactants)
S_DfHr = sum(R(:,1) .* R(:,2)); % ΣνΔfH° (reactants)
S_DfHp = sum(P(:,1) .* P(:,2)); % ΣνΔfH° (products)
H = S_DfHp - S_DfHr; % ΔrH°, standard reaction enthalpy
end
```

The balanced reaction equation for the combustion of table sugar (sucrose, $C_{12}H_{22}O_{11}$) is

$$C_{12}H_{22}O_{11}(s) + 12 O_2(g) \rightarrow 12 CO_2(g) + 11 H_2O(l)$$

with standard formation enthalpies (which you can find in databases and/or tabulated appendices to most Physical Chemistry textbooks)

compound	$C_{12}H_{22}O_{11}(s)$	$O_2(g)$	$CO_2(g)$	$H_2O(l)$
$\Delta_f H^\circ$/kJ mol^{-1}	-2226.1	0	-393.5	-285.8

we call our function from the *Command Window* as

```
>> sucrose_st_comb_H=DrH_st([1 -2226.1; 12 0],[12 -393.5;
11 -285.8])
sucrose_st_comb_H =
            -5639.7
```

Note that our `DrH_st()` function works like any other mathematical function in MATLAB. It reads in some numbers provided as input arguments and returns a number. It can be called from the *Command Window* or from any scripts and functions.

Another common type of calculation that can be done using user-defined functions is acid–base titrations. To find the unknown concentration of an acid (or base) solution, we transfer a certain volume of the solution to a flask, along with a small amount of a suitably chosen indicator, and fill a burette with a base (or acid) solution of known concentration. We slowly keep adding small volumes of the solution in the burette until the indicator changes colour at the equivalence (or stoichiometric) point, where the reaction between the acid and base has just gone to completion.

For example, if 38.5 cm^3 of 0.200 mol dm^{-3} HCl solution was added to 50.0 cm^3 of NaOH solution before the colour change occurred, we can calculate the concentration of the sodium hydroxide solution based on the chemical equation

$$NaOH + HCl \rightarrow H_2O + Na^+ + Cl^-$$

which gives us the NaOH to HCl molar ratio (1 : 1) we require for the calculation. This means that the number of moles of HCl added (n_{HCl}) is equal to the amount of NaOH (n_{NaOH}) at the stoichiometric point. Therefore, the number of moles of sodium hydroxide is

$$n_{NaOH} = n_{HCl} = c_{HCl} \times V_{HCl} = 0.200 \text{ mol dm}^{-3} \times 0.0385 \text{ dm}^3 = 7.70 \times 10^{-3} \text{ mol},$$

and from this we can calculate the concentration of the NaOH solution as

$$c_{NaOH} = n_{NaOH}/V_{NaOH} = 7.70 \times 10^{-3} \text{ mol}/0.0500 \text{ dm}^3 = 0.154 \text{ mol dm}^{-3}.$$

Exercise 2.5

Write a user-defined function to solve the following acid–base titration problem. To allow for flexibility, ensure that the function reads in the stoichiometric coefficients of the acid and base as input arguments.

In a titration, 29.7 cm^3 of 0.1 mol dm^{-3} H$_2$SO$_4$ was added to neutralise 25 cm^3 of KOH solution. Calculate the concentration of the KOH solution.

Answer:

We start by writing the general chemical equation applicable to any acid–base neutralisation reaction

$$b\text{B(OH)}_x + a\text{H}_y\text{A} \rightarrow bx\text{H}_2\text{O} + b\text{B}^{x+} + a\text{A}^{y-}$$

with constraint $bx = ay$ resulting from the atom and charge balances. Thus, for the reaction equation representing neutralisation in the question

$$2\text{KOH} + \text{H}_2\text{SO}_4 \rightarrow 2\text{H}_2\text{O} + 2\text{K}^+ + \text{SO}_4^{2-}$$

B and A correspond to "K" and "SO$_4$", respectively, with $b=2$, $x=1$, $a=1$, $y=2$.

With b and a being the stoichiometric coefficients of the base and acid, respectively, at the equivalence point

$$\frac{b}{a} = \frac{n_{\text{base}}}{n_{\text{acid}}} = \frac{c_{\text{base}} V_{\text{base}}}{c_{\text{acid}} V_{\text{acid}}}$$

where n_{base} and n_{acid} are the numbers of moles of the base and acid which can be calculated from their concentrations (c_{base}, c_{acid}) and volumes (V_{base}, V_{acid}). Rearrangements yield the formulae for calculating the concentration of the base and the acid

$$c_{\text{base}} = \frac{b}{a} \frac{c_{\text{acid}} V_{\text{acid}}}{V_{\text{base}}}$$

$$c_{\text{acid}} = \frac{a}{b} \frac{c_{\text{base}} V_{\text{base}}}{V_{\text{acid}}}$$

Given their similarities, we can merge the equations above into one by introducing compound U, whose concentration is unknown, and compound S, in standard solution, whose concentration we know. With u, c_{U}, V_{U} and s, c_{S}, V_{S} being the stoichiometric coefficient, concentration and volume of U, and the stoichiometric coefficient, concentration and volume of S, respectively, the unknown concentration can be calculated as

$$c_{\text{U}} = \frac{u}{s} \frac{c_{\text{S}} V_{\text{S}}}{V_{\text{U}}}$$

irrespective of whether we are looking for the concentration of the acid or the base. Notice that as long as the volumes have the same units, the unknown concentration will be given in the same unit as the concentration of the standard solution. Turning this equation into a user-defined function will look like this.

```
function CU=titration(VU,u,VS,CS,s)
% titration function for "uU + sS→..." type reactions
% U: solution with unknown concentration (Titrand)
% S: standard solution with known concentration (Titrant)
% VU: volume of U solution; u: stoichiometric coefficient of U
% CU: concentration of U solution
```

```
% VS: volume of S solution; s: stoichiometric coefficient of S
% CS: concentration of S solution
CU=u/s*CS*VS/VU;
end
```

With the function saved, the problem is solved as

```
>> c=titration(25,2,29.7,0.1,1)
c =
    0.2376
```

The units for the KOH concentration are $\text{mol}\,\text{dm}^{-3}$ (the same as the units for the concentration of H_2SO_4).

2.4 Appendix: Figure Formatting Using Arrays

Without changing `analyte_disp_on_map()`, we now set an arbitrary pin colour and size for each sample. These do not reflect the level of the analyte, as we are only interested in learning how to pass arrays where we previously used single-valued input arguments. In Exercise 2.2, `PinSize` and `PinColor` were assigned `1.5` and `'yellow'` or `'y'`, which we would like to replace by arrays. These arrays will have to

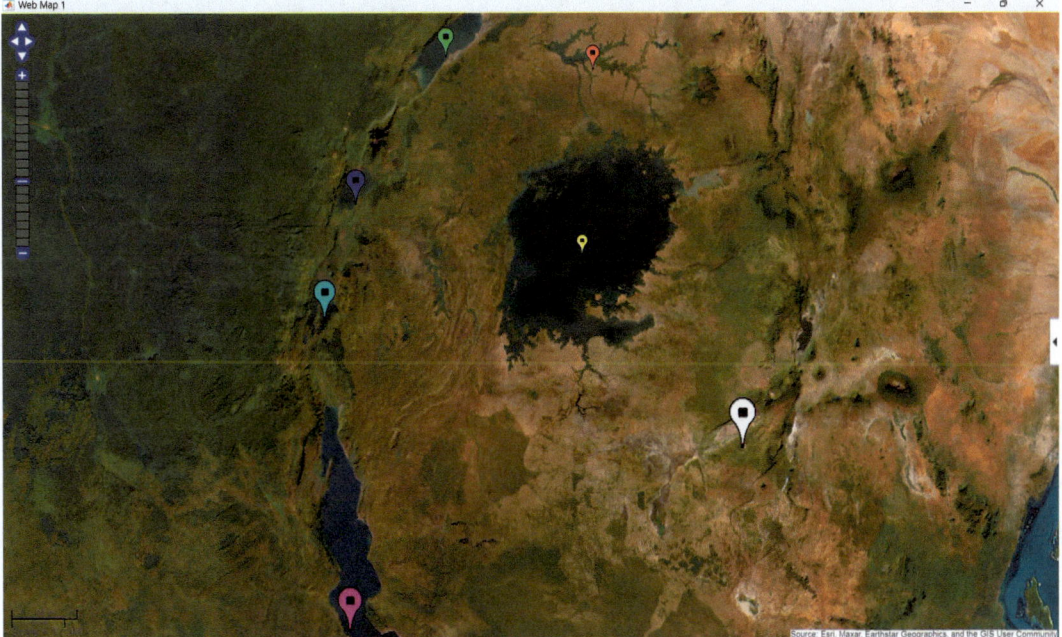

Figure 2.5 The appearances of all pins on a map can be set in one step by passing all colours and sizes as vectors to `wmmarker()`.

have the same number of elements as the number of samples in order to set the pin colour and size for each pin individually.

```
>> ps = linspace(0.8,2,7);
```

will generate a 7-member array containing a series of equally-spaced numbers ranging from 0.8 to 2, which will serve as pin sizes. Individual pin colours are provided in a cell array (given in curly brackets)

```
>> pc = {'y','r','g','b','c','m','w'};
```

where the letter codes stand for yellow, red, green, blue, cyan, magenta and white, respectively. Executing

```
>> analyte_disp_on_map(env_data,'nitrate (mM)',ps,pc,'World Imagery')
```

will create the map seen in Figure 2.5.

3 Presenting Data in Chemistry: Plots and Charts

Working in a lab involves collecting data. If you are synthetizing a chemical, you record the reaction conditions so that you can reproduce your results later, or to allow others to synthetize the compound using your procedure. You write down the initial composition of the reaction mixture, including the amounts of solvents, temperature, the times you took samples from the reaction mixture for analysis and the amounts of the substances the samples contained. Often, we re-run reactions multiple times, each time slightly differently so that we can gain an understanding of what factors influence the outcome of the reaction and by how much. For example, we run a reaction at different temperatures and/or vary the ratio of solvents in the reaction mixture. Each time after work-up, we determine the amount of our target compound and usually begin to see some trends after just a few runs. Instead of only making notes on how the yield changed in response to varying the reaction conditions, we typically visualise this information to more easily see any emerging trends.

MATLAB offers a wide variety of data display options. Those often used in Chemistry include 2D scatter and line plots, histograms and more complex data display arrangements, such as stacked multiple line plots to present the progress of change in chemical composition over time.

(Graphic outputs in this chapter were primarily generated with MATLAB Online and may look slightly different to those created by the desktop version. There are some minor differences between the layouts of MATLAB's online and desktop versions, which means that accessing some of the features discussed below are not done in exactly the same ways. You can access MATLAB Online through your MathWorks account.)

A First Look at Coding in Chemistry: Solving Problems Using MATLAB
By Tamas Bansagi
© Tamas Bansagi 2025
Published by the Royal Society of Chemistry, www.rsc.org

3.1 Scatter and Line Plots

If we have a small amount of data, the quickest way to display them is by selecting the array we wish to visualise in WORKSPACE and choosing a diagram type from the PLOTS tab.† For example, the simple array of data set as

```
>> data = [1.00  0.39;  2.00  0.63;  3.00  0.77;  4.00  0.86;  5.00  0.91;
6.00  0.95; 7.00  0.96; 8.00  0.98; 9.00  0.98; 10.00  0.99];
```

can be quickly plotted using MATLAB's graphic user interface by the following steps: click on data in WORKSPACE (□ 1), then click on the scatter plot icon in the PLOTS tab (□ 2), as shown in Figure 3.1, which will result in the plot shown in Figure 3.2.

To annotate the diagram (add a title, axis labels, grid lines and so on) go to the FIGURE tab (available in MATLAB Online (shown in Figure 3.3), in the desktop version open the "Insert" dropdown menu in the figure window for annotation) and select the annotations you wish to add. For changing fonts and other figure settings click on the "Inspector" icon in the Toolstrip (or on the "Property Inspector" icon in the figure window in the desktop version).

Most of the time, however, we need to handle much more data and create several diagrams, which typically takes longer using the graphic user interface than using commands directly. Especially if you need to set sub- and super-scripts, which happens very often in chemistry.

We can recreate the figure above using the following commands:

```
>> scatter(data(:,1),data(:,2))
>> xlabel('time/min')
>> ylabel('[Product]/mol dm^{-3}')
>> title('Product formation')
>> grid on
```

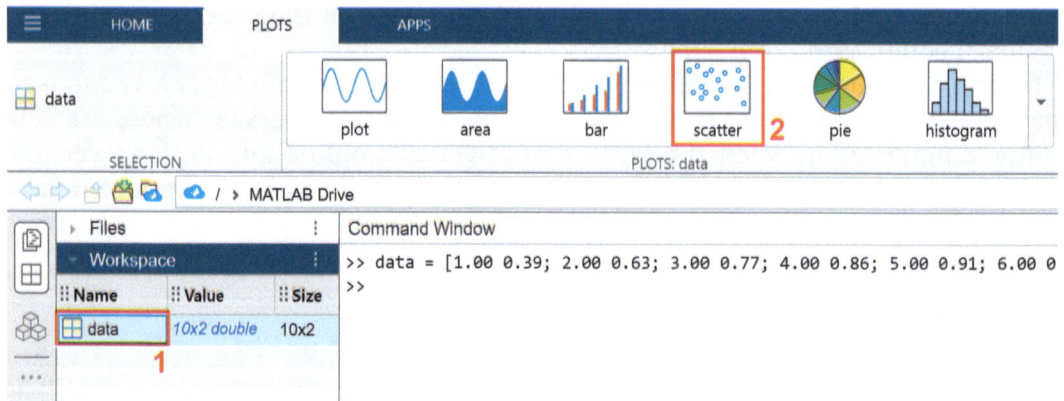

Figure 3.1 Creating a scatter plot in MATLAB interactively by two clicks.

---

†Octave also offers data visualization through Workspace. Double-click on an array and then click on the "Plot selected data" icon in Variable Editor.

Presenting Data in Chemistry: Plots and Charts

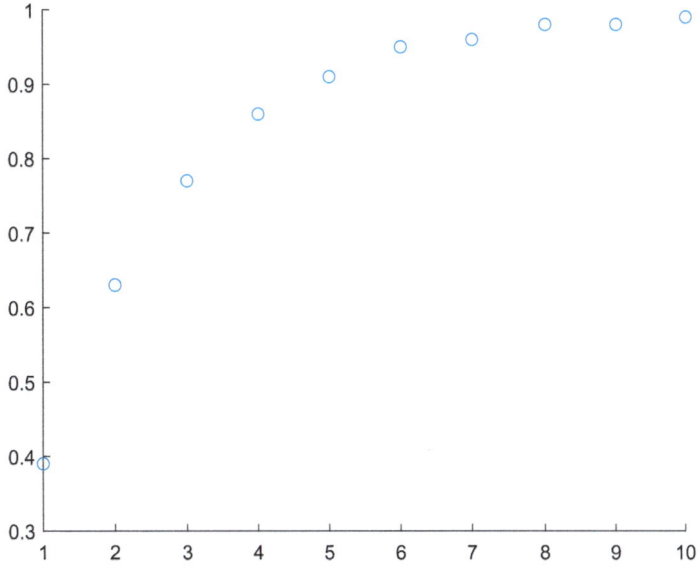

Figure 3.2 Default scatter plot generated by MATLAB.

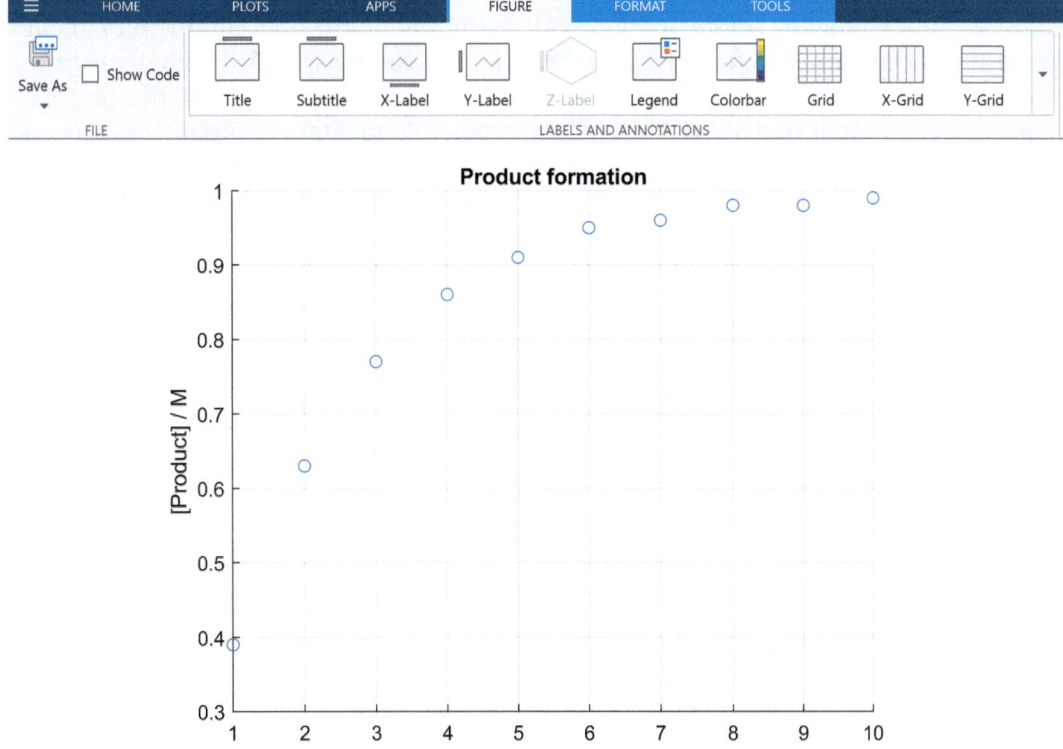

Figure 3.3 Figure properties you can set interactively on the FIGURE tab in MATLAB online. (In the desktop version, these can be accessed from the "Insert" dropdown menu in the figure window.)

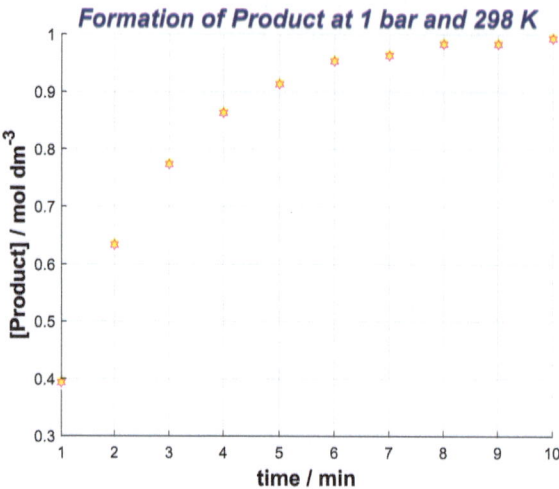

Figure 3.4 Figure properties set using commands *via* property-value pairs.

Even when we want to style our figure, commands are often quicker than using the graphic user interface. Let us say we would like to have red squares for markers, large bold fonts in the axis labels and blue italic fonts in the title. We can conveniently create this plot by specifying the marker format as `'rs'` (r for 'red' and s for 'square') directly within the `scatter()` command (more options at https://uk.mathworks.com/help/matlab/ref/scatter.html). Similarly, the font size, weight and angle can be specified within `xlabel()`, `ylabel()` and `title()` by using *property name–value pairs*. In general, the process of generating the output of a MATLAB function is possible to alter *via* property name–value pairs. In Figure 3.4, the font size and weight properties of the text within the objects created by `xlabel()`, `ylabel()` and `title()` are changed from their default values by setting `'fontsize'` to `14` and `'fontweight'` to `'bold'`. (For the list of available property names and values visit the respective commands on the MathWorks website.)

```
>>scatter(data(:,1), data(:,2), 'hexagram', 'markeredgecolor', 'r',
'markerfacecolor', 'y')
>> xlabel('time/min','fontsize',14,'fontweight','bold')
>> ylabel('[Product]/mol dm^{-3}','fontsize',14,'fontweight','bold')
>> title('Formation of Product at 1 bar and 298 K', 'color', 'b',
'fontangle', 'italic', 'fontsize',16)
>> grid on
```

When we have a scientific law, theory or model describing the relationship between two quantities or a large amount of data points to present we prefer not to use markers and create line graphs instead of scatter plots. The function to create line graphs in MATLAB is called `plot()`. For example, drawing the line graph for the expression describing the build-up of product, P, in a reaction A→P from reactant A, $[P] = [A]_0 - e^{-kt}$ where $[A]_0 = 1$ mol dm^{-3} and $k = 0.5$ min^{-1} (with t representing time) can be done as follows (Figure 3.5)

```
>> clear data % clear array "data" before redefining it with different numbers
>> data(1,:) = 0:0.01:10; % generate first row of data points
```

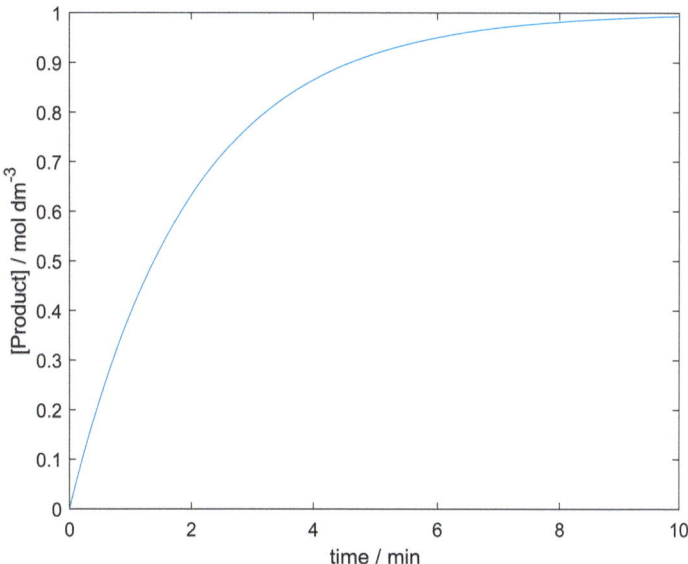

Figure 3.5 By default, the box property of MATLAB figures is on.

```
>> data(2,:) =1−exp(−0.5.*(0:0.01:10)); % generate second row of data
points using elementwise multiplication (.*)
>> plot(data(1,:),data(2,:))
>> xlabel('time/min')
>> ylabel('[Product]/mol dm^{−3}')
```

It is good practice to set axis limits so that line graphs are clearly visible and not merged with the edges of the box around the plot area, which could be conveniently achieved by executing `axis padded`. Line graphs starting from the origin and/or closely approaching the x axis are usually appropriate. We can also set the axis limits manually in a single command as

```
>> axis([0  10  0  1.1]) % x axis limits: 0  10; y axis limits: 0  1.1
```

and/or we turn the box around the plot area – automatically generated by `plot()` – off by

```
>> box off
```

Having executed the latter to commands, you should obtain the improved diagram shown in Figure 3.6.

Markers can be added to graphs generated with `plot()` by simply defining the marker type – for example `'rs'` – after the array storing the values of the dependent variable. Line styles can be defined *via* special symbols: `'-'` for solid, `'--'` for dashed, `':'` for dotted and `'-.'` for dash-dotted lines. Marker types, colours and line styles can be conveniently combined into one short special symbol set. For example, shorthand `'m-.d'` results in magenta diamond markers connected by a dash-dotted line.

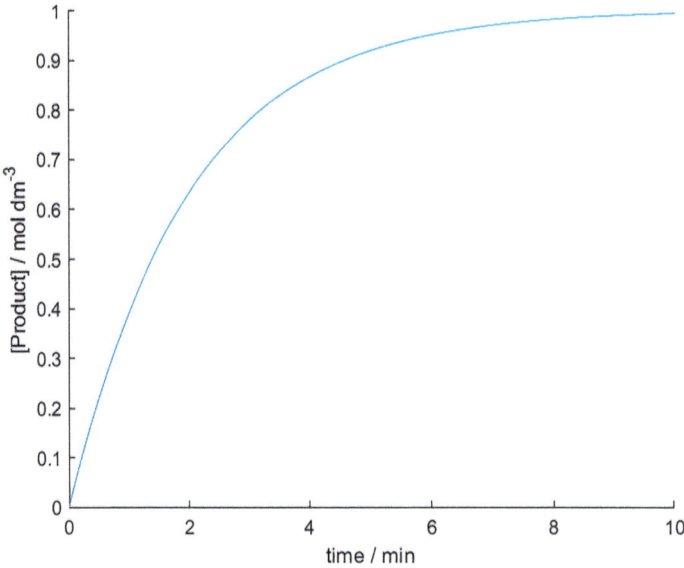

Figure 3.6 A MATLAB figure with the box property turned off.

Exercise 3.1

Write a MATLAB script to create the line graph of an isotherm for an ideal gas. Isotherms are diagrams showing the pressure of a gas as its volume is changed at constant temperature. The expression for isotherms directly comes from the ideal gas equation ($pV = nRT$). For one mole of ideal gas ($n = 1$ mole), the isotherm at a particular temperature is given by

$$p = \frac{RT}{V}$$

Generate two thousand equally spaced volumes between 0.01 and 0.1 m³, set the temperature to 298 K and use the expression p = R*T./V to compute the corresponding pressure values. Use special symbols to set the line style to blue dashed. Label the axes and add title to the diagram.

Answer:

```
R=8.3145; % Gas constant in J/K/mol
T=298; % Temperature in K
V=linspace(0.01,0.1,2000); % generate array of volumes in m^3
p=R*T./V;% calculate pressure using elementwise division (./)
plot(V,p,'b--') % create plot
xlabel('V/m^3') % label x axis
ylabel('p/Pa') % label y axis
title('Isotherm of 1 mole of ideal gas at 298 K')
grid on
box off
axis([0.01 0.1 0 2.5e5])
```

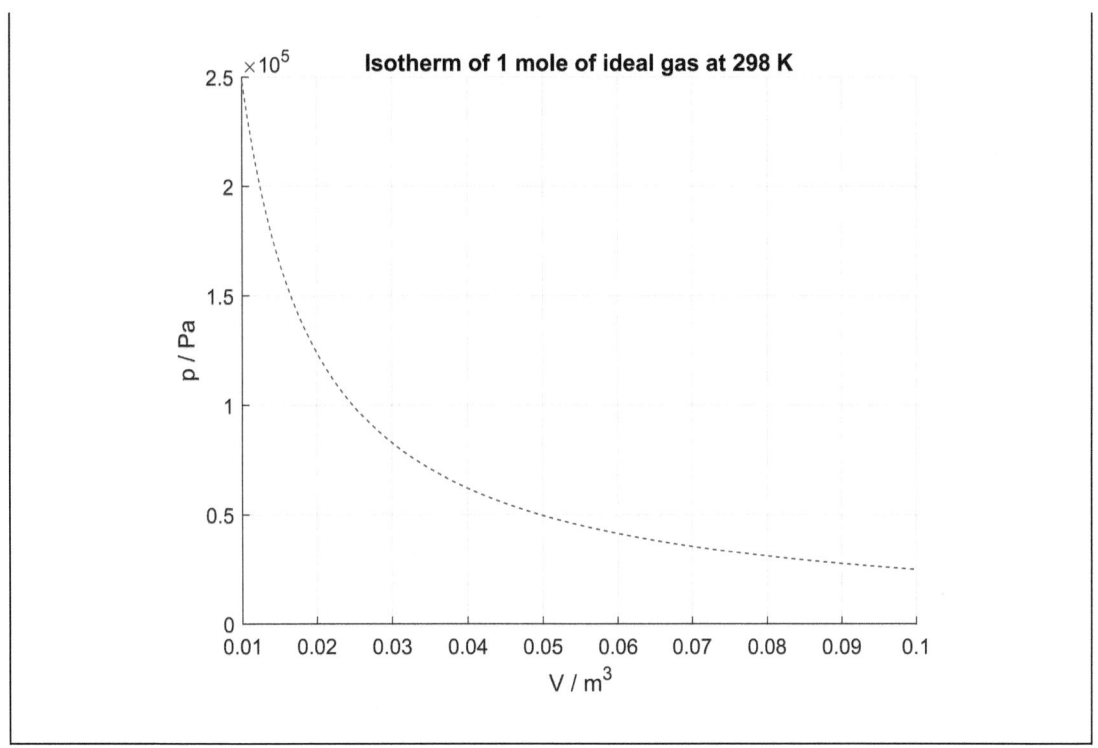

At the end of this section on producing basic graphs, it is worthwhile reviewing the general guidelines for producing scientific figures:

- axes are labelled (including the symbol or name of the quantity plotted on each axis being followed by its unit(s) separated by "/" which indicates that the unitless tick labels result from dividing the corresponding values of the physical quantity by the unit(s) given in the axis label – Guggenheim notation)
- the plot has an informative title or figure caption (whichever is appropriate but not both, including the name of the substance or system and relevant conditions: temperature, pressure, solvent, pH, sample preparation method, *etc.*)
- data points are *not* connected (unless the number of data points is so large that markers would otherwise overlap, or the graph of an expression is plotted)
- axis limits are set tight or slightly padded (using `axis tight` or `axis padded` to override limits MATLAB set automatically) to minimise empty spaces
- grid lines are optional

3.2 Log Plots

When a data set spans across orders of magnitudes, logarithmic diagrams are often best suited to display the data points. Logarithmic plots are also useful when we suspect a logarithmic relationship between the quantities, which we can simply confirm by obtaining a linear trend in a logarithmic diagram. Here, we create a semi-log plot by displaying the concentration data on a logarithmic y axis (Figure 3.7).

```
t=0:10; % time in minutes
c=[10.26  6.52  4.67  3.08  1.91  1.58  1.14  0.81  0.45  0.26
0.13]; % concentration of reactant in mM
semilogy(t,c,'rd') % create semi-log plot (only y axis is
logarithmic)
xlabel('t/min') % label x axis
ylabel('c/mmol dm^{-3}') % label unchanged: axis transformed,
not data
axis([0  10  0.1  13])
grid on % display grid lines
```

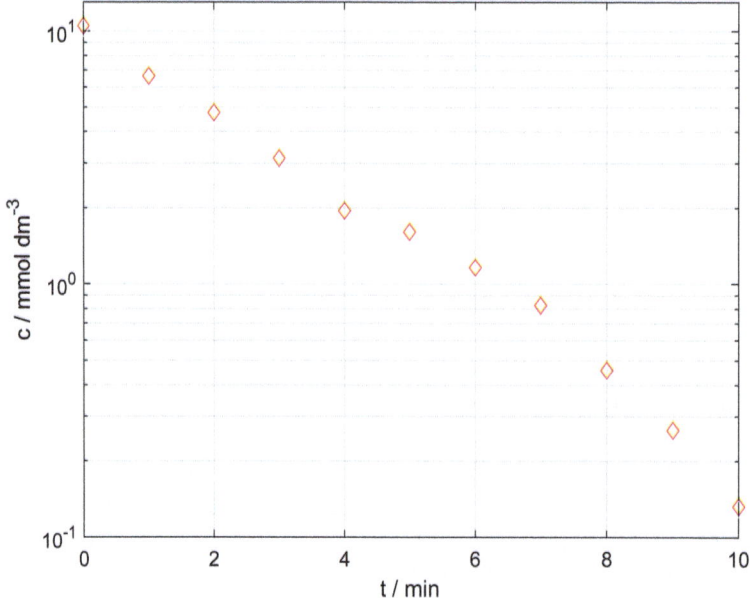

Figure 3.7 MATLAB figure with logarithmic y-axis generated by `semilogy()`.

MATLAB also supports a logarithmic x-axis, `semilogx()`, and log–log plots, `loglog()`, where both axes are logarithmic. Based on Figure 3.7, we can draw the conclusion that the reaction we were following is likely first order because we obtained a linear trend by plotting the reactant concentration on a logarithmic axis against time.

3.3 Combining Graphs and Plots

It is good practice to display data sets intended for comparison in a single diagram. MATLAB makes this easy; the above commands to create diagrams can be executed one after another without deleting the previous diagram, as long as we instruct MATLAB to hold the current diagram before new content is added. This can be achieved by executing the `hold on` command between adding visual elements to an existing figure.

Now, we display three sets of reactant concentrations measured in three experiments for a reaction at different temperatures.

```
t=0:9; % time points for measuring the reactant concentration
c1=[3.000, 1.819, 1.103, 0.669, 0.406, 0.246, 0.149, 0.090, 0.054, 0.033];
c2=[2.000, 1.213, 0.735, 0.446, 0.270, 0.164, 0.099, 0.060, 0.036, 0.022];
c3=[1.000, 0.606, 0.367, 0.223, 0.135, 0.082, 0.049, 0.030, 0.018, 0.011];
```

Then, we use `plot()` to display the data sets one after the other in the same diagram (Figure 3.8).

```
plot(t,c1,'-k') % plot 1st data set
hold on % keep figure open for more content
plot(t,c2,'-r') % plot 2nd data set
hold on % keep figure open for more content
plot(t,c3,'-g') % plot 3rd data set and create figure displaying 3rd
data set along with the previously supressed 1st and 2nd data sets
xlabel('t/min'); ylabel('[A]/mol dm^{-3}') % add axis labels
```

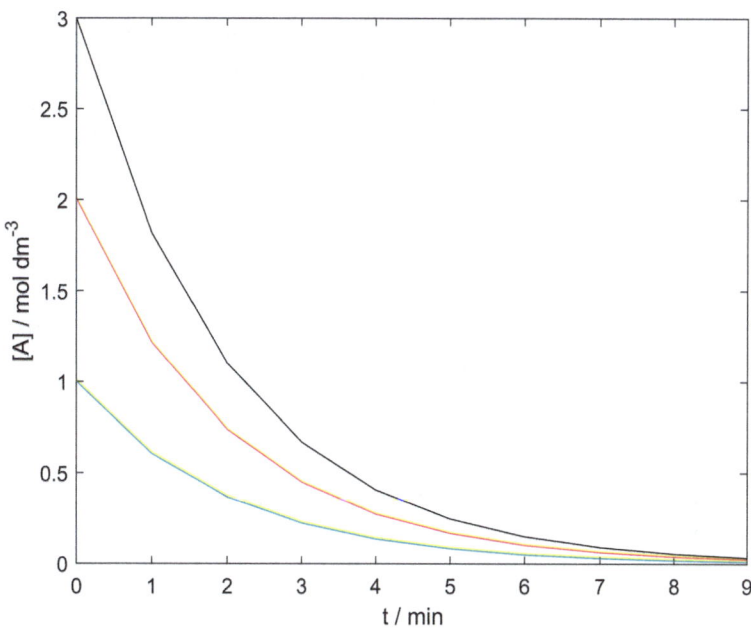

Figure 3.8 Multiple data sets in one figure created by executing `hold on` between `plot()` commands.

The same result can be efficiently achieved by calling `plot()` only once. Passing the three data sets to `plot()` in one sequence shortens code considerably.

```
>> plot(t,c1,'-k',t,c2,'-r',t,c3,'-g') % this single command replaces
calling plot() three times
```

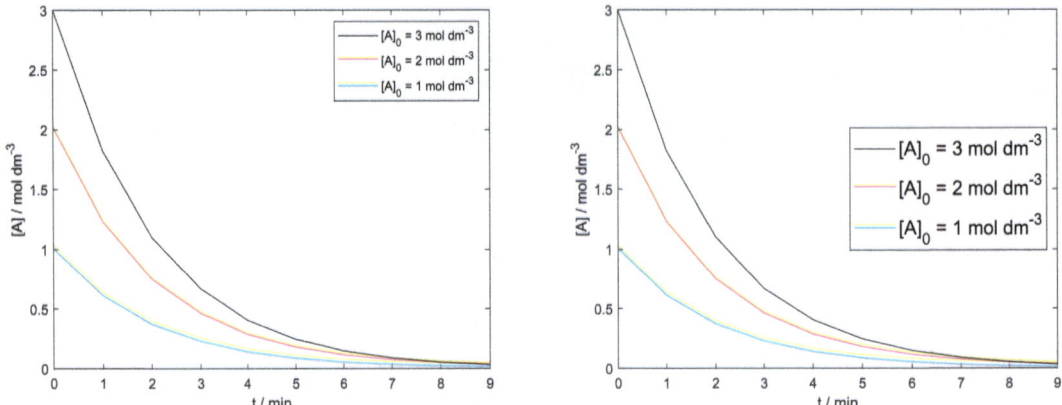

Figure 3.9 Adding, styling and positioning figure legends with the legend() function.

Having multiple graphs in one diagram often requires adding labels. The function for inserting labels is called legend() and is straightforward to use. We only need to provide the texts in the order of displaying the curves and MATLAB does the rest. The command below results in the left diagram in Figure 3.9.

```
>> legend('[A]_0=3 mol dm^{-3}','[A]_0=2 mol dm^{-3}','[A]_0=1 mol dm^{-3}')
```

The format and location of legends can be conveniently set within the argument of the legend() function by adding formatting instructions *via* property name–value pairs introduced in Section 3.1. On the right of Figure 3.9 we have the same legend content, but with increased font size displayed in the middle on the right hand side of the diagram. The command generating this legend is as follows:

```
>> legend('[A]_0=3 mol dm^{-3}','[A]_0=2 mol dm^{-3}','[A]_0=1 mol dm^{-3}','fontsize',14,'location','east')
```

Sometimes we prefer presenting multiple data sets in composite figures made up of multiple diagrams. The process of creating such figures can be streamlined by using the subplot() function. All we have to do is specify the layout of panels within the figure and which panel the next diagram should slot in before it is drawn. Within subplot(), always preceding a command creating a specific graphic output, we need to define how many rows and columns of panels the figure will contain and the position for the next diagram created within that layout. The script below displays the kinetic data seen in Section 2.2 on linear (top) and logarithmic scales (bottom) arranged in a 2 by 1 layout (Figure 3.10).

```
t=0:10;
c=[10.26  6.52  4.67  3.08  1.91  1.58  1.14  0.81  0.45  0.26  0.13];
subplot(2,1,1) % display figures in 2 rows and 1 column, draw 1st diagram
plot(t,c,'bs') % create plot with linear axes
xlabel('t/min')
ylabel('c/mmol dm^{-3}')
```

Presenting Data in Chemistry: Plots and Charts 39

```
axis([0  10  0  11])
grid on

subplot(2,1,2) % display figures in 2 rows and 1 column, draw 2nd diagram
semilogy(t,c,'rd') % create plot with logarithmic y-axis and linear x-axis
xlabel('t/min')
ylabel('c/mmol dm^{-3}') % label unchanged: axis transformed, not data
axis([0  10  0.1  13])
grid on
```

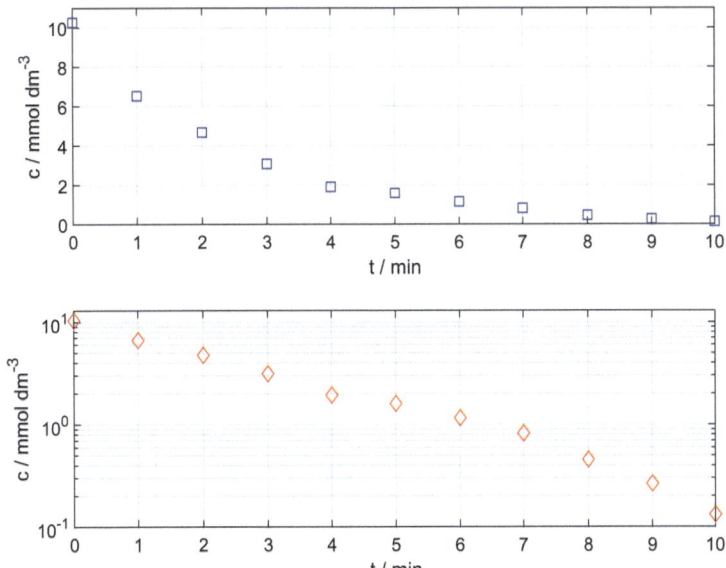

Figure 3.10 Subplots are another way to present more information in a figure. They are used often in science and engineering and are easy to create with MATLAB.

To reduce redundancy in composite figures, common axis labels and tick labels can be removed. These will allow for expanding subplots into the widened gaps for improved readability.

Exercise 3.2

Write a MATLAB script that creates an altered version of the figure above with the upper subplot having an unlabelled *x*-axis and no x-tick labels. Reduce the resulting gap in the middle of the figure by vertically expanding both subplots by 15%.

Answer:
The way we tackle this task is by removing `xlabel('t/min')` from line 5 and letting MATLAB create the rest of the figure as before; but now we will alter the figure properties to our requirements once it has been generated. To be able to conveniently change subplots, we will first load their properties into *axes objects* `fig1` and `fig2` when they are created, for these will be serving as handles through which we can access and modify

their properties. To make the automatically-generated *x*-axis tick labels of the upper subplot disappear, we set the `fig1.XTickLabel` property to an empty array.

For expanding the subplots to reduce the gap in the middle of the figure, we need to first read out their MATLAB-assigned positions, stored in `fig1.Position` and `fig2.Position`, and load them into arrays `L` (left), `B` (bottom), `W` (width) and `H` (height). The meaning of the numbers in the `fig1.Position` and `fig2.Position` properties are displayed in the diagram below. With the bottom left and top right corners of the figure window being the (0,0) and (1,1) reference points, respectively, the first two elements (loaded into `L` and `B`) define the location of the bottom left corners of the subplots, whereas the third and fourth elements (loaded into `W` and `H`) define the widths and heights of subplots. Extending the height of the upper subplot (`fig1`) into the empty space in the middle requires lowering its bottom left corner (*via* decreasing `B`) and increasing its height (*via* making `H` bigger) by the same amount, `H*0.15`. To vertically stretch the lower subplot (`fig2`) into the gap in the middle, we only need to increase its height (*via* making `H` larger) by the same increment.

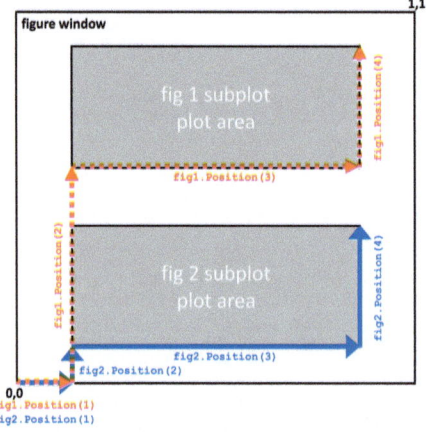

```
t=0:10;
c=[10.26  6.52  4.67  3.08  1.91  1.58  1.14  0.81
0.45  0.26  0.13];
ve=0.15; % vertical expansion in subplot size (corresponds to 15%)
fig1=subplot(2,1,1); % display figures in 2 rows and 1 column,
draw 1st diagram, add handle (fig1) to subplot object for accessing its
properties
plot(t,c,'bs') % create plot with linear axes
ylabel('c/mmol dm^{-3}')
axis([0  10  0  11])
grid on
fig1.XTickLabel=[]; % set tick labels on x-axis to empty array
% access and load the automatically-assigned position of fig1 subplot into
4 variables: L(left) B(bottom) W(width) H(height)
L=fig1.Position(1); B=fig1.Position(2);
W=fig1.Position(3); H=fig1.Position(4);
```

```
fig1.Position=[L B-H*ve W H+H*ve]; % alter vertical position
(lower original by 15% of height) and height of fig1 (by 15% of original)
fig2=subplot(2,1,2); % display figures in 2 rows and 1 column,
draw 2nd diagram, add handle (fig2) to subplot object for accessing its
properties
semilogy(t,c,'rd')   %  create  plot  with  logarithmic  y-axis  and
linear x-axis
xlabel('t/min')
ylabel('c/mmol dm^{-3}') % label unchanged: axis transformed, not data
axis([0  10  0.1  13])
grid on
% access and load the automatically-assigned position of fig2 subplot into
4 variables: L(left) B(bottom) W(width) H(height)
L=fig2.Position(1); B=fig2.Position(2);
W=fig2.Position(3); H=fig2.Position(4);
fig2.Position=[L B W H+H*ve]; % increase height of fig2 (by 15% of
original)
```

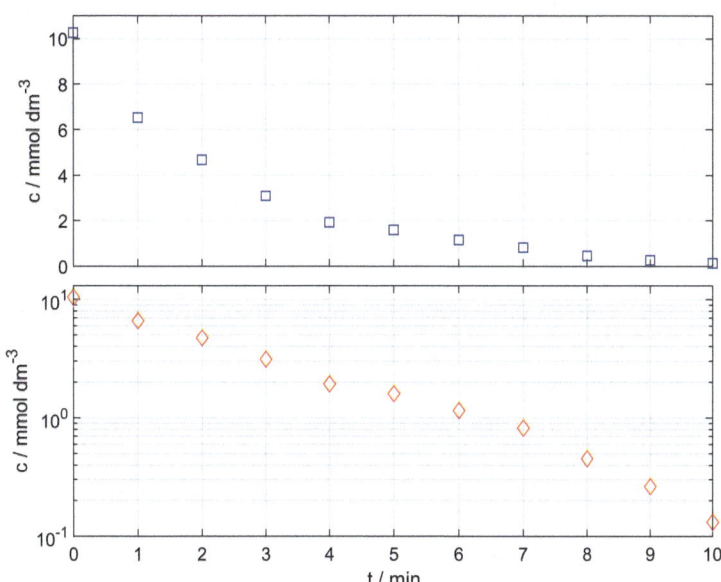

By adjusting the position properties of subplots[‡] (or by interactively resizing and dragging them in the figure window), we can create overlapping subplots, which is useful in making one subplot the inset of another.

[‡] In Octave, adjusting these properties is done slightly differently (which also works in MATLAB). Adjust plot properties through the `gca` axes object directly after a plot is drawn instead of creating the `fig1` and `fig2` objects for the subplots. To remove the tick labels from the *x*-axis of the upper subplot, use `set(gca,'XTickLabel',[])`. Changing its position and size can be done *via* `P=get(gca,'Position'); L=P(1); B=P(2); W=P(3); H=P(4); set(gca,'Position',[L B-H*ve W H+H*ve])`.

Sometimes we would like to create subplots that span over multiple subplot areas. This can be achieved by listing the required subplot areas in square brackets as the third input argument of `subplot()`.

Exercise 3.3

A simple decomposition reaction was studied to determine its rate constant. Because both reactant (A) and product (B) absorb light in the UV and visible regions, the experimental runs were proposed to be monitored using a spectrophotometer. The below spectral data for the two species were recorded between wavelengths 200 and 800 nm at every 10 nm:

Absorbance of A: 0.064 0.125 0.224 0.352 0.475 0.548 0.551 0.507 0.457 0.437 0.455 0.501 0.561 0.624 0.682 0.731 0.769 0.793 0.803 0.803 0.801 0.812 0.844 0.891 0.920 0.890 0.784 0.624 0.457 0.319 0.222 0.159 0.116 0.085 0.062 0.044 0.031 0.022 0.015 0.010 0.006 0.004 0.003 0.002 0.001 0.001 0.000 0.000 0.000 0.000 0.000 0.000 0.000 0.000 0.000 0.000 0.000 0.000 0.000 0.000 0.000

Absorbance of B: 0.000 0.001 0.003 0.005 0.007 0.008 0.010 0.011 0.013 0.016 0.019 0.024 0.030 0.038 0.048 0.061 0.077 0.098 0.127 0.166 0.224 0.302 0.397 0.488 0.554 0.581 0.578 0.566 0.560 0.564 0.572 0.578 0.577 0.567 0.548 0.523 0.491 0.457 0.422 0.392 0.379 0.393 0.446 0.526 0.599 0.620 0.569 0.466 0.355 0.269 0.220 0.197 0.190 0.189 0.190 0.192 0.194 0.196 0.197 0.198 0.199

Following recording the UV-Vis spectra of species A and B, the two wavelengths at which their concentrations can be independently monitored were selected from the regions of no spectral overlap. A preliminary experiment was then conducted to collect absorbances at these wavelengths which were subsequently used to calculate the concentrations of A and B as the reaction progressed.

Create two composite figures to present the results. In the first figure, place the two separate spectra in the top row and the concentrations spanning across the bottom. In the second figure, place the separate spectra in the left column and the concentrations spanning across the right column. Label all subplots.

time/min:	0	1	2	3	4	5	6	7	8	9	10
[A]/mmol dm^{-3}:	10.26	6.52	4.67	3.08	1.91	1.58	1.14	0.81	0.45	0.26	0.13
[B]/mmol dm^{-3}:	0.01	3.71	5.62	7.21	8.32	8.61	9.19	9.35	9.68	9.93	10.03

Answer:

```
wl = 200 : 10 : 800;  % wavelength in nm
Abs_A = [0.064 0.125 0.224  0.352 0.475 0.548 0.551 0.507 0.457  0.437
0.455 0.501 0.561 0.624 0.682 0.731 0.769  0.793 0.803 0.803 0.801 0.812
0.844 0.891 0.920  0.890 0.784 0.624 0.457 0.319 0.222 0.159 0.116  0.085
0.062 0.044 0.031 0.022 0.015 0.010 0.006  0.004 0.003 0.002 0.001 0.001
```

```
0.000 0.000 0.000 0.000 0.000 0.000 0.000  0.000 0.000 0.000 0.000 0.000
0.000  0.000 0.000]; % absorbance of species A
Abs_B=[0.000 0.001 0.003 0.005 0.007 0.008 0.010  0.011 0.013 0.016
0.019 0.024 0.030  0.038 0.048 0.061 0.077 0.098 0.127  0.166 0.224 0.302
0.397 0.488 0.554 0.581 0.578  0.566 0.560 0.564 0.572 0.578 0.577 0.567
0.548 0.523 0.491 0.457 0.422 0.392 0.379 0.393 0.446  0.526 0.599 0.620
0.569 0.466 0.355 0.269 0.220 0.197 0.190 0.189  0.190 0.192 0.194 0.196
0.197 0.198 0.199]; % absorbance of species B
t=0:10; % time in minutes
% concentrations of A and B in mM at times given in t:
c_A=[10.26 6.52 4.67 3.08 1.91 1.58 1.14 0.81 0.45 0.26 0.13];
c_B=[0.01 3.71 5.62 7.21 8.32 8.61 9.19 9.35 9.68 9.93 10.03];
subplot(2,2,1)
plot(wl,Abs_A,'-b')
xlabel('wavelength/nm')
ylabel('absorbance')
legend('A','Location','southeast')
subplot(2,2,2)
plot(wl,Abs_B,'-r')
xlabel('wavelength/nm')
ylabel('absorbance')
legend('B','Location','southeast')
subplot(2,2,[3,4]) % subplot 3: horizontally spanning across
the bottom
plot(t,c_A,'cd',t,c_B,'m^')
xlabel('time/min')
ylabel('concentration/mmol dm^{-3}')
axis([0 10 0 11])
legend('[A]', '[B]','Location','east')

figure % create a new figure window (while keeping previous ones
unchanged)
subplot(2,2,1)
plot(wl,Abs_A,'-b')
xlabel('wavelength/nm')
ylabel('absorbance')
legend('A','Location','southeast')
subplot(2,2,3)
plot(wl,Abs_B,'-r')
xlabel('wavelength/nm')
ylabel('absorbance')
legend('B','Location','southeast')
subplot(2,2,[2,4]) % subplot 3: vertically spanning across the right
plot(t,c_A,'cd',t,c_B,'m^')
xlabel('time/min')
ylabel('concentration/mmol dm^{-3}')
axis([0 10 0 11])
legend('[A]', '[B]','Location','east')
```

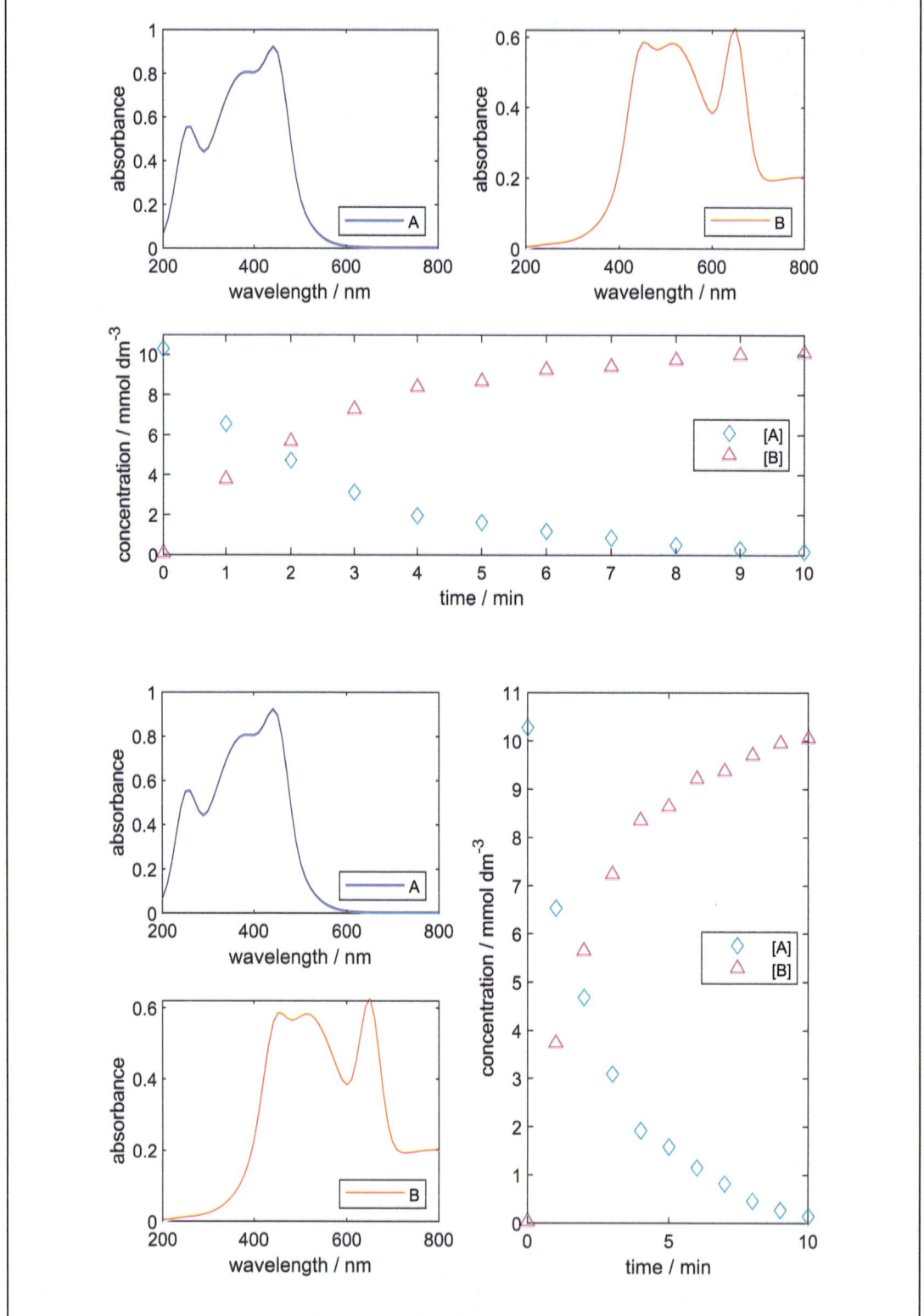

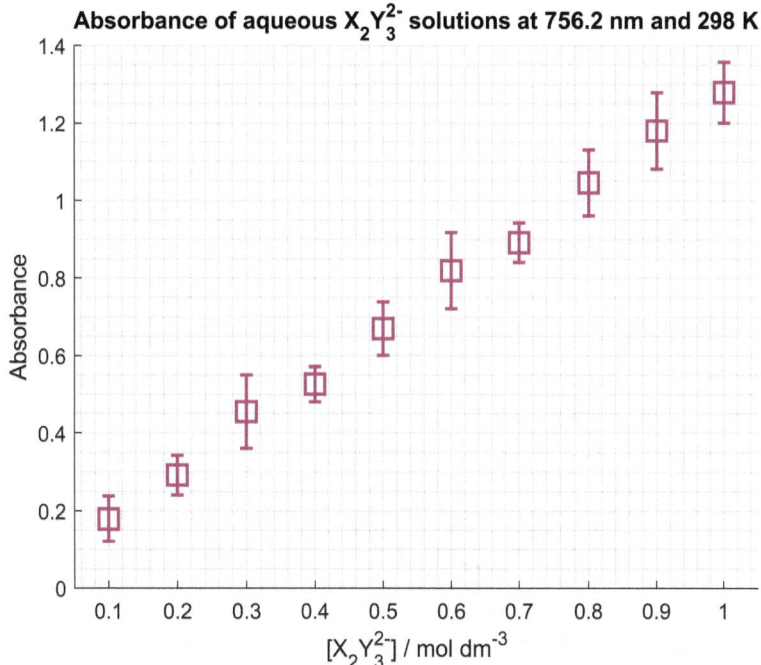

Figure 3.11 Show the uncertainty in your data. MATLAB makes it easy to add error bars to your data points.

3.4 Data Uncertainty in Plots

Graphical expression of uncertainty in data is usually done by adding error bars to data points in diagrams. For example, imagine that you have measured the absorbances in Exercise 3.1 multiple times and found that the absorbances for each concentration varied and spread around their average value. The extent of this variability can be effectively captured visually by drawing error bars at some estimated bounds below and above each data point. (We will discuss how these uncertainties are estimated in Chapter 4. Now, we are only concerned with presenting them on graphs.)

For displaying uncertainties in data points, the errorbar() function is used in MATLAB. It works very similarly to the plot() function. In our example, the uncertainties in the absorbance values will be stored in array abs_err which is passed to errorbar() as the third input array, following arrays conc (concentrations, x-values) and abs (absorbances, y-values). By default, error bars are drawn by MATLAB as vertical line sections centred on the data points as seen in Figure 3.11.

```
conc = [0.1  0.2  0.3  0.4  0.5  0.6  0.7  0.8  0.9  1.0];      %
concentrations in M
abs = [0.1785  0.2911  0.4545  0.5255  0.6688  0.8185  0.8908  1.0448
    1.1790  1.2785]; % Average absorbances
abs_err = [0.0584, 0.0511, 0.0945, 0.0455, 0.0687, 0.0985, 0.0507, 0.0847,
0.0990, 0.0785]; % must be the same length as the data it corresponds to
errorbar(conc,abs,abs_err,'ms','markersize',12,'linewidth',1.5)
```

```
xlabel('[X_2Y_3^{2-}]/mol dm^{-3}')
ylabel('Absorbance')
title('Absorbance of aqueous X_2Y_3^{2-} solutions at 756.2 nm and 298 K')
axis([0.05  1.05  0  1.4])
grid minor
box off
```

Exercise 3.4

Write a MATLAB script to create two calibration curves in one composite figure using the data below.

concentration = [0.01 0.02 0.03 0.04 0.05 0.06 0.07 0.08 0.09 0.1];
abs_1 = [0.16 0.24 0.40 0.52 0.62 0.75 0.85 0.98 1.13 1.21];
uncertainty_abs_1 = [0.02 0.03 0.02 0.02 0.02 0.03 0.02 0.03 0.04 0.02];
abs_2 = [0.11 0.15 0.21 0.27 0.33 0.37 0.43 0.50 0.55 0.63];
uncertainty_abs_2 = [0.03 0.01 0.02 0.02 0.03 0.01 0.02 0.04 0.01 0.02];

Add legends to the top left corners for "Chemical 1" and "Chemical 2". Use the same scales on both diagrams to emphasize the difference in absorbances between the two compounds. Because the y-axis label and tick labels in the subplot on the right are redundant, remove them and reduce the gap between the two subplots.

In case the ticks on the horizontal axes are too sparce, manually set the x-axis tick positions to 0, 0.02, 0.04, 0.06, 0.08 and 0.1 by overwriting the XTick properties of the subplots.

Answer:

```
concentration = [0.01  0.02  0.03  0.04  0.05  0.06  0.07  0.08  0.09  0.1];
abs_1 = [0.16  0.24  0.40  0.52  0.62  0.75  0.85  0.98  1.13  1.21];
uncertainty_abs_1 = [0.02  0.03  0.02  0.02  0.02  0.03  0.02  0.03  0.04  0.02];
abs_2 = [0.11  0.15  0.21  0.27  0.33  0.37  0.43  0.50  0.55  0.63];
uncertainty_abs_2 = [0.03  0.01  0.02  0.02  0.03  0.01  0.02  0.04  0.01  0.02];

he = 0.12; % horizontal increase in subplot size (corresponds to 12%)

fig1 = subplot(1,2,1);
errorbar(concentration,abs_1,uncertainty_abs_1,'rs','markersize',6,'linewidth',1)
xlabel('c/mol dm^{-3}')
ylabel('Absorbance')
axis([0  0.105  0  1.4]) % set scales manually to ensure they are the same in both plots
legend('Chemical 1','Location','northeast')
grid minor
% access and load the automatically-assigned position of fig1 subplot
% into 4 variables: L(left) B(bottom) W(width) H(height)
L = fig1.Position(1); B = fig1.Position(2);
```

```
W=fig1.Position(3); H=fig1.Position(4);
fig1.Position=[L B W+W*he H]; % increase width of fig1 (by 12% of original)
fig1.XTick=[0:0.02:1]; % set tick positions manually

fig2=subplot(1,2,2);
errorbar(concentration,abs_2,uncertainty_abs_2,'bo','markersize',6,
'linewidth',1)
xlabel('c/mol dm^{-3}')
axis([0 0.105 0 1.4]) % set scales manually to ensure they are the
same in both plots
legend('Chemical 2','Location','northeast')
grid minor
fig2.YTickLabel=[]; % set tick labels on y-axis to empty array
% access and load the automatically-assigned position of fig2 subplot
into 4 variables: L(left) B(bottom) W(width) H(height)
L=fig2.Position(1); B=fig2.Position(2);
W=fig2.Position(3); H=fig2.Position(4);
fig2.Position=[L-W*he B W+W*he H]; % increase width of fig2 (by 12% of
original)
fig2.XTick=[0:0.02:1]; % set tick positions manually
```

3.5 Plots with Two *y*-Axes

Convenient comparison between some data sets can be effectively aided by presenting them on charts with two different *y*-axes. MATLAB makes it very easy to create such arrangements. The set of commands generating graphic outputs relating to the *left y-axis* simply need to be preceded by the `yyaxis left` command,[§] while the group of commands generating graphic outputs associated with the *right y-axis* need to be issued after the `yyaxis right` command. Use the `colororder()` function to pre-define in one step the two colours that will be applied by MATLAB to the two separate groups of graphic objects corresponding to the different axes.

Exercise 3.5

Rault's Law predicts the partial vapour pressures of the components of a liquid mixture as $p_i = x_i p_i^*$, where p_i is the estimated partial vapour pressure of component i, x_i is its mole fraction in the liquid phase and p_i^* denotes the vapour pressure of pure liquid i. The law applies to ideal mixtures only, yet it is useful in assessing non-ideal behaviour of real mixtures. For that, measured partial vapour pressures are plotted together with pressures predicted by Rault's Law, which visually captures the deviation from ideality.

Calculate the ideal partial vapour pressures for compounds of A and B in a binary liquid mixture at the compositions (mole fractions of A) given below if the vapour pressures of pure liquid A and B are 85.2 Pa and 109.7 Pa, respectively. Create a diagram with two *y*-axes to display the predicted (ideal) and measured (real) partial vapour pressures together for each compound on a different axis. Use red colour for compound A (ideal: solid line, real: circles) and blue colour for B (ideal: solid line, real: upward-pointing triangles).

x_A: 0.00 0.05 0.10 0.15 0.20 0.25 0.30 0.35 0.40 0.45 0.50 0.55 0.60 0.65 0.70 0.75 0.80 0.85 0.90 0.95 1.00;
p_A/Pa: 0.7 9.2 16.1 25.4 33.5 40.1 46.2 53.9 59.1 64.1 67.2 71.9 76.2 79.4 81.1 81.8 84.6 83.9 85.4 86.2 86.5;
p_B/Pa: 111.4 108.0 106.6 102.3 101.3 97.9 93.4 87.3 82.9 77.1 74.0 68.7 62.8 55.1 48.5 39.6 35.2 25.7 16.6 10.9 1.6;

Answer:

```
% measured (real) partial pressures for A and B in Pa
pa_m = [0.7 9.2 16.1 25.4 33.5 40.1 46.2 53.9 59.1 64.1 67.2 71.9 76.2 79.4
81.1 81.8 84.6 83.9 85.4 86.2 86.5];
pb_m = [111.4 108.0 106.6 102.3 101.3 97.9 93.4 87.3 82.9 77.1 74.0 68.7
62.8 55.1 48.5 39.6 35.2 25.7 16.6 10.9 1.6];

xa = 0:0.05:1; % create mole fractions for A in liquid phase
xb = 1 - xa;   % mole fraction of B in liquid phase
pa0 = 85.2;    % vapour pressure of pure liquid A in Pa
pb0 = 109.7;   % vapour pressure of pure liquid B in Pa
pa_RL = xa*pa0; % predicted partial pressures of A using Rault's Law
pb_RL = xb*pb0; % predicted partial pressures of B using Rault's Law

% pre-assign colours for the two y-axes and their associated plotted
objects; make 1st/2nd axis and their associated plotted objects blue/red
colororder({'b','r'}) % colours are passed together as a cell array
```

[§] As of mid-2024, `yyaxis` is not yet implemented in Octave.

```
yyaxis left % left y-axis (for compound B)
plot(xa,pb_RL,'-') % colour already set (blue), only define line style
hold on
scatter(xa,pb_m,'o') % colour already set (blue), only define marker
type
ylabel('p_B/Pa') % xa=0 => pure B

yyaxis right % right y-axis (for compound A)
plot(xa,pa_RL,'-') % colour already set (red), only define line style
hold on
scatter(xa,pa_m,'^') % colour already set (red), only define marker type
ylabel('p_A/Pa') % xa=1 => pure A
xlabel('x_A')
```

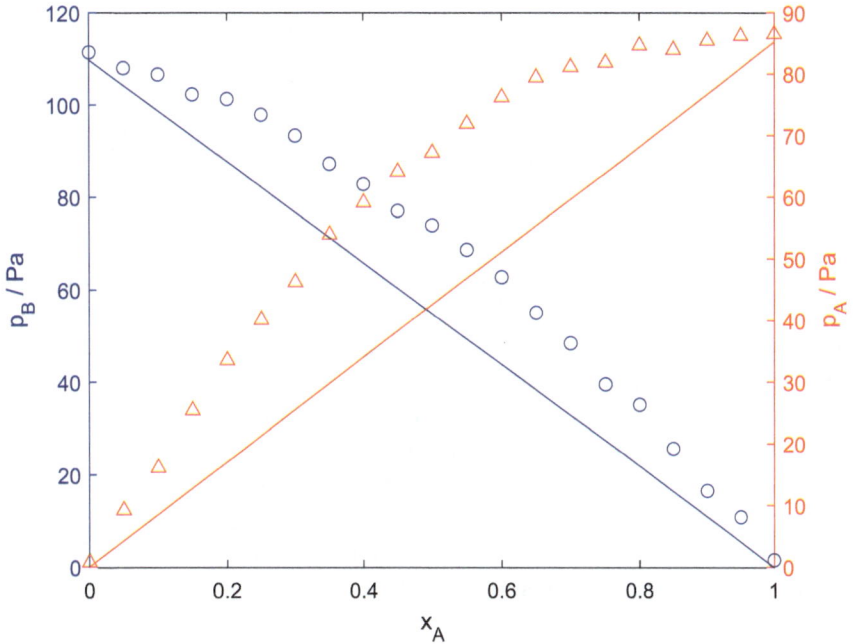

By comparing the predicted linear trends to the corresponding measured partial pressures, we observe that ideal behaviour is limited to the vicinities of $x_A = 0$ (where $x_B = 1$) and $x_A = 1$ (where $x_B = 0$); *i.e.*, when the liquid mixture is composed almost entirely of either substance B or A, respectively.

3.6 Histograms

When concerned about the reproducibility of a measurement or an experiment, we determine the same quantity or repeat the experiment under the same conditions multiple times and look at the spread of the resulting data. To best visualise them, we display the distribution of our data in histograms. These are bar charts concisely

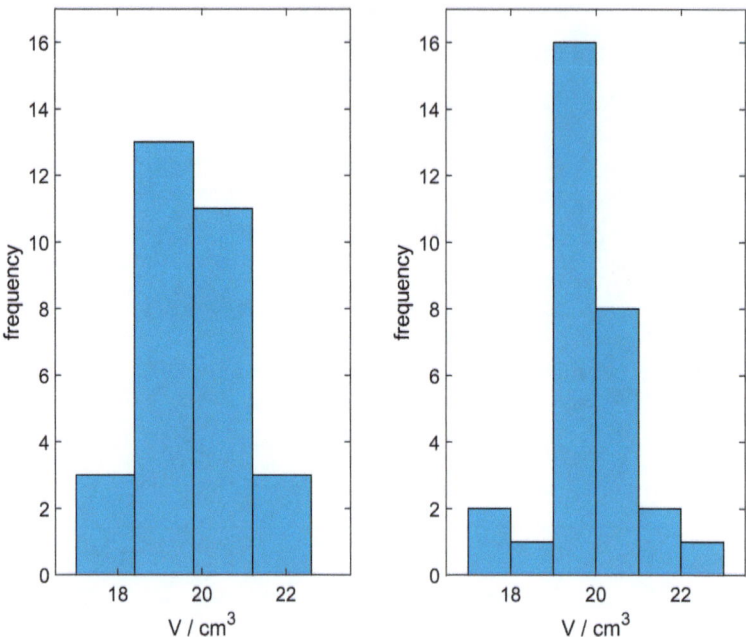

Figure 3.12 Histograms of titration end points. The `histogram()`[¶] function allows users to flexibly bin their data.

showing the number of data points laying within intervals, called bins, across the spread of our data set. Let's say a first-year cohort of chemistry students have pooled their synthesis yields or titration end points and they would like to display the distributions of these results. To establish the distribution, they partition the spread of the data points (yields or end points) into small domains and count how many data points fall into each domain, which is called data binning. A bar chart composed of rectangles, where the bases of the rectangles represent the binning domains and the heights of the rectangles represent the number of data points held within the bins.

In this example, we have 30 titration end points and we are drawing two histograms (Figure 3.12). By setting the number of bins, MATLAB will automatically define the bin boundaries for us (left). Manually setting bin boundaries is also possible (right).

```
v=[19.5  19.6  20.7  19.7  20.8  21.6  20.1  19.5  18.0  19.9  17.9  19.1
   19.1  20.5  19.8  19.9  19.1  19.0  17.1  19.1  20.9  20.2  19.0  19.1
   20.4  21.9  19.5  20.5  19.2  22.3];
% titration end points determined by the students

subplot(1,2,1)
histogram(v,4) % Create 4 bins, boundaries set by MATLAB
xlabel('V/cm^{3}')
ylabel('frequency')
axis([16.5 23.5 0 17])
subplot(1,2,2)
```

[¶] Use `hist()` in Octave.

Presenting Data in Chemistry: Plots and Charts 51

```
histogram(v,17:1:23) % Set bin boundaries manually:
17↔18↔19↔20↔21↔22↔23
xlabel('V/cm^{3}')
ylabel('frequency')
axis([16.5  23.5  0   17])
```

3.7 Appendix: Plot Stacking

Several data series that belong together are sometimes best displayed as stacked 2D graphs. For example, if we take spectroscopic measurements at regular time intervals during a reaction and we would like to display how the spectrum of the reaction mixture changes over time (to show the concurrent removal of reactants and formation of products).

In that case, the absorbance data collected at multiple wavelengths at multiple time points can be compactly stored in a 2D array, where each row (separated by semicolons) represents a spectrum taken at some time during the experiment (Figure 3.13). The accompanying wavelength data and time points are stored in separate 1D arrays. We first use `repmat()` to create a 2D wavelength and a time point array (having the same dimensions as the absorbance array) from the respective 1D arrays. We transpose the 2D arrays, so that each subsequent line plot in the final diagram will draw out a spectrum. As this overall diagram will be displaying data in 3D we will be using the function `plot3()`, the 3D version of `plot()`, with the 2D input arrays to be plotted column-wise. Often, we need to rotate the final diagram to make sure that key features are clearly visible. Here, the diagram shows how a species absorbing at 480 nm was removed while another species (initially not present) absorbing at 620 nm was produced in the reaction.

```
wl=[400  430  460  490  520  550  580  610  640  670]; % wavelength
A=[1.23e-04  1.83e-02  3.67e-01  1.05e+00  3.67e-01  1.83e-02  1.23e-02
1.12e-02  1.38e-02  2.31e-03;  9.87e-05  1.46e-02  2.94e-01  7.84e-01
2.94e-01  1.83e-02  7.36e-02  2.06e-01  7.35e-02  3.66e-03;  7.40e-05
1.09e-02  2.20e-01  6.03e-01  2.20e-01  1.83e-02  1.47e-01  4.02e-01
1.47e-01  7.32e-03;  4.93e-05  7.32e-03  1.47e-01  4.10e-01  1.47e-01
1.83e-02  2.20e-01  5.92e-01  2.20e-01  1.09e-02]; % absorbances
t=(0:10:30)'; % generate 1D time domain (make it a column vector by
transposing - using ()' - the inner 1D row array of time points)
TimePoz=repmat(t,1,10); % create 2D time array from 1D time points: repeat
column array t 10 times and load result into 2D array TimePoz
WLPoz=repmat(wl,4,1); % create 2D wavelength array from 1D wavelength
data: repeat row array wavelength 4 times and load result into WLPoz
% Now WLPoz, TimePoz and A are all arrays of 4 rows and 10 columns (4×10)
plot3(WLPoz',TimePoz',A') % plot3() reads in the 2D arrays as 10×4 arrays
thus they need to be transposed using (')
view([13,21]) % rotate 3D plot [azimuth angle, elevation angle]
xlabel('wavelength/nm')
ylabel('time/min')
zlabel('A')
```

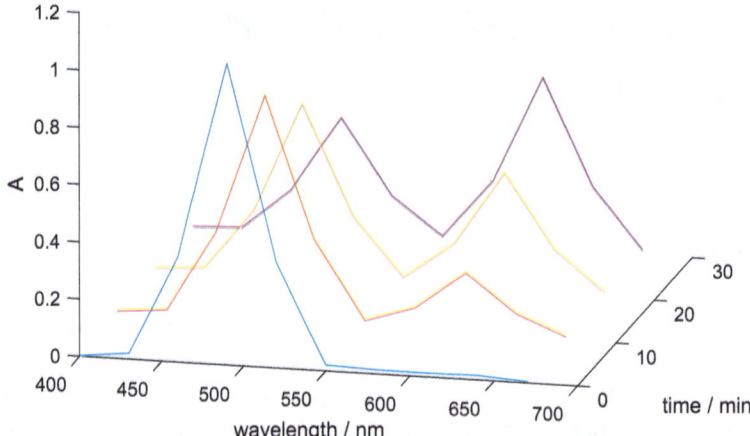

Figure 3.13 Stacked graphs can concisely and effectively capture the progression of reactions. It only takes a few lines of MATLAB code to generate them.

4 Curve Fitting in Chemistry

When carrying out experiments or synthetizing compounds, we usually have expectations regarding the outcome. These expectations are based on natural laws, theories and models discussed in textbooks and scientific articles, information available in databases, our previous observations and calculations, or a combination of these. Having completed an experiment or reaction, with all relevant data representing the outcome collected, we always critically examine our data, which includes comparing them with these expectations. When the results are hard to explain, we usually vary the conditions slightly to see some trends emerge in the outcome, which help us hypothesise what could have caused the puzzling observations. To test these hypotheses, we again need to compare predicted outcomes with the trends seen in experiment.

Comparing data collected with predicted results is at the heart of science and serves as the foundation of the *scientific method*. Curve fitting is a way of looking at what trend best describes our data and/or how closely the data follow what we expected based on established scientific laws, theories and/or hypotheses. Theoretical predictions and the laws of science most often come in the form of mathematical expressions. When we want to look at how closely our data match the predicted trends, we plot the data points along with the curves computed from the mathematical expressions known (or suspected) to play a key role in the phenomenon investigated. Often these expressions contain some parameters that we can freely adjust to make the predicted curves fit the data. These parameters are physical quantities or constants that determine, or are hypothesised to affect, the outcome of our experiments, but their values are specific to the particular experiment, chemical system or compound we are investigating.

For example, consider an experiment where we pass light through a solution of a light-absorbing substance (Figure 4.1). As the concentration (c) of the substance and the size of the container (ℓ) increase more light is observed to be absorbed by the solution.

The law that describes this phenomenon is called the Beer–Lambert Law and it is expressed mathematically as

$$A = \varepsilon c \ell$$

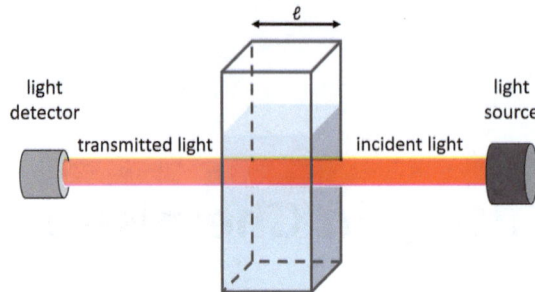

Figure 4.1 The intensity of light going through the cuvette decreases in the presence of light-absorbing substances in the cuvette.

where A is the absorbance of the solution, given as the intensity of the incident light (I_0) divided by the intensity of the transmitted light (I); $A = I_0/I$, and ε is called the molar absorptivity or molar extinction coefficient. At a constant container width, the linear dependence of A on the concentration is predicted to take place through parameter ε whose value is not pre-determined by the Beer–Lambert Law; every light-absorbing substance has a specific molar absorptivity at certain wavelengths (colour) of incident light. In order to establish the value of ε for a substance we prepare solutions containing different concentrations of the substance and measure the absorbance for each solution at a fixed wavelength and container width. Subsequently, we plot the absorbance values against the concentration values in a diagram, which reveals that the points are distributed along a line, as predicted by the Beer–Lambert Law. If we wish to obtain the best estimate of ε, we need to find the line which runs closest to the data points. The sum of the lengths of the vertical line segments connecting the linear curve and the data points can be taken as a measure of how close the absorbances are to the line with slope ε, as long as ℓ remains constant (Figure 4.2). Thus, the best estimate for the molar absorptivity will yield the linear curve with the vertical line segments above and below overall cancelling out.[†]

4.1 Fitting Curves to Data: *Lines*

What we were doing in the above example was – what is called – fitting a line to our data to establish the best estimate for the gradient which directly yielded the value of the molar absorptivity, ε, of the substance in question at the chosen wavelength of light. MATLAB makes it easy to fit lines and curves to data. One of the functions for fitting is fittingly called `fit()`.[‡] It requires the independent and dependent data sets as columns of numbers, as well as specifying the type of curve we wish to fit. (The arrays holding the values of the independent and dependent variables can be called whatever we prefer – we do not have to name them x and y.)

[†] This is one possible way to approach the problem. This can be thought of as the line of best fit ensuring that the sum of differences (positive and negative) between the data points and the line of best fit are equal to zero. Another approach, the Method of Least Squares, most often used to obtain linear fits (discussed in Chapter 5), requires the line of best fit to yield the smallest possible sum of squared differences.

[‡] As of mid-2024, `fit()` is not yet implemented in Octave. For fitting lines while calculating uncertainties in the slope and intercept (often used in undergraduate lab classes) see Section 6.7 or use `polyfit()`.

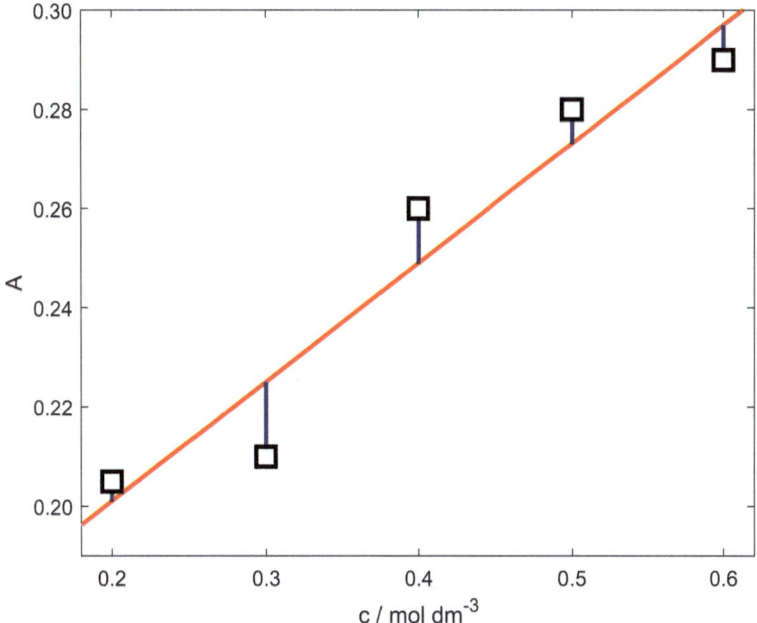

Figure 4.2 Data points scattered around a linear trend line.

To look at how the `fit()` function works, imagine that we would like to determine the activation energy of a reaction by using the linearised Arrhenius equation.

$$\ln k = -\frac{E_a}{R}\frac{1}{T} + \ln A$$

$$y = m\,x + c$$

We first run the reaction at multiple temperatures (T) and determine the rate constant (k) for each temperature which we load into arrays T and k as

```
>> T=[200  220  240  260  280  300  320  340  360  380  400];
>> k=[2.0e-04  6.5e-04  3.2e-03  4.7e-02  3.8e-01  8.8e-01  2.1e+00
1.2e+01  1.9e+01  7.9e+01  1.3e+02];
```

The equation above yields a line if we plot the natural logs of the rate constants ($\ln k$) against the inverse of the temperatures ($1/T$), therefore, in the next step we compute these as:

```
>> inv_T=1./T;
>> ln_k=log(k);
```

where we used the (./) element-wise division operator to instruct MATLAB to take each value in array T and compute its inverse. Now we call the `fit()` function to fit a line to the ln_k *vs.* inv_T data points and return the best estimates for the gradient and intercept. Please note that `fit()` requires column array inputs; therefore, we need to turn inv_T and ln_k, which are currently rows, into columns by transposing them using the (') operator. The third input argument `'poly1'` tells MATLAB to fit a line (a first order polynomial, hence the name `'poly1'`) to the data. The result of the fitting will be loaded into object `arrh_fit`

```
>> arrh_fit = fit(inv_T',ln_k','poly1')
arrh_fit =
    Linear model Poly1:
    arrh_fit(x)=p1*x+p2
    Coefficients (with 95% confidence bounds):
        p1 = -5683   (-6170, -5196)
        p2 = 18.92   (17.18, 20.66)
```

MATLAB has fitted a line described by the expression `p1*x+p2`, where `p1` and `p2` are the two parameters MATLAB estimated to produce the best possible linear fit to our data. Their values are listed under "`Coefficients (with 95% confidence bounds):`", `p1 = -5683` is the best estimate for the *gradient* and `p2 = 18.92` is the best estimate for the *intercept*. (Please note that the unit of the intercept is Kelvins, K, and that we must write it after the numeric value of the intercept in all our scientific communications.) The numbers in parentheses are their so-called *confidence bounds*. They specify a range for each fitting parameter within which the true value of that parameter is 95% likely to be found. Confidence bounds have the same units as the parameter they belong to. The domains between the confidence bounds are called *confidence intervals* and they are very useful and widely used, because their widths are indicative of how uncertain each estimated parameter is. In contrast, R^2 (coefficient of determination) often seen associated with fits provides only some generic information about the goodness of a fit. It only measures how well our data follow the proposed trend, *i.e.*, the fitted expression; therefore, it has limited relevance to our main objective: finding the fitting parameters and their uncertainties.

We now display the data points and the linear fit in a single command as the `arrh_fit` object can be directly passed to the `plot()` function along with the data points.

```
>> plot(arrh_fit,'-b',inv_T,ln_k,'rs')
>> xlabel('1/T/K^{-1}')
>> ylabel('ln k')
>> axis([2.4e-3  5.1e-3 -10   5.5])
```

The activation energy is calculated from the gradient of the fitted line seen in Figure 4.3. The gradient, fitting parameter `p1`, is a property of the object `arrh_fit` and can be accesses as `arrh_fit.p1` to be used to obtain the activation energy (gradient = $-E_a/R$, thus $E_a = -$gradient $\times R$) as

```
>> Ea =-arrh_fit.p1 * 8.3145
Ea =
    4.7252e+04
```

Given the units of the gradient, $(K^{-1})^{-1} = K$, and the gas constant, $J\,mol^{-1}\,K^{-1}$, the units of E_a are $KJ\,mol^{-1}\,K^{-1} = J\,mol^{-1}$. Hence, the sought activation energy is 4.7252×10^4 $J\,mol^{-1}$. (The number of decimal places fitting parameters should be rounded to will be discussed in Section 4.5.)

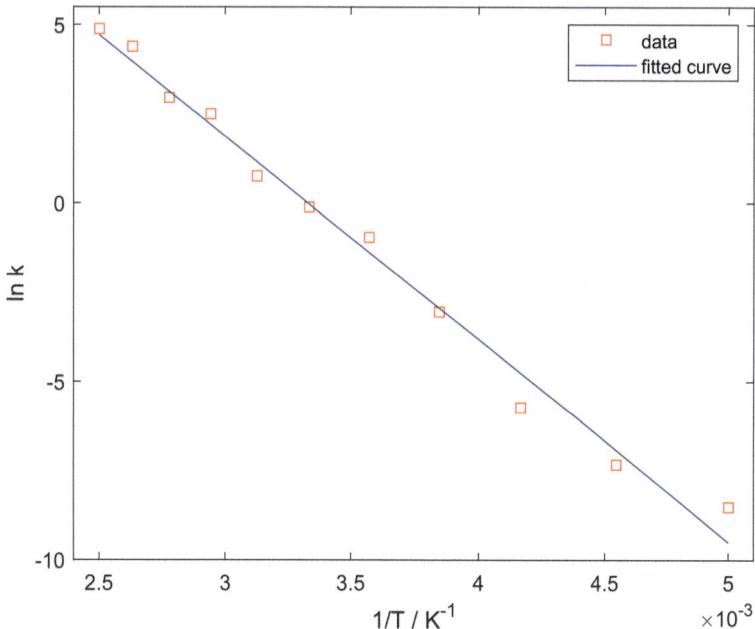

Figure 4.3 Arrhenius plot for a reaction whose activation energy can be calculated from the slope.

Note that in reality we are unable to calculate the true values of physical parameters, like the activation energy of a reaction. We can only estimate their values based on whatever data we have. 4.7252×10^4 J mol^{-1} is our best estimate only, and this value on its own (even with R^2 provided) tells us nothing about how good this estimate is. To have some idea about how close this estimated activation energy may be to the true value we use the confidence bounds or confidence intervals. We will discuss how to deal with uncertainties in more detail later, now we apply the above expression for calculating E_a to the confidence bounds of `p1` by substituting -6170 and -5196 in a single command into the expression in place of `p1` as

```
>> -[-6170 -5196] * 8.3145
ans =
    1.0e+04 *
    5.1300    4.3202
```

which means that, based on our temperature and rate constant data, the true value of E_a is 95% likely to be between 4.3202×10^4 J mol^{-1} and 5.1300×10^4 J mol^{-1}.

Fitting lines to experimental data is common in introductory chemistry laboratory classes. In the following example, we will apply what we have learnt so far to a spectrophotometry experiment.

Exercise 4.1

We measured the absorbances for a set of manganese concentrations at $\lambda = 585$ nm to create a calibration curve (using a cuvette with $\ell = 1.000$ cm). Fit a line to the data given below.

$c/10^{-4}$ mol dm^{-3}: 1.00　2.00　3.00　4.00　5.00　6.00　7.00　8.00　9.00　10.0

Absorbance:　0.2336　0.3396　0.5889　0.7654　0.9079　1.0838　1.2503　1.3832　1.5775　1.6971

Answer:

```
c=(1:10)*1e-4; % manganate ion concentrations
A=[0.2336 0.3396 0.5889 0.7654 0.9079 1.0838 1.2503 1.3832 1.5775
1.6971]; % Absorbances measured
f=fit(c',A','poly1') % fit linear calibration curve to Absorbances
plot(f,'-c',c,A,'rs')
legend('[MnO_4^-]','cal. curve',
'fontsize',12,'location','northwest')
axis padded
xlabel('[MnO_4^-]/mol dm^{-3}')
ylabel('A')
```

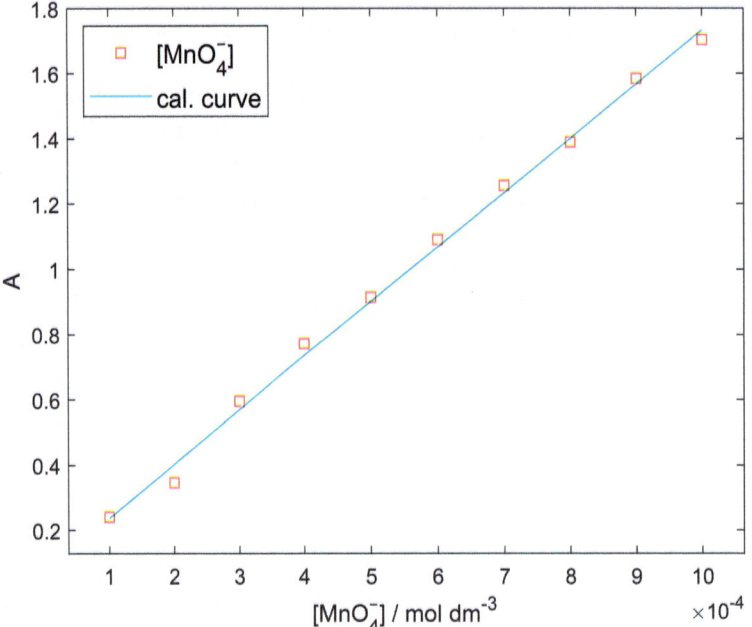

```
Linear model Poly1:
f(x)=p1*x+p2
Coefficients (with 95% confidence bounds):
  p1=1663 (1585, 1741)
  p2=0.06807 (0.01979, 0.1163)
```

The slope (p1) directly yields the molar absorption coefficient of manganese (ε = 1663 dm^3 mol^{-1} cm^{-1}) at 582 nm. Based on our measurements, the true value of ε is 95% likely to be between 1585 and 1741 dm^3 mol^{-1} cm^{-1}.

Curve Fitting in Chemistry

In the next example, we will look at how linear fits can be used to determine the position of a step change, which is useful in finding the stoichiometric point of a titration.

Exercise 4.2

0.010 M NaOH was added in 1.0 cm^3 amounts from a burette to a conical flask containing 50 cm^3 of acetic (or ethanoic) acid solution. The pH of the solution in the conical flask was monitored with a pH electrode. The following pH values were recorded for 0 to 50 cm^3 NaOH solution added:

pH: 3.21 3.39 3.54 3.68 3.75 3.86 3.96 4.07 4.14 4.21 4.29 4.36
4.39 4.46 4.54 4.57 4.61 4.68 4.75 4.86 4.93 5.04 5.14 5.29 5.46
5.82 6.71 8.57 10.14 10.43 10.54 10.64 10.75 10.79 10.86 10.89 10.93
10.96 11.00 11.04 11.07 11.11 11.11 11.14 11.14 11.18 11.18 11.21
11.21 11.25 11.25;

Write a MATLAB script that fits a line to the above pH data points in each of the three distinctly different linear regimes: slowly changing pH at low (1) and high pH (2), and the region characterised by rapid pH change in between them (3). From the resulting three equations, determine the stoichiometric or equivalence point (the volume of NaOH solution added for which the number of moles of CH$_3$COOH and NaOH in the flask are equal).

Answer:

At the stoichiometric point, the pH increase is fastest upon the addition of NaOH (as there is no more unreacted CH$_3$COOH left to react with it while buffering the pH increase). Determining the precise location of that point on the horizontal axis would require a smooth titration curve which we do not have. Thus, we can only estimate the location of the stoichiometric point. One strategy is to fit a line to the linearly-changing low pH values and another one to the linearly-changing high pH values, followed by fitting a third line to the series of most rapidly increasing pH values. After a quick visual inspection of our data (assign the values given above to array `pH`, then select the array in *Workspace* and click "Plot" in the PLOTS tab), we could decide to fit the first line to pH values 10 through 20, the second line to values 40 through 51, and the third one to values 26 through 29. Once the lines have been fitted, we extract the fitting parameters (the gradients and intercepts of the fitted lines) from the fit objects. Using the `num2str()` function, we convert these numerical values to their respective text characters, to be displayed in the figure legend incorporated into the expressions of the lines fitted to the pH values in the three different domains. Then, we display all results and expressions fitted.

In the last step, we algebraically work out the location of the equivalence point in the *Command Window* from the fitted equations displayed in the figure legend. We first calculate the two locations (V_L and V_H) on the horizontal axis where the third fitted line intersects the first and the second fitted lines, then we take the average of these two locations, which yields our estimate for the position of the stoichiometric point of the titration (V_{ST}).

```
pH=[3.21  3.39  3.54  3.68  3.75  3.86  3.96  4.07  4.14  4.21  4.29
4.36  4.39  4.46  4.54  4.57  4.61  4.68  4.75  4.86  4.93  5.04  5.14
5.29  5.46  5.82  6.71  8.57  10.14  10.43  10.54  10.64  10.75  10.79
10.86  10.89  10.93  10.96  11.00  11.04  11.07  11.11  11.11  11.14
11.14  11.18  11.18  11.21  11.21  11.25  11.25];
v=0:50; % generate NaOH solution volumes added to the flask
f1=fit(v(10:20)',pH(10:20)','poly1'); % fit line to flat low pHs
f2=fit(v(40:end)',pH(40:end)','poly1'); % fit line to flat high pHs
f3=fit(v(26:29)',pH(26:29)','poly1'); % fit line to steep mid pHs
% generate character strings in the format of 'pH=p1 V + p2' where p1
% and p2 are the fitting parameters (for slopes & intercepts). Unit of p1:
% cm^{-1}
eq1_expr=['pH= ',num2str(f1.p1),' cm^{-3} V + ',num2str(f1.p2)];
eq2_expr=['pH= ',num2str(f2.p1),' cm^{-3} V + ',num2str(f2.p2)];
eq3_expr=['pH= ',num2str(f3.p1),' cm^{-3} V - ',num2str(-f3.p2)];
plot(v,pH,'ks') % plot measured pHs
hold on
plot(f1,'-r') % plot line fitted to low pHs
hold on
plot(f2,'-b') % plot line fitted to high pHs
hold on
plot(f3,'-y') % plot line fitted to mid pHs
ylim([2.5  12])
xlabel 'V/cm^3' % short for xlabel('V/cm^3')
ylabel 'pH' % short for ylabel('pH')
grid on
legend('measured pH',eq1_expr,eq2_expr,eq3_expr,'Location',
'southeast')
```

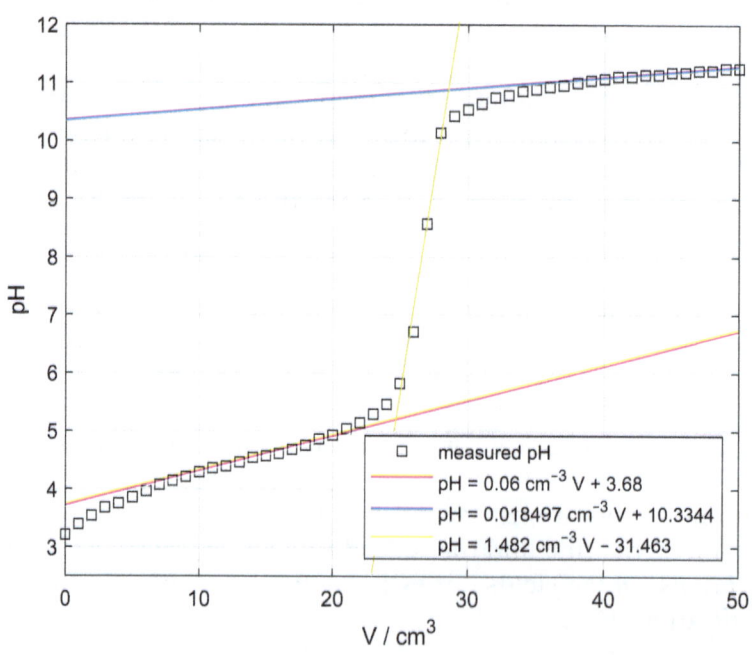

```
>> V_L=(3.68+31.463)/(1.482-0.06)
V_L =
    24.7138
>> V_H=(10.3344+31.463)/(1.482-0.018497)
V_H =
    28.5598
>> V_ST=mean([V_L V_H])
V_ST =
    26.6368
```

Computing the estimated stoichiometric point can be done in a single command (which we could include in our script) by taking the fitting parameters directly from the fit objects as

```
V_ST=mean([(f1.p2-f3.p2)/(f3.p1-f1.p1) (f2.p2-f3.p2)/(f3.p1-f2.p1)])
V_ST =
    26.6368
```

The same method can be used in calorimetry experiments to determine the size of a temperature step change – in the presence of temperature fore- and after-drifts – which is then used to calculate the enthalpy change of the process investigated.

4.2 Fitting Curves to Data: *Polynomials*

MATLAB can fit not only lines but also higher order polynomials. When data points are not distributed along a line or when theory suggests that the independent variable (x) should be raised to the second, third or higher powers to obtain the values of the dependent variable ($y = f(x)$), we instruct MATLAB to fit higher order polynomials to the data. In this example we have some data we would like to fit a linear, quadratic and cubic expression to and see which tracks the data points best.

```
x=[0.100  0.200  0.300  0.400  0.500  0.600  0.700  0.800  0.900  1.000];
y=[0.740  1.084  1.237  1.113  0.729  0.270  -0.251  -0.620  -0.758
-0.683];
f1=fit(x',y','poly1') % linear fit
f2=fit(x',y','poly2') % quadratic fit
f3=fit(x',y','poly3') % cubic fit
plot(f1,'-r',x,y,'k^') % plot results of linear fit
hold on
plot(f2,'-g') % plot results of quadratic fit
```

```
hold on
plot(f3,'-b') % plot results of cubic fit
legend('Data','linear fit','quadratic fit','cubic fit')
axis([0  1.1  -1  1.5]) % set axes limits to best scale data and fits in
figure
```

We find that the linear (first order polynomial: $y = p_1 x + p_2$) and the quadratic (second order polynomial: $y = p_1 x^2 + p_2 x + p_3$) expressions produce poor fits, while the cubic curve (third order polynomial: $y = p_1 x^3 + p_2 x^2 + p_3 x + p_4$) closely traces the data points (Figure 4.4).

4.3 Fitting Curves to Data: *Splines*

Occasionally, we encounter data sets that do not follow any distinct trend, yet we would like to add a curve to the diagram to aid the eye. In situations like this we use *splines* that are flexible curves that nicely follow the trend in the data whatever it may be. Please note that it is bad practice to use splines

- if an expression is available for fitting,
- just to connect your data points without explaining.

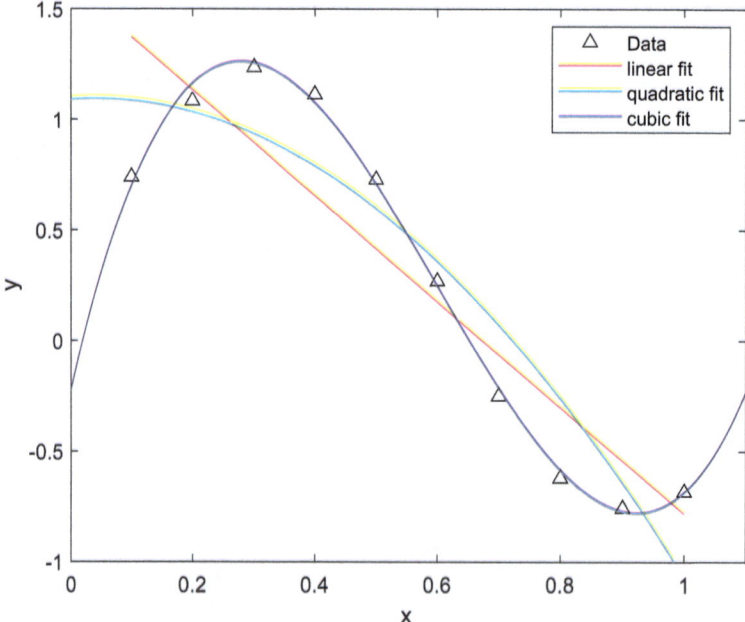

Figure 4.4 Different polynomials fitted to a set of data. The third-order polynomial (cubic) curve follows the data points closest.

Curve Fitting in Chemistry

Acknowledging the fact that splines can be useful sometimes (and that no doubt you see them often even when they are probably not appropriate) we look at how to generate them in MATLAB.

```
x=[0.100  0.200  0.300  0.400  0.500  0.600  0.700  0.800  0.900  1.000];
y=[0.047  0.112  0.151  0.184  0.154  0.026  −0.140  −0.344  −0.426
−0.431];
f3=fit(x',y','poly3') % cubic fit
fsp=fit(x',y','smoothingspline') % spline fit
plot(f3,'-b',x,y,'k^') % plot results of cubic fit
hold on
plot(fsp,'-m') % plot results of spline fit
legend('Data', 'cubic fit', 'spline')
axis([0  1.1  −0.5  0.3]) % set axes limits to best scale data & fits in
figure
```

The cubic curve poorly captures the trend of the data; the spline fitted, however, nicely tracks the data points (Figure 4.5). Note that while splines look appealing to the eye, they signal to the reader that you do not have a deep understanding of your data; otherwise, you would have fitted a specific expression.

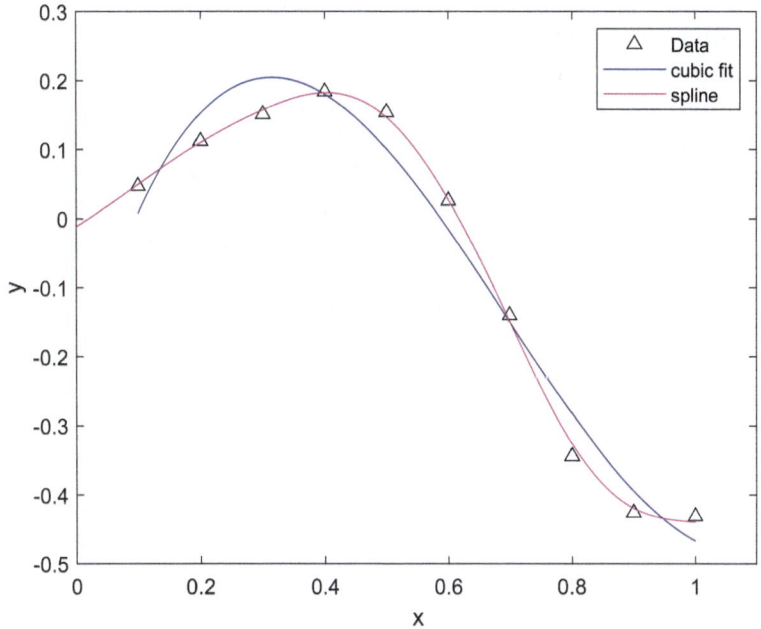

Figure 4.5 Spline fitted to data points. It nicely traces the data points aiding the eye; however, offering no understanding on what parameters determine the trend in the data and how.

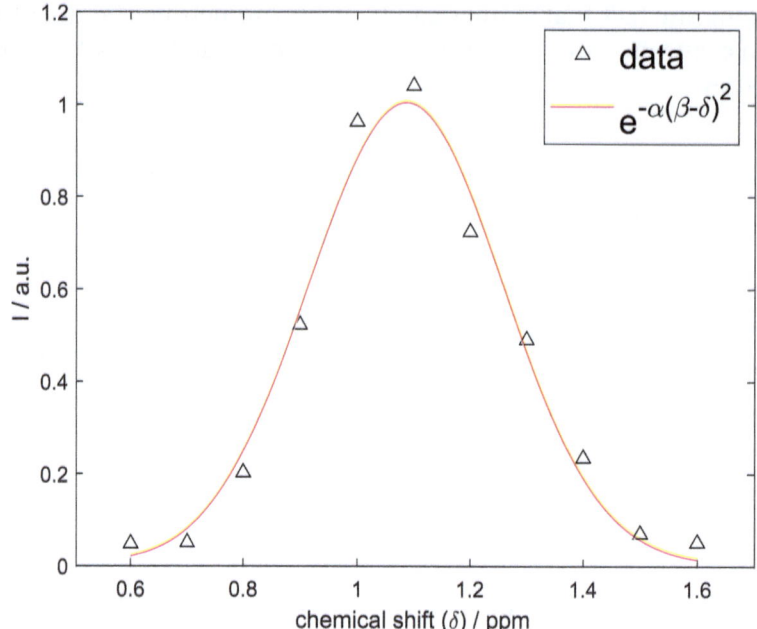

Figure 4.6 Data points with a bell-shaped trend. Peaks are common features of spectra and chromatograms.

4.4 Fitting Curves to Data: *Custom Expressions*

When we do have an expression that fits our data but it is not a polynomial, we can still use the `fit()` function.[§] All we need to do is supply the expression to MATLAB. In the following example we fit an expression to data points that trace out a peak, which we very often encounter in chemistry on chromatograms and spectra (Figure 4.6). Mathematically, peaks can be captured by exponential functions. Thus, we choose here to fit the expression $y = e^{-\alpha(\beta - x)^2}$ to our Nuclear Magnetic Resonance (NMR) intensities.

When fitting custom expressions, it is often useful to provide guidance to MATLAB on where to begin the parameter search. Therefore, after defining the custom function, we give MATLAB the starting values for α and β as `'startpoint', [1  1]`, which instructs MATLAB to launch the search for α and β at $\alpha_0 = 1$ and $\beta_0 = 1$. (This is by no means an educated guess, as we will see below; yet 1 is used often for initiating parameter searches for convenience.)

```
S=[0.600  0.700  0.800  0.900  1.000  1.100  1.200  1.300  1.400  1.500
1.600];
I=[0.049  0.052  0.203  0.523  0.963  1.041  0.725  0.491  0.235  0.071
0.052];
```

[§] In Octave, use `nlinfit()` which requires user-defined functions discussed in Chapter 9.

```
fc = fit(S',I','exp(-a*(b-x)^2)','startpoint',[1 1]) % custom function
fit
plot(fc,'-r',S,I,'k^') % plot results of custom function fit
legend('data','e^{-\alpha(\beta-\delta)^2}','fontsize',16)
axis([0.5 1.7 0 1.2]) % set axes limits to best scale data & fits
in figure
xlabel('chemical shift (\delta)/ppm'); ylabel('I/a.u.') % label axes
```

```
fc =
    General model:
    fc(x) = exp(-a*(b-x)^2)
    Coefficients (with 95% confidence bounds):
      a = 17.08 (14.41, 19.76)
      b = 1.087 (1.07, 1.103)
```

This script also shows how Greek letters can be accessed in MATLAB. In line 5, where we labelled the diagram we wrote \alpha, \beta and \delta to output "α", "β" and "δ", respectively. The list of Greek letters and special characters are available at: https://uk.mathworks.com/help/matlab/creating_plots/greek-letters-and-special-characters-in-graph-text.html

The horizontal axes of NMR spectra are reversed, which we can achieve by the command

```
>> set(gca,'XDir','reverse')
```

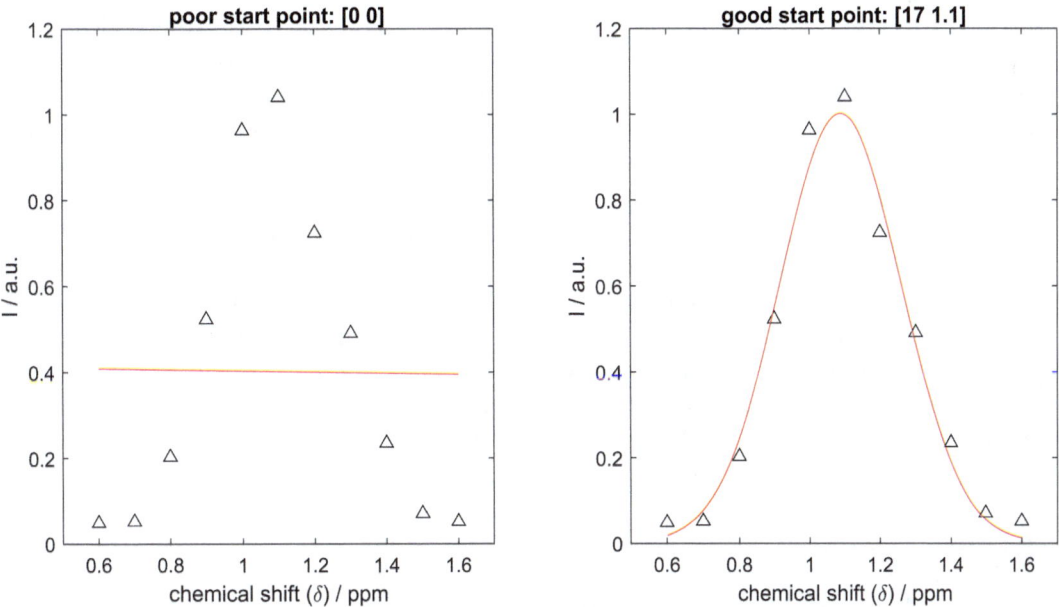

Figure 4.7 Bad fit, good fit. When fitting a non-polynomial, user-defined expression, a random parameter guess often results in a poor fit. This can be viewed as a great example of the RIRO (or GIGO) principle of computer science standing for "Rubbish In, Rubbish Out" (or "Garbage In, Garbage Out").

When the parameter search results in a poor fit, improved starting points are required. Instead of blind guessing, we need to carefully examine how each parameter affects the shape of the curve and set starting points close to their best-estimate values. In this expression, β directly determines the location of the maximum on the horizontal axis. The role of α is less straightforward. It is related to the broadness of the peak and can be given in terms of the width of the peak at half maximum (h) as $\alpha_0 = 4\ln 2/h^2$. Therefore, a quick visual inspection of the distribution of data points in the diagram is enough to roughly estimate α and β which then can be used as a suitable start point $[\alpha_0\ \beta_0]$ ($h \approx 1.3 - 0.9 = 0.4$, thus $\alpha_0 \approx 17$).

As Figure 4.7 shows, the success of finding the best estimates for fitting parameters is very sensitive to choosing the starting point for the parameter search. [1 1] was a lucky start point. As seen above, a slightly different start point, [0 0], results in a failed search.

Exercise 4.3

Using MATLAB, fit the expression

$$y = a + \frac{b}{1+e^{(c-x)}}$$

which is used for describing sigmoid step changes, to the titration data in Exercise 4.2 to determine the position of the equivalence point.

Answer:

In this expression, a controls where the base of a step change is on the vertical axis, b determines the size of the step change and c positions the midpoint of the step change along the horizontal axis which corresponds to the equivalence point. Because this expression produces a smooth transition between two horizontal lines, we should expect the fitted curve to only loosely follow the measured values in the linear low and high pH regimes. It should, however, trace the pH values closely in the vicinity of the stoichiometric point, and that is what matters most for titrations.

We pass the above expression to `fit()` as `'a+b/(1+exp(c-x))'`. Because our measured values span across approximately 7 pH units, the midpoint of the low pH domain is around pH 4 and the equivalence point is about 27 cm^3, we set the initial guesses for fitting parameters `a,b,c` to `[7 4 27]`. Then, as in Exercise 4.2, using the values of the fitting parameters we construct the character string of the equation for the legend, followed by displaying the results in a diagram. We also output the estimated stoichiometric point to the *Command Window*.

```
pH=[3.21  3.39  3.54  3.68  3.75  3.86  3.96  4.07  4.14  4.21  4.29
4.36  4.39  4.46  4.54  4.57  4.61  4.68  4.75  4.86  4.93  5.04  5.14
5.29  5.46  5.82  6.71  8.57  10.14  10.43  10.54  10.64  10.75  10.79
10.86  10.89  10.93  10.96  11.00  11.04  11.07  11.11  11.11  11.14
11.14  11.18  11.18  11.21  11.21  11.25  11.25];
v=0:50; % generate NaOH solution volumes added to the flask
```

Curve Fitting in Chemistry

```
f4 = fit(v(10:end)',pH(10:end)','a+b/(1+exp(c-x))','startpoint',
[7 4 27]);
eq4_expr = ['pH = ',num2str(f4.a),' + ',num2str(f4.b),'/
[1 + exp(',num2str(f4.c),' - V/cm^3)]'];
plot(v,pH,'ks')
hold on
plot(f4,'--m')
ylim([2.5 12])
xlabel 'V/cm^3'
ylabel 'pH'
grid on
legend('measured pH',eq4_expr,'Location','southeast')
V_ST = f4.c
```

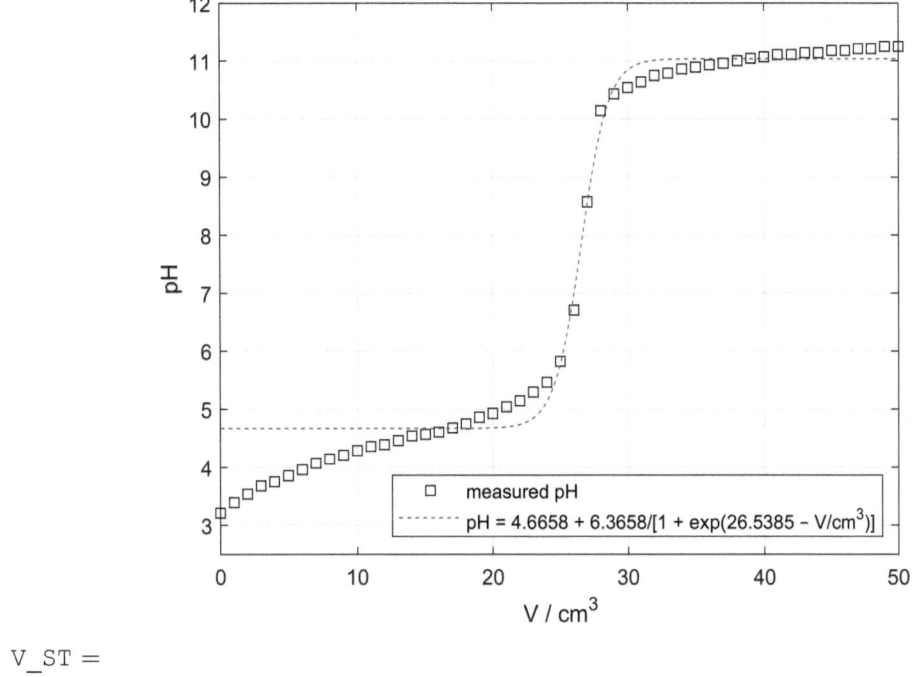

```
V_ST =
    26.5385
```

The difference between this and the equivalence point estimated in Exercise 4.2 is less than 0.1 cm³, which is an acceptable level of uncertainty given the uncertainties present in titrations (primarily steaming from the slight variabilities in glassware volumes).

4.5 Quoting Fitting Parameters

We often need to report the parameters obtained through fitting as they are usually some physical constants or quantities that describe some important aspects of a chemical system or phenomenon we are investigating. Consider β which determines what is called peak position, an important quantity across many areas of chemistry. We quote peak positions (and all fitting parameters and quantities we measure/determine) in the format of:

$$\textit{best estimate} \pm \textit{uncertainty of best estimate} \quad \textit{units (if applicable)}$$

where

- the *uncertainty* is rounded up to 1 significant figure (unless we have at least a hundred data points, in which case we round up to 2 significant figures; but this rarely occurs in undergraduate labs; we round up to avoid underestimating uncertainties)
- the *best estimate* is rounded to the same number of decimal places as the uncertainty.

For quoting the peak position, we need the best estimate of β and its uncertainty which we obtain from the confidence bounds. First, we load the confidence bounds into an array we name here `fit_conf_bounds` using the `confint()` function as

```
>> fit_conf_bounds = confint(fc)
fit_conf_bounds =
    14.4082    1.0703
    19.7616    1.1031
```

where the numbers in the *second* column, `1.1031` and `1.0703`, are the confidence bounds for β (the *second* parameter specified in the custom expression argument of `fit()` above). To give the uncertainty of the peak position in standard scientific format, we need half the range of the confidence interval. Therefore, we load in the numbers stored in column 2 of `fit_conf_bounds` and compute the size of the confidence interval as the modulus of the difference of the confidence bounds using the `abs()` function. We subsequently divide the range of the interval by 2 and load the result into an array we call here db. Writing all steps concisely in one line:

```
>> db = abs(fit_conf_bounds(2,2) - fit_conf_bounds(1,2))/2
```

yields $\Delta\beta$ as

```
db =
    0.0164
```

which we round to one significant figure by using the `round()` function

```
>> db_1_sf =round(Db,1,'significant')
db_1_sf =
    0.0200
```

Note that the `round()` function rounds to the nearest required digit, thus we need to double check that the result of rounding is correct, *i.e.* the uncertainty has been and if

MATLAB has rounded down we must round up. Because here rounding $\Delta\beta$ to 1 s.f. meant rounding to the second decimal place, we need to round β to 2 d.p.

```
>> b_dp=round(b,2,'decimal')
b_dp=
      1.0900
```

We can now quote the chemical shift in the format of "$\delta = \beta \pm \Delta\beta$ ppm" using b_dp and db_1_sf:

$$\delta = 1.09 \pm 0.02 \text{ ppm}$$

Exercise 4.4

We measured the concentration of compound A multiple times to monitor its decomposition. The process can be described mathematically with the expression $[A]=[A]_0\, e^{-kt}$, where $[A]$ is the concentration of A at any given time (t) through the course of the reaction, $[A]_0$ is the initial concentration of A at the start, $t=0$, and k is the rate constant in $\min^{-1}$. (i) Fit the above function to the tabulated values below and estimate the initial concentration of A and the rate constant of the decomposition reaction (2-parameter fit). (ii) Fix the initial concentration of A in the fitting expression and only estimate the value of the rate constant (1-parameter fit).

t/min:	0	1	2	3	4	5	6	7	8
	9	10	11						
[A]/mM:	1.568	1.124	0.858	0.747	0.475	0.459	0.339	0.313	0.168
	0.107	0.091	0.022						

Answer:

```
time=[0:11];
A=[1.568 1.124 0.858 0.747 0.475 0.459 0.339
0.313 0.168 0.107 0.091 0.022];
fa=fit(time',A','A0*exp(-k*x)','startpoint',[A(1) 1]) % 2-parameter
fit
fb=fit(time',A','1.568*exp(-k*x)','startpoint',[1]) % 1-parameter
fit
figure('Position',[100 100 800 400]) % set position and size of
figure on screen [left bottom width height]
subplot(1,2,1)
plot(fa,'-b',time,A,'>m')
xlabel('time/min')
ylabel('[A]/mmol dm^{-3}')
axis padded
legend('[A]','[A]_0e^{-kt}')
title('2-parameter fit')
subplot(1,2,2)
plot(fb,'-b',time,A,'>m')
xlabel('time/min')
```

```
ylabel('[A]/mmol dm^{-3}')
axis padded
legend('[A]','[A]_0e^{-kt}')
title('1-parameter fit')
```

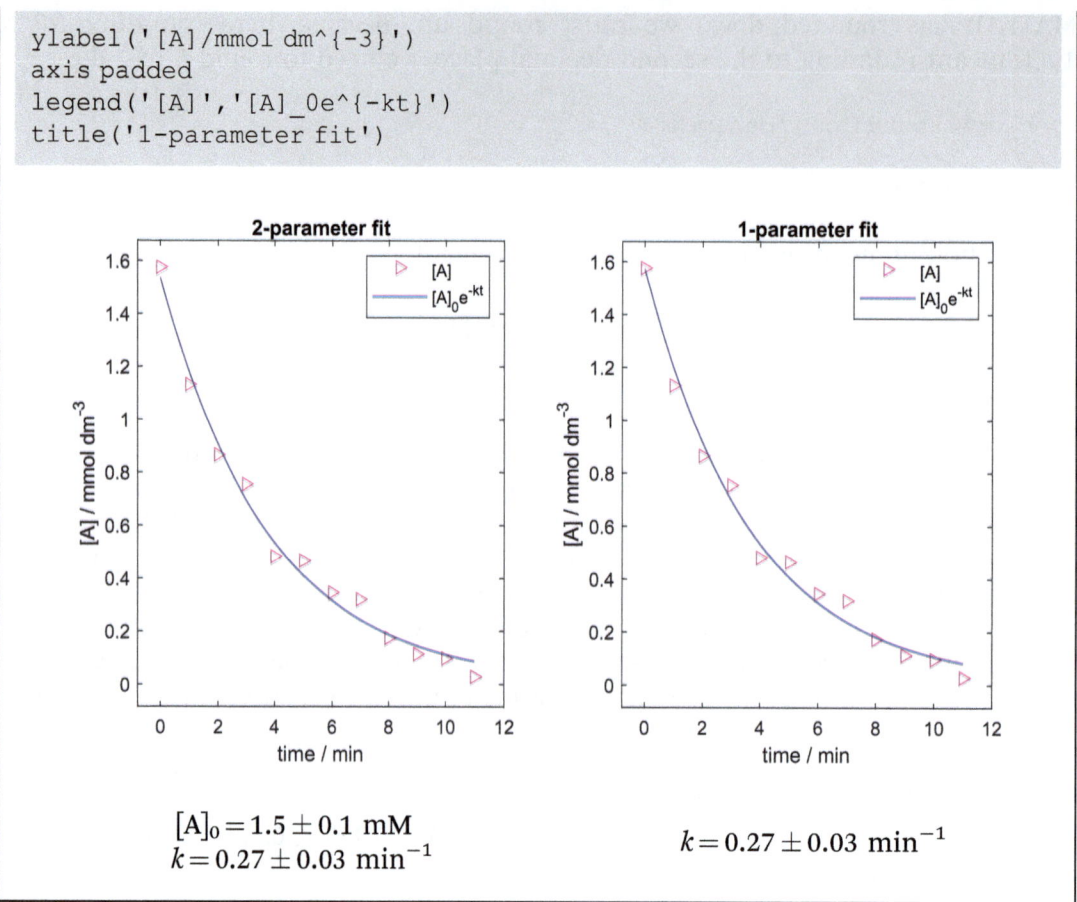

$[A]_0 = 1.5 \pm 0.1$ mM
$k = 0.27 \pm 0.03$ min^{-1}

$k = 0.27 \pm 0.03$ min^{-1}

Exercise 4.5

Using the data in Exercise 4.4, the natural logarithm of [A] plotted against t yields a linear trend. Fit a line to the ln([A]) vs. t points and determine the value of the rate constant *without* fixing [A]$_0$ (a) and *with* fixing [A]$_0$ (b).

Answer:

```
time=[0:11];
A=[1.568  1.124  0.858  0.747  0.475  0.459  0.339  0.313  0.168  0.107
0.091  0.022];
ln_A=log(A);
fa=fit(time',ln_A','poly1') % (a) without fixing ln[A]0 (2-parameter
fit)
fb=fit(time',ln_A','0.4495-k*x','startpoint',[1]) % (b) With ln[A]0
fixed; ln[A]0=0.4495 (1-parameter fit)
figure('Position',[100 100 800 400]) % set position and size of
figure on screen [left bottom width height]
subplot(1,2,1)
plot(fa,'-b',time,ln_A,'>m')
xlabel('time/min')
```

```
ylabel('ln[A]')
axis padded
legend('[A]','ln[A]_0 - kt')
title('2-parameter fit')
subplot(1,2,2)
plot(fb,'-b',time,ln_A,'>m')
xlabel('time/min')
ylabel('ln[A]')
axis padded
legend('ln[A]','ln[A]_0 - kt')
title('1-parameter fit')
```

$\ln[A]_0 = 0.6 \pm 0.5$
$k = 0.33 \pm 0.07 \text{ min}^{-1}$

$k = 0.30 \pm 0.04 \text{ min}^{-1}$

Notice that the uncertainties are bigger than those previously obtained when fitting custom exponential expressions. This is because linearising our data by taking the natural logarithm of the concentrations enlarges the differences between the line fitted and the data points at low concentrations, which affects the fit.

Fitting parameters obtained *via* non-linear fitting usually have smaller uncertainties than those attained by linear fitting to the corresponding linearised data.¶

¶ Here, for example, if the measured absorbances were 1.1 and 0.2, and the corresponding predicted absorbances (on the fitted line) were 1.0 and 0.1, the differences would be 0.1 in both cases; however, linearisation by taking natural logs would yield differences $\ln(1.1) - \ln(1.0) = 0.0953$ and $\ln(0.2) - \ln(0.1) = 0.6931$. This can be compounded by the low measured values falling close to the detection limit of the instrument where the relative scatter of data points usually sharply increases.

Exercise 4.6

In their research article, Koe, J.R.; Fujiki, M.; Motonaga, M.; Nakashima, H. *Macromolecules* 2001, **34**, 1082, the authors report the following potential energies as a function of dihedral angle for a particular polysilane.

dih. ang/°:	139	149	155	160	165	169	175
	180	185	194	200	204	210	219
E_{pot}/kJ mol^{-1}:	−879	−2322	−2054	−2151	−2201	−2054	−1322
	−1247	−1711	−2322	−2582	−2565	−2540	−1125

Fit a fourth order polynomial (quartic) curve to the data set and estimate the parameters of the double-well potential.

Answer:

```
ang=[139 149 155 160 165 169 175 180 185 194 200 204 210 219];
E=[-879 -2322 -2054 -2151 -2201 -2054 -1322 -1247 -1711 -2322 -2582 -2565 -2540 -1125];
f=fit(ang',E','poly4')
plot(f,'-r',ang,E,'bd')
xlabel('dihedral angle/deg')
ylabel('E_{pot}/kJmol^{-1}')
legend('data','Quartic fit')
```

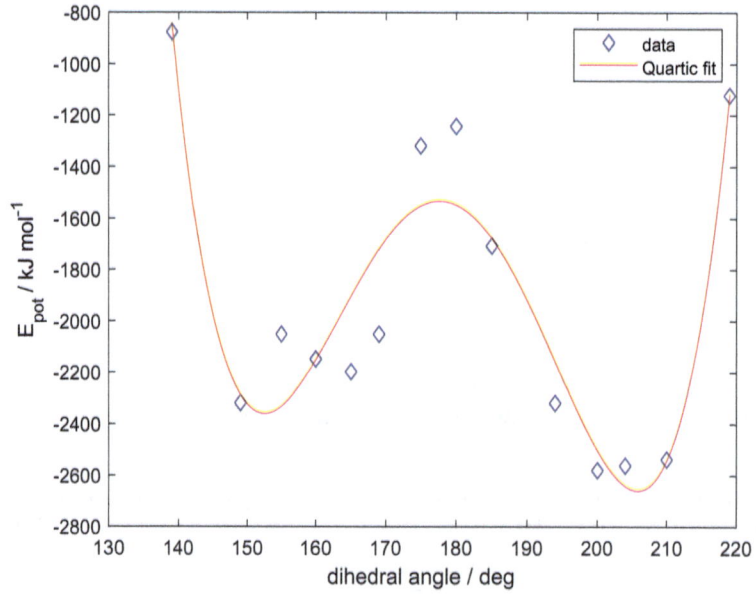

4.6 Appendix: Displaying Data Points Neglected from Fitting

Sometimes we would like to display all our data, including even those data points we decided to disregard during the fitting. With MATLAB, we can conveniently exclude data points from the fit while still displaying them on the same diagram using a different marker to distinguish them from the data points used for the fitting (Figure 4.8). To improve the fit in Exercise 4.4, we could exclude data points 8 and 12 from the fitting by extending the previous argument of `fit()` with " ,'exclude', [8 12] " as

```
fa=fit(time',A','A0*exp(-k*x)','startpoint',[A(1) 1],'exclude',[8 12])
```

For distinguishing the neglected data points from the rest we add " ,[8 12],'k*' " to the argument of `plot()` as

```
plot(fa,'-b',time,A,'>m',[8 12],'k*')
```

As a result, we obtain a slightly improved estimate for $[A]_0$ (1.53 ± 0.09 mM, notice the decreased uncertainty/narrowed confidence interval) and the diagram below.

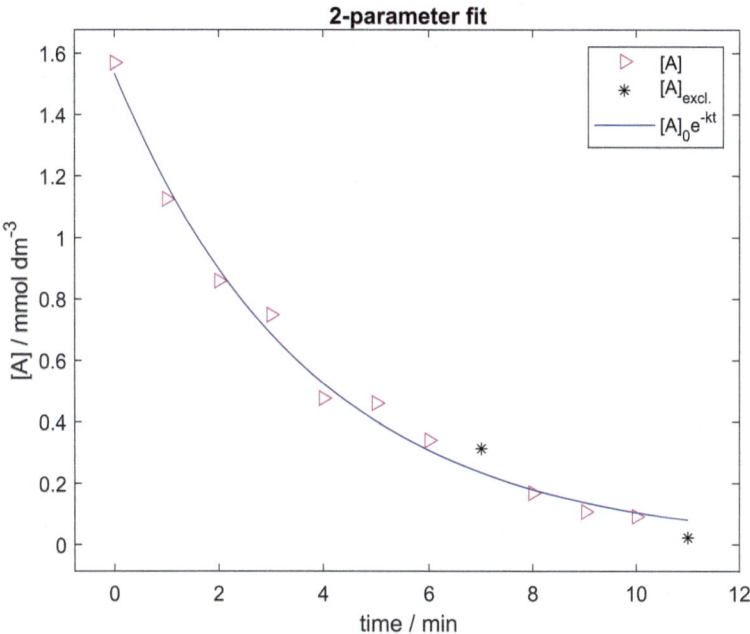

Figure 4.8 Data points excluded from fitting (*) displayed with the rest of the data. This can give readers a better sense of the variability in your experimental method and your data analysis choices. It is quick and easy in MATLAB to assign different markers to selected data points in a data set.

5 Measurements and Their Uncertainties

Imagine that we are tasked with investigating a reaction where bispyrenes of various lengths form. We have decided to analyse the results using atomic force microscopy because this powerful imaging technique is able to resolve the structure of single molecules. (This imaging technique is able to directly scan molecules trapped on a surface using electrostatic and magnetic forces between the molecules and a slowly moving sharp tip only a few atoms wide.)

Given that we are able to directly count the number of CH_2 units in the hydrocarbon linkers between the two pyrene ends in each product molecule, we could in theory establish the exact composition of the product mixture. All we need to do, once the reaction has reached completion, is trap all product molecules on a surface and scan them (Figure 5.1). Let us say a total of 3482 molecules were produced in a nano droplet reactor. When we count the number of CH_2 units in each molecule on the AFM scan of the entire surface, we obtain 3482 numbers. Each number, $n_{(CH2)}$, represents how many CH_2 units were found in one particular molecule within the population of 3482 molecules produced in the reaction. For this nano batch, we find that $n_{(CH2)}$ ranges between 2 and 22. Despite the wide range in the number of CH_2 units in the aliphatic linkers, we find that most often the product molecules have hydrocarbon chains between 8 and 15 CH_2 units long. To quantitatively display the length distribution of the aliphatic linkers within the population of our 3482 product molecules we draw a histogram, as shown in Figure 5.2. The horizontal axis shows the number of CH_2 units and on the vertical axis we have how many molecules containing that number of CH_2 units were found amongst the product molecules.

Not impressed with this result, we decide to re-run the reaction more carefully hoping to obtain a narrower spread in the number of CH_2 units. After the second, more controlled nano batch, we scan the entire population of product molecules and find that we have produced not only more (4159) molecules, but also that the length range of aliphatic linkers has narrowed as seen in Figure 5.3.

The shapes of the two histograms appear to be very similar, both bell-shaped. In fact, as we are about to see, the same mathematical expression can be fitted to both histograms.

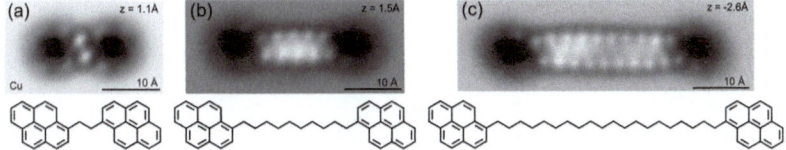

Figure 5.1 AFM images and the corresponding structures of bispyrene products with aliphatic linkers composed of 2 (a), 10 (b), and 20 (c) CH_2 units (adapted from *Chem. Sci.* 2017, 8, 2315). By counting the white spots representing CH_2 units in the AFM scans we can directly determine the number of CH_2 units making up the hydrocarbon linker in each bispyrene molecule scanned.

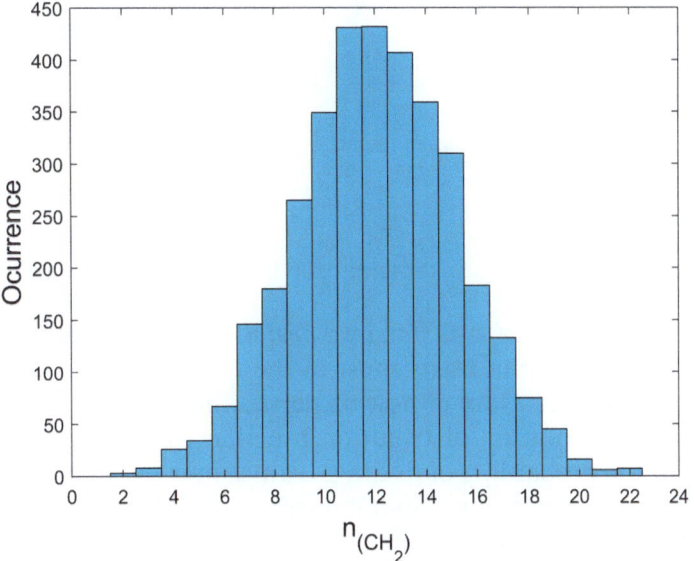

Figure 5.2 Distribution of aliphatic linker lengths in the first nano batch reaction.

5.1 Normal Distribution

Histograms like the ones above are ubiquitous in science. They crop up everywhere, which can be explained by what we are already familiar with from our chemistry studies. Just like chemical reactions, all phenomena are influenced by a wide range of factors which we are unable to control to an arbitrarily high degree. This lack of absolute control over every conceivable parameter of processes (for example, chemical reactions, observations of natural phenomena or even simple measurements) results in unavoidable random interferences from a multitude of sources which affect the outcome of experiments (and observations). Because these random fluctuations tend to have bell-shaped distributions, the quantities we measure in relation to processes we design or observe become similarly distributed under the influences of these fluctuations. This observation is formalised in the *central limit theorem* which underpins modern data analysis.

All bell-shaped distributions, also called *normal* or *Gaussian* distributions, can be mathematically described by the expression

$$f(x) = \frac{1}{\sigma\sqrt{2\pi}} e^{-\frac{(x-\mu)^2}{2\sigma^2}}$$

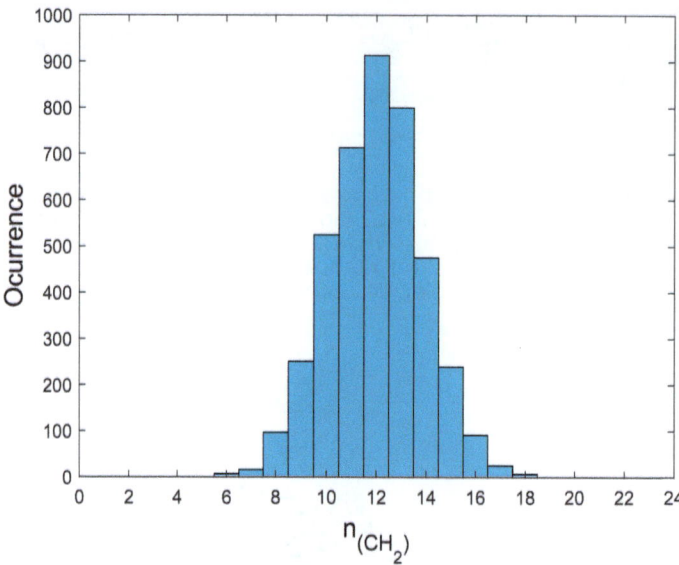

Figure 5.3 Distribution of aliphatic linker lengths in the second nano batch reaction.

where μ and σ are called *mean* and *standard deviation*, respectively. The mean defines the location of the peak on the x-axis most of the measurements cluster around. The standard deviation is the measure of how dispersed the measured values are about the mean. The graph of the normal distribution function traces out the frequency of the occurrences of – or the probability of finding – values in measurements or experimental outcomes. Values are most frequently obtained in the vicinity of the mean, and the probability of encountering a value quickly drops moving away from the mean.

An important property of the normal distribution is that about 2/3 of the obtained/observed values (68%) are within the $\mu \pm \sigma$ interval. In our example, if scanning all the bispyrene molecules produced in a nano batch leads to the number of CH_2 units narrowly clustering about the mean of the number of CH_2 units, the peak of the bell-shaped curve fitted to the histogram of CH_2 units will be narrow and tall.[†] As a result, σ for that batch will be small as 2/3 of the observed number of CH_2 units will lie very close to the mean value of CH_2 units. However, if the observed number of CH_2 units scatterers widely about the mean, the peak of the bell-shaped curve fitted to the histogram of CH_2 units will be broad and small (less sharp), resulting in a larger σ. A high degree of control of a reaction will lead to a small σ, whereas a large σ will indicate lack of control. This makes the mean and standard deviation ($\mu \pm \sigma$) for a run suitable for capturing the characteristics of the outcome in terms of the distribution of linker lengths. In Figure 5.4, the effect of σ is shown; as the standard deviation grows the distribution broadens and the height of the peak reduces.

It is standard practice to report the characteristics of the distribution of a quantity measured across the entire population of entities in the format of $\mu \pm \sigma$, once the mean and

---

[†] The outcome of a reaction in our scenario is described by a set of whole numbers each representing the number of CH_2 units linking the pyrene ends of a particular product molecule. This means that $n_{(CH_2)}$ can only take discrete values. Despite being a continuous function, the normal distribution can be (and is often) used to describe the distribution of discrete outcomes.

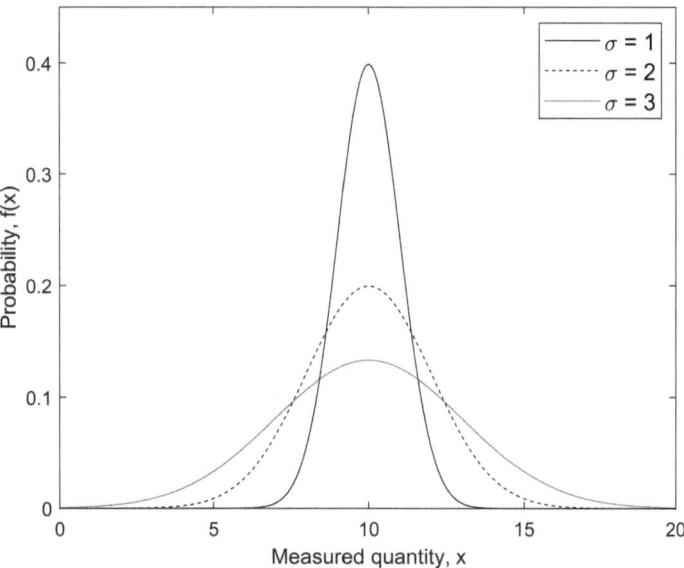

Figure 5.4 Bell-shaped distributions around the same mean with different standard deviations. The areas under the curves are all equal to 1, which is an important property of probability distribution functions.

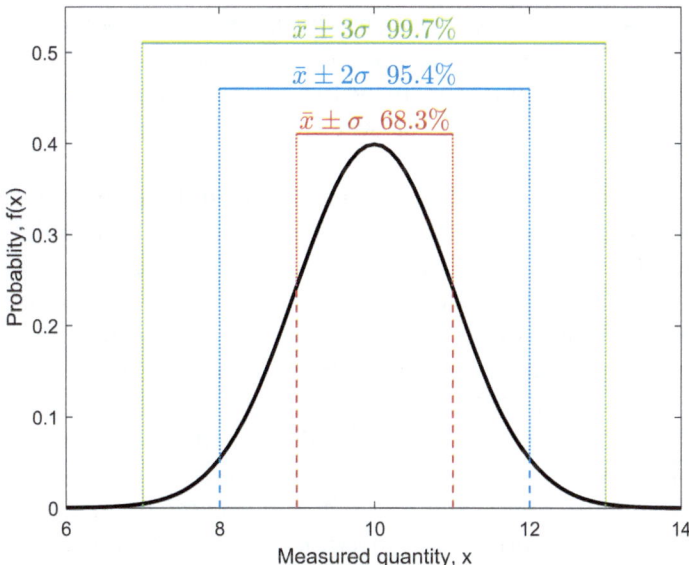

Figure 5.5 Probability of finding a value within domains centred around the mean (areas under the curve between the dashed lines) for an item randomly selected from a population with a bell-shaped distribution.

the standard deviation are determined. These immediately tell us the value about which roughly 2/3 of the observed values cluster. (It also enables us to draw the graph of the bell-shaped distribution that best fits the histogram of the data.) The $\mu \pm 2\sigma$ or $\mu \pm 3\sigma$ formats are also commonly used. In the former case about 95% of the data are found within the specified range, whereas in the latter case the given interval contains over 99% of the data as shown in Figure 5.5 for the solid curve ($\mu = 10, \sigma = 1$) in Figure 5.4. These percentages

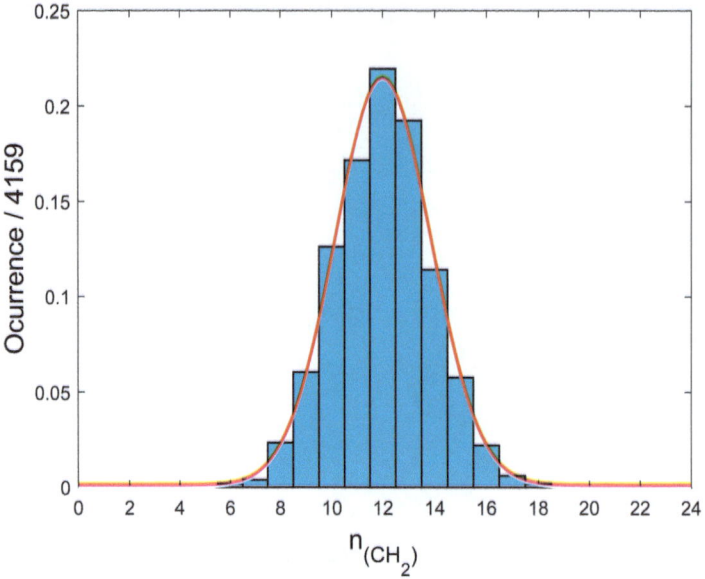

Figure 5.6 Normal distribution curve fitted to the results of the second nano batch. Each total count of molecules with a given aliphatic linker length, n_{CH2}, was divided by the total number of molecules produced (4159).

also mean that if we were to randomly pick a value (x) from the pool of data collected across the entire population of entities, the value would be roughly 68% likely to be between $\mu - \sigma$ and $\mu + \sigma$ ($\mu - \sigma < x \leq \mu + \sigma$). Similarly, this value would be about 95% likely to be between $\mu - 2\sigma$ and $\mu + 2\sigma$ ($\mu - 2\sigma < x \leq \mu + 2\sigma$) or over 99% likely to be between $\mu - 3\sigma$ and $\mu + 3\sigma$ ($\mu - 3\sigma < x \leq \mu + 3\sigma$).

In our experiment, we saw that the number of CH_2 units widely varied, yet most often we observed on our AFM scans molecules having around 12 CH_2 units (408 out of 3842 molecules in batch 1 and 951 out of 4159 molecules in batch 2.) Let us look at the second nano batch and see what happens when we divide the number of occurrences by the total number of molecules synthesized. What we obtain is a diagram displaying the occurrences relative to the total number of molecules, which shows that the molecules with 12 CH_2 units account for about 22% (0.22) of the entire population of product molecules. (This also means that if we were to randomly take one molecule from the product mixture it would be 22% likely a bispyrene molecule with a linker 12 CH_2 units long.) The length distribution of the aliphatic linkers in the second nano batch can be approximated with a normal distribution with *mean*, $\mu = 12$, and *standard deviation*, $\sigma = 1.8$ as indicated by the red curve superimposed on the relative occurrences in Figure 5.6.

5.2 Population Mean and Standard Deviation

For a complete data set with normal distribution (like the lengths of aliphatic linkers within the entire population of product molecules) the *mean* (μ) of the values measured is calculated as

$$\mu = \frac{1}{N} \sum_{i=1}^{N} x_i$$

where the N is the size of the data set (the total number of bispyrene molecules produced in a particular reaction) with x_i representing an individual data point (a single bispyrene molecule in the product mixture). To calculate the mean length of aliphatic linkers, we need to add up the number of CH_2 units found in each molecule and divide the sum by the total number of molecules produced. We should obtain a length very close to 12.

The standard deviation for the entire population of data points (a measure of how spread the aliphatic linker lengths are around the mean aliphatic linker length; or more specifically, the average deviation of the linker lengths from the population mean linker length) is calculated as

$$\sigma = \sqrt{\frac{\sum_{i=1}^{N}(x_i - \mu)^2}{N}}$$

(The population mean and standard deviation calculated using these formulae are the same as those obtained by fitting the normal distribution expression to the histogram produced from the data collected.)

In MATLAB, the **population mean** and **population standard deviation** are calculated by using the mean() and std() functions. *For the sake of simplicity let us assume that we had a particularly poor yield in one nano batch resulting in only 20 bispyrene molecules.* We look at the sparse AFM scan and count the number of CH_2 units in all the 20 molecules formed. We then calculate the mean and standard deviation of the aliphatic linker length for the population of bispyrenes as shown below.

```
n_CH2 =[13   15   8   14   13   10   11   13   16   17   10   15   13   12   13   12
12   15   11   13];
mp_n_CH2 =mean(n_CH2) % population mean
sp_n_CH2 =std(n_CH2,1) % population standard deviation


mp_n_CH2 =
    12.8000
    sp_n_CH2 =
    2.1354
```

Please note that we used std(n_CH2,1) to calculate the population standard deviation, specifically that the name of the row of numbers (n_CH2) is followed by ",1". This will be important in the next section.

5.3 Sample Mean and Standard Deviation

Unfortunately, scanning all bispyrenes formed in a reaction – so that we can count the number of CH_2 units in each molecule directly – would take a very long time. Instead of trapping every single molecule on a large surface and scanning them all, we could just use a much smaller sticky surface and trap only a small fraction of the molecules for scanning. We expect that the mean and standard deviation of the length of aliphatic linkers in the molecules trapped on the small sticky surface will be roughly the same as the mean and standard deviation of the length of aliphatic linkers for the entire population of bispyrenes produced in a nano batch. Taking small samples for analysis sounds much more feasible and would considerably speed up the process; however, we would lose the ability to obtain the exact values of the mean and the standard deviation and we will only be able to *estimate* them from a sample.

At this point we need to assess our priorities and decide how much uncertainty we are willing to tolerate in return for speeding up the data analysis. In fact, this is a question scientists and engineers commonly face: to avoid wasting effort, time and money, what is the minimum amount of data required to yield sufficiently certain estimates for the population mean and standard deviation.[‡] The level of certainty usually depends on the requirements of what we wish to use the mean and standard deviation for.

For a sample from a complete data set with normal distribution the *mean* ($\bar{x}$) is calculated as

$$\bar{x} = \frac{1}{n} \sum_{i=1}^{n} x_{s,i}$$

where n is the sample size (the number of bispyrene molecules on the small sticky plate) with $x_{s,i}$ representing an individual data point within the sample (the number of CH_2 units found in the ith bispyrene molecule on the small sticky plate scanned). This formula instructs us that for calculating the mean length of aliphatic linkers within the sample, like before, we need to add up the numbers of CH_2 units found in the molecules in the sample and divide the sum by the number of molecules in the sample (*i.e.* the number of bispyrenes on a small sticky sample plate). We should obtain again a length close to 12.

The sample standard deviation of data points in a sample (a measure of how spread the aliphatic linker lengths are around the mean aliphatic linker length in the sample) is calculated as

$$s_{n-1} = \sqrt{\frac{\sum_{i=1}^{n} (x_{s,i} - \bar{x})^2}{n-1}}$$

Note that for samples, we need to divide by $(n-1)$, as this yields the un-biased estimate for the population standard deviation (σ).

In the denominators of the formulae for calculating σ and s_{n-1}, N and $(n-1)$, are the *degrees of freedom*, the number of values that can freely vary in a calculation. In the expression for the sample standard deviation, each linker length is a result of a draw from the population, thus each can freely vary; but because $\bar{x}$ depends on them, we lose

---

[‡] Pollsters face the same problem when surveying the population. For example, asking everyone eligible to vote about their political preferences would have an enormous cost. Instead, only a small group of them are interviewed (sample) for establishing the popularities of parties across the entire electorate (population). Taking a sample randomly is far more complicated than it sounds. A sample must be representative of the population: pollsters must ensure the voters selected for a sample reflect the demographic characteristics (gender, income, education, religion, *etc.*) of the electorate; chemists must make sure a mixture is homogeneous before sampling, or when this is not possible or practical mix aliquots taken from locations representative of the inhomogeneities present in the chemical system investigated.

one degree of freedom. (Every time we take a sample we will have a different set of linker lengths, which will result in different mean linker length from one sample to the next.) However, in the formula for the population standard deviation, the degrees of freedom is N, because μ cannot vary freely, for it is a property of the population.

In MATLAB, the **sample mean** and **sample standard deviation** are also calculated by using the `mean()` and `std()` functions. From a usual nano batch containing thousands of bispyrenes we take a sample made up of 10 product molecules, look at the AFM scan and count the number of CH_2 units in all of the 10 molecules the sample contains. We then calculate the sample mean and sample standard deviation of the aliphatic linker length for the molecules in the sample as follows.

```
n_CH2 = [13  15  8  14  13  10  11  13  18  17];
mean_n_CH2 = mean(n_CH2) % sample mean
s_n_CH2 = std(n_CH2) % sample standard deviation s_(n-1) of n_CH2
```

```
mean_n_CH2 =
   13.2000
s_n_CH2 =
   3.0478
```

Please note that now we used `std(n_CH2)` to calculate the sample standard deviation, specifically that the name of the row of numbers (`n_CH2`) is now *not* followed by "`,1`" unlike before when we were calculating the population standard deviation.

5.4 Standard Error

Let us use again a sticky plate small enough to trap only 10 bispyrene molecules out of the thousands produced in a typical nano batch. As before, after the AFM scan is taken, we count the number of CH_2 units within a sample. Then, we put the sticky plate back to the product mixture and purge the plate by sonicating the mixture. Subsequently, we wait until the plate becomes fully covered again, remove it and take another AFM scan to count the number of CH_2 units within this second sample. After re-sampling a few more times, we realise that each sample gave a slightly different sample mean and sample standard deviation. How can we decide which sample mean and sample standard deviation were the closest to the (exact) population mean and population standard deviation? Unfortunately, we cannot. Worse still, we also realise that just by looking at samples we have no way of ever knowing for sure what the exact values of the population mean and population standard deviation are. What we instinctively know, however, is that the larger the sample size, the more likely it is that the sample mean and sample standard deviation are close to the (exact) population mean and population standard deviation. Similarly, taking more samples of the same sample size and averaging the sample means and sample standard deviations will result in better estimates of the population mean and population standard deviation.

So let us start taking samples and see what happens as we increase the sample size. First, we take small samples (use a sticky plate that is so small it collects only 10 bispyrene molecules) and find that the sample mean values greatly scatter within a wide range. Then, we take larger samples (by using a bigger sticky plate that collects 50 molecules), which results in the sample mean values scattering in a narrower range. By taking even larger samples (say, 100 molecules), we realise that the sample mean values start to cluster even more and the range of the values further reduces. In other words, what we are seeing is that the standard deviation of the sample means decreases as the sample size is increased. To qualitatively capture this observation we introduce the *standard deviation of the (sample) mean* (SDOM) which is often called the **standard error**, α, in statistics. The standard error is calculated as

$$\alpha = \frac{s_{n-1}}{\sqrt{n}}$$

where s_{n-1} is the sample standard deviation and n is the sample size. The standard error is a measure of how uncertain our estimate of the (exact) population mean is based on the sample size. The larger the sample (as n increases), the better the estimate of the (exact) population mean becomes (α decreases). The expression above we could use to quote our *estimate of the (exact) population mean in standard scientific format* as

$$\mu \pm \alpha = \bar{x} \pm \frac{s_{n-1}}{\sqrt{n}}$$

from a *large sample*. It is important that we realise that in reality we are extremely rarely able to determine a population mean (μ) directly. In laboratory experiments trillions of molecules form which we cannot scan individually, therefore we have no choice but to merely estimate μ from samples. Also note that s_{n-1} calculated for a sample is an unbiased estimator of σ, which is a measure how much x_i values vary from μ. $\pm \alpha$ represents the width of the interval within which μ is likely to be found based on the sample we took. The larger the sample size the smaller α becomes and if the sample size is large enough μ will be 68% likely to be in the interval between $\bar{x} - \alpha$ and $\bar{x} + \alpha$. In other words, we can be 68% confident that $\bar{x} - \alpha < \mu \leq \bar{x} + \alpha$. For 95% certainty – only one in twenty chance that μ falls outside the specified domain – we would need to increase the size of the interval to $\bar{x} \pm 1.96\alpha$.

A consequence of the central limit theorem is that we can estimate the mean (μ) and standard deviation (σ) of a population from samples. If we take many large enough samples (of sample size n) from a population, the means of the individual samples $(\bar{x}_1, \bar{x}_2, \ldots, \bar{x}_i, \ldots, \bar{x}_k)$ will approximate a normal distribution. The mean of these sample means, $\mu_{\bar{x}} = (\sum \bar{x}_i)/k$, will approach the population mean

$$\mu_{\bar{x}} \approx \mu$$

and the standard deviation of the sample means, $\sigma_{\bar{x}} = \left[\sum (\bar{x}_i - \mu_{\bar{x}})^2/k\right]^{1/2}$, will tend toward $\sigma/\sqrt{n}$, thus

$$\sigma_{\bar{x}} \approx \sigma/\sqrt{n}$$

as the number of samples increases. The latter is harder to see but it can be qualitatively understood by using our intuition that the means of samples spread much less about μ than individual measured values; therefore $\sigma_{\bar{x}}$ ought to be smaller than σ.

We would also expect the sample means to scatter less about μ as we increase the sample size. Taking all possible samples of size n from the population – which would usually be very impractical or even impossible to do – would result in $\mu_{\bar{x}} = \mu$ and $\sigma_{\bar{x}} = \sigma/\sqrt{n}$. Therefore, the variable

$$z = \frac{\bar{x} - \mu}{\sigma_{\bar{x}}} = \frac{\bar{x} - \mu}{\sigma/\sqrt{n}}$$

where $\bar{x}$ denotes the mean of a large enough sample, typically $n > 30$ as a rule of thumb, would be normally distributed with mean $= 0$ and standard deviation $= 1$. The special normal distribution with such parameters plays a central role in statistics and is called the *z-distribution* or *standard normal distribution*; or for short: N(0,1).

5.5 Small Sample Sizes and Student's *t*-distribution

As we have seen a sample mean ($\bar{x}$), which we can determine, differs from the population mean (μ) we would ideally want to find. We can think of the difference between them ($\bar{x} - \mu$) as a quantity that would keep changing from sample to sample. The values of ($\bar{x} - \mu$) we expect to have a bell-shaped distribution centred on zero and that the quantity

$$t = \frac{\bar{x} - \mu}{\alpha} = \frac{\bar{x} - \mu}{s_{n-1}/\sqrt{n}}$$

will have a normal distribution with mean $= 0$ and standard deviation $= 1$. However, it can be shown that this is not the case because the sample standard deviation (s_{n-1}) and the population standard deviation (σ) are different when the sample size is small (which is typically the case). If this fact is taken into account we find that t follows the so-called *Student's t-distribution*. This distribution is wider than the standard normal distribution, but as the sample size increases the width gradually narrows and approaches the width of the normal distribution with mean $= 0$ and standard deviation $= 1$. This is important because this means that we need to adjust how we quote the uncertainties in measured quantities. The range (confidence bounds) within which the (exact) population mean is expected to be found strongly depends on the sample size, especially when the sample size is small, which is almost always the case when we measure something in undergraduate chemistry labs. From small samples (*i.e.*, from a few measurements or reactions run under the same conditions) the estimated value of the (exact) population mean should be given as

$$\mu = \bar{x} \pm t(p, \nu) \frac{s_{n-1}}{\sqrt{n}}$$

for which *t-values*, $t(p,\nu)$, are available in tables. *t-Values* depend on ν, the degrees of freedom which is calculated as $\nu = n - 1$, and the confidence limits (p) we set depending on how certain we would like to be about our estimate of the population mean. There is a different *t-value* for every sample size and every confidence limit. As the sample size grows, *t-values* decrease and approach 1 for 68%, and 1.96 for 95% confidence.

Now we have a way to address the question: what is the minimum number of measurements/experiments we need to take/conduct? It depends on the level of uncertainty required by what we intend to use the estimated population mean (μ) for. If we need to estimate μ for an application that is very sensitive to the uncertainty in μ, we need $t(p, \nu)s_{n-1}/\sqrt{n}$ to be below a small threshold value, which we can achieve by improving our measurements (increasing precision), which will reduce s_{n-1}, and/or increasing the number of measurements to reduce $t(p, \nu)$ while also increasing $\sqrt{n}$.

We do not need to remember or look up *t*-values, because MATLAB can compute them for us. All we need to do is call the `tinv()` function[§] and obtain the *t*-value for the confidence level we require and for our sample size.

Exercise 5.1

The melting temperature of a substance has been measured 11 times. What is the melting temperature of the substance at 95% and 68% confidence levels? Comment on the uncertainties of the melting temperature.

T_m/°C: 42.70   42.60   42.78   42.83   42.58   42.68   42.65   42.76   42.73   42.71   42.59

Answer:

```
Tm=[42.70  42.60  42.78  42.83  42.58  42.68  42.65  42.76  42.73
    42.71  42.59];
m_Tm=mean(Tm) % sample mean
s_Tm=std(Tm); % sample standard deviation
cl=0.95; % confidence limit (true Tm is 95% likely to be in conf. int.)
p=(1+cl)/2; % convert cl=> p for MATLAB (MATLAB doesn't use cl
directly)
n=length(Tm); % sample size (each measurement is taking one sample)
nu=n-1; % degrees of freedom
t=tinv(p,nu); % obtain t-value
dTm=t*s_Tm/sqrt(n) % uncertainty in Tm
% quote melting temperature as Tm=m_Tm±dTm °C
% round dTm up to 1 S.F. and m_Tm to same D.P. as dTm
```

m_Tm =
    42.6918
dTm =
    0.0548

After rounding, at 95% confidence level: $T_m = 42.69 \pm 0.06$ °C and at 68% confidence level: $T_m = 42.69 \pm 0.03$ °C. The reported uncertainty in the melting temperature is greater at the 95% confidence level, because it is more likely to find the true value of the melting point in a wider temperature domain.

[§] In Octave, the Statistics package needs to be loaded in before `tinv()` can be used. For that, add "`pkg load statistics`" to the top of your code.

5.6 Size of Error Bars

What we have discussed in this chapter can be applied to the lengths of error bars. Repeat or replicate measurements carried out under the (supposedly) same conditions have been found to produce slightly varying results. Similarly, there is always some variance in the yield of replicate reactions. The uncertainty in the result of a measurement (or the yield of a reaction) is similar to the unpredictability in randomly selecting a value from a large pool of values. Thus, a set of measured values (or yields) can be thought of as a sample taken from a population of values. The set of measured values can be concisely presented by the value they cluster about, the sample mean ($\bar{x}$), and the extent of their scatter, which can be shown by drawing an error bar between one standard deviation (s_{n-1}) below and above the mean. However, instead of displaying the spread of the measured values (which does not change with n, the number of repeat/replicate measurements), error bars could also be justifiably used to indicate the estimated bounds of the domain within which the population mean (true value) is 95 or 68% likely to be found. In this case, the size of an error bar depends on n as its end points are drawn at $\bar{x} \pm 1.96 s_{n-1}/\sqrt{n}$ or $\bar{x} \pm s_{n-1}/\sqrt{n}$, respectively for a large number of repeat/replicate measurements. If only a few repeat/replicate measurements were carried out, $\bar{x} \pm t(p,\nu)s_{n-1}/\sqrt{n}$ should be used for the error bar positions instead. Irrespective of the number of repeat/replicate measurements, standard error ($\bar{x} \pm s_{n-1}/\sqrt{n}$) is also a popular choice for drawing error bars.

As we have seen error bars can be ambiguous and easily misinterpreted without checking and stating what they represent. In the next exercise we will compute these different quantities commonly used when drawing error bars.

Exercise 5.2

The carbohydrate content of a glycoprotein is determined in repeat measurements to be 12.6, 11.9, 13.0, 12.7, 12.5, 13.3, 11.6, 11.3, 13.4, 10.2, 12.7, 11.8 g of carbohydrate per 100 g of protein. What is the carbohydrate content of the glycoprotein? Use standard deviation, standard error, 68 and 95% confidence levels to indicate the uncertainty in the carbohydrate content. Amend the above code to calculate confidence intervals in one step and output results in standard format ("*mass* ± Δ*mass* g" without rounding).

Answer:

```
mass = [12.6  11.9  13.0  12.7  12.5  13.3  11.6  11.3  13.4  10.2
12.7  11.5];
n = length(mass); % sample size (number of repeat measurements)
m_mass = mean(mass); % sample mean
s_mass = std(mass); % sample standard deviation (SD)
se_mass = std(mass)/sqrt(n); % standard error (SE)
cl = [0.68  0.95]; % assign confidence limits in one step (68% & 95% CI)
p = (1+cl)/2; % convert cl => p for MATLAB (MATLAB doesn't use cl
directly)
nu = n-1; % degrees of freedom
t = tinv(p,nu); % retrieve t-value
```

```
dmass=t*se_mass % uncertainties in mass given as confidence intervals
disp(['SD: m=',num2str(m_mass),'±',num2str(s_mass),' g'])
disp(['SE: m=',num2str(m_mass),'±',num2str(se_mass),' g'])
disp(['68% CI: m=',num2str(m_mass),'±',num2str(dmass(1)),' g'])
disp(['95% CI: m=',num2str(m_mass),'±',num2str(dmass(2)),' g'])
```

SD: m=12.225±0.94304 g
SE: m=12.225±0.27223 g
68% CI: m=12.225±0.28353 g
95% CI: m=12.225±0.59918 g

After rounding they become (SD) 12 ± 1 g, (SE) 12.2 ± 0.3 g, (68% CI) 12.2 ± 0.3 g and (95% CI) 12.2 ± 0.6 g.

5.7 Comparing Values: Are They Different or Not?

Here we only look at the simplest case, comparing a set of experimentally determined values against the accepted literature value. This often arises in undergraduate laboratories where we have a set of repeat/replicate measurements and would like to establish if their mean is in agreement with the literature value or not. This can be then used to ascertain how well our method performs and whether it requires improvements.

Knowing the literature value (μ_0), which we take as the true value (or exact population mean), enables us to determine how far our estimate of the true value (μ) falls from the accepted value. This will be used to judge whether the literature value is within the confidence bounds we calculated. Recall that our claim that the true value is $\mu = \bar{x} \pm t(p,\nu)s_{n-1}/\sqrt{n}$ carries a probability statement as well. At 95% confidence level, we assert that according to our investigation the true value is 95% likely to be within those bounds. This also means that the literature value is only 5% likely to fall outside the domain we specified, which could be visualised by drawing the t-distribution corresponding to the number of measurements we made[¶] (Figure 5.7).

The red dashed line segments represent the bounds of the confidence interval with the slider on the slide bar at the bottom showing the scaled position of the literature value, $t = (\mu - \mu_0)/\alpha$, relative to our estimate μ. If $\mu = \mu_0$, the slider is in the middle at 0. If the *literature value is between the confidence limits*; i.e., t is within the confines of the dashed line segments, we have no reason to reject our claim that $\mu = \bar{x} \pm t(p,\nu)s_{n-1}/\sqrt{n}$. In this case we can say that *our estimate is in agreement with the literature value*. However, if the *literature value is outside the confidence interval*, we must reject our claim and state instead that *our estimate is not in agreement with the literature value*.[∥] At 95% confidence level, we

---

[¶] With increasing number of repeat/replicate measurements the peak of the t-distribution narrows and becomes taller while the sides flatten as it approaches the standard normal distribution. For a large number of repeat/replicate measurements we use the z-distribution to find the confidence limits to compare the value of $(\mu - \mu_0)/\alpha$ against.

[∥] When you do research and show using this type of analysis that your result is different from the generally accepted, it might mean that you have discovered a shortcoming of an established method or theory. The progress of science is made through steps where new theories/methods providing significantly more accurate predictions than established ones take over.

Measurements and Their Uncertainties

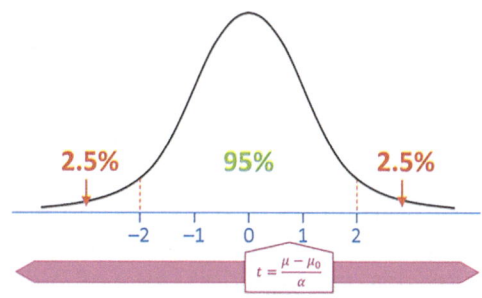

Figure 5.7 *t*-Value representing how many scaled standard deviations the true value falls from our estimate. If it is outside the [−2,2] domain, there is less than 5% chance that our estimate is in agreement with the true value.

are giving only 5% chance of the literature value to fall outside the confidence limits. If that happens we need to reassess the accuracy of our method.**

Given how the shape of the *t*-distribution depends of the number of repeat/replicate measurements, the more measured values you have, the narrower the confidence interval becomes at the same confidence level. As long as the literature value continues to fall within the confidence limits, this will mean you can be more certain about your method producing accurate results. Increasing the confidence level for the same number of repeat/replicate measurements will result in the broadening of the confidence interval, which will leave the literature value less chance of falling outside. At 99% confidence level, we would be giving this possibility only 1-in-100 chance.

For example, in Exercise 5.2 ($n=12$, $\bar{x}=12.225$ g and $s_{n-1}=0.94304$ g), we made a statement at the 68% confidence level that – based on our repeat measurements – the true value is within the 12.2 ± 0.3 g interval. The *t-value* required for making that claim is 1.0415. Raising the confidence level to 95% (for which the *t-value* is 2.2010) resulted in an increase in the domain size to 12.2 ± 0.6 g. These claims are equivalent to stating that the odds of the true value being outside the 12.2 ± 0.3 g and 12.2 ± 0.6 g intervals are respectively 32 and 5% likely. (Notice how increasing the confidence level resulted in a less useful claim. In an extreme case, one could claim with complete certainty that the true value is within the $12.2 \pm \infty$ g interval, which would be a completely useless statement.)

Exercise 5.3

We carefully weight out 0.1000 mol of (6.5380 g) of Zn and transfer it into a vessel fitted with a sensitive pressure sensor. Inside the vessel, which can be sealed and evacuated, is a glass dish containing concentrated HCl in stoichiometric amount. Given that the volume of the vessel is 10.000 litre, we can estimate the value of the gas constant (R), assuming the reaction

$$\text{Zn} + 2\text{HCl} \rightarrow \text{ZnCl}_2 + \text{H}_2$$

goes to completion, by first evacuating the vessel, then shaking it lightly to tip over the glass cup and subsequently measuring the pressure of the H_2 gas at 298.15 K

** Comparing results is not only important in natural sciences, engineering and medicine but also in social sciences and many industries. In statistics, answering the question of whether our sample mean is different from the established value is discussed as part of *hypothesis testing* under a *two-tailed one-sample t-test*.

once the bubble formation has stopped. We repeat the experiment several times and measure the pressures given below.

$p/10^4$ Pa: 2.5285   2.4659   2.7212   2.5302   2.6497   2.7104   2.6913   2.6405

a. We first assume that the volumes of the glass cup and HCl solution are negligible compared to the volume of the vessel ($V = 1.0000 \times 10^{-2}$ m^3). Substituting the pressures into the ideal gas equation ($pV = nRT$; thus $R = pV/nT$) yields the following values for the universal gas constant:

$R/\text{J mol}^{-1}\text{K}^{-1}$: 8.4806   8.2707   9.1269   8.4863   8.8871   9.0907   9.0267   8.8563

b. Realising that the combined volumes of the glass cup and HCl solution is 0.3415×10^{-3} m^3 and therefore they cannot be neglected ($V = 9.6585 \times 10^{-3}$ m^3), substituting the same pressures into the ideal gas equation results in the following values for the universal gas constant:

$R/\text{J mol}^{-1}\text{K}^{-1}$: 8.1910   7.9882   8.8153   8.1965   8.5836   8.7803   8.7184   8.5538

Are our estimates of R in agreement with the literature value?

Answer:

The above literature value of the gas constant, recommended by the Committee on Data for Science and Technology (CODATA), is regarded as the internationally accepted (true) value of R. The two sets of R values we produced are two samples; thus, we can compute the sample means and standard deviations for them. From these, we can estimate the bounds of the domains expected to contain the literature value with some probabilities which are usually set to 95% for statistical significance. Thus, the literature value will be 95% likely to be within the $\bar{x} \pm t(p,\nu)s_{n-1}/\sqrt{n}$ interval. We cannot tell if our sample means are equal to the literature value, but we can estimate the likelihood of our sample means being significantly different from the true value. For case (a)

```
>> m_R=mean(R);
>> s_R=std(R);
>> cl=0.95;
>> p=(1+cl)/2;
>> nu=length(R)-1;
>> t=tinv(p,nu);
>> dR=t*s_R/sqrt(n)
>> disp(['R=',num2str(m_R),'±',num2str(dR),'J/mol/K'])
R=8.7782±0.27004 J/mol/K
```

which after rounding becomes $R = 8.8 \pm 0.3$ J mol^{-1} K^{-1}. Based on our measurements, we claim that the literature value of R is 95% likely to be between 8.5 and 9.1 J mol^{-1} K^{-1}. As the internationally accepted literature value falls outside this region, we must reject our statement and conclude that our method yields values of R *not in agreement* with the literature value. In other words, our method yields R values *significantly different* from the literature value.

> For case (b), the commands above generate the output
>
> ```
> R=8.4784±0.26082 J/mol/K
> ```
>
> which after rounding becomes $R = 8.5 \pm 0.3$ J mol^{-1} K^{-1}. Therefore, we now claim that the literature value of R is 95% likely to be between 8.2 and 8.8 J mol^{-1} K^{-1}. As the literature value recommended by CODATA lies within this region, our statement is justified and we conclude that our revised method yields R values *in agreement* with the literature value; *i.e.*, our updated method yields R values which are *not significantly different* from the literature value.

5.8 Identifying Outliers

We sometimes have data points that fall far away from the trend of the majority of data points or lie far outside the cluster of data points. Whether to exclude such outlying data points from further analysis typically requires careful consideration and there are a number of developed methods to help us decide. Despite their usefulness, they do not eliminate ambiguity around outliers. For that reason, the best way to deal with outliers is by knowing our experiments. Paying attention to detail – to the experimental conditions, specifics of sample preparation, settings of the instruments used – taking notes regularly and repeating measurements/experiments we suspect to be erroneous will probably help us identify outliers far more than applying statistical methods.

Here, we are going to look at two typical scenarios: repeating a measurement under the same conditions, or varying a parameter and measuring a quantity in response. In both cases we can apply intuitively what we have learnt in this chapter about standard deviation (σ), namely that about 99.7% of data points are within $\pm 3\sigma$ of the mean (μ) for a quantity with a bell-shaped distribution. This leaves approximately 0.3% chance, on average about 1 in 370 repeat measurements, for obtaining a value smaller than $\mu - 3\sigma$ or greater than $\mu + 3\sigma$ due to unavoidable random fluctuations present in all experiments and measurements.[††]

If we were to remeasure a quantity under the same conditions thousands of times we should expect to see a few of these extreme values naturally occur. We, however, at best make only a few repeat measurements in undergraduate (or even research) laboratories; therefore, if we happen to encounter a value outside the $\mu \pm 3\sigma$ domain we can consider that as an outlier.

The same is true for experiments where a quantity is repeatedly measured while an experimental parameter is varied. In these cases, we usually fit curves to our data points as part of the analysis. We can also use these curves to quantify how much the data points

---

[††] Note that with increasing number of measurements/experiments, the uncertainties in the quantities estimated decrease; however, it has no influence on their standard deviations, which are characteristic to the population. For example, taking larger and larger samples of bispyrenes will reduce the uncertainty in the estimated mean linker length (μ) for a batch. The standard deviation of linker length (σ estimated as s_{n-1}) will not change in the batch, no matter how large the sample is. Encountering a linker length outside the $\mu \pm 3\sigma$ range will likely be an outlier, perhaps caused by some slight vibration shifting the AFM tip, which you can be more certain of (and more inclined to exclude that linker length) if it is corroborated by your notes saying someone walked past the instrument while the scan was taken.

scatter. The distance between each data point and the fitted curve, called *regression residual*, can be regarded as a quantity whose values keep changing from one data point to the other. When a data point is above the fitted curve, the residual is positive, when it is below the curve the residual is negative and the residual is zero when a data point is right on the curve. Because fitted curves are constructed to ensure the residuals cancel out (the sum of residuals is zero) we expect the residuals to have a bell-shaped distribution centred on zero. Using the spread of residuals around their zero mean can be used to calculate the *standard deviation of residuals* (s_e) which, in turn, will help us decide how unlikely a certain observed value is in our data series. As before, if it is below -3σ (under the curve) or over $+3\sigma$ (above the curve) it is very likely to be an outlier.

Exercise 5.4

Decide if there are any likely outliers amongst the temperatures in Exercise 5.1.

Answer:

We take the data, calculate the mean and standard deviation, followed by generating the critical temperature values ($\bar{T} - 3s_{n-1}$ and $\bar{T} + 3s_{n-1}$) for identifying outliers which we will then output into the *Command Window*.

```
Tm = [42.70  42.60  42.78  42.83  42.58  42.68  42.65  42.76  42.73
42.71  42.59];
m_Tm = mean(Tm); % sample mean
s_Tm = std(Tm); % sample standard deviation
Tcv = [m_Tm - 3*s_Tm  m_Tm + 3*s_Tm]; % data point(s) outside this domain
are likely outlier(s)
disp(['Likely outliers: T < ',num2str(Tcv(1)),' °C or T > ',num2str
(Tcv(2)),' °C'])
```

Likely outliers: T < 42.4471 °C or T > 42.9366 °C

We conclude that none of temperatures are outliers.

Exercise 5.5

Draw the boundaries at $\pm 3\sigma$ around the fitted curves in Exercise 4.5 and Exercise 4.6 to allow convenient identification of outlying data points.

Answer:

In both cases, we first evaluate the fitted expression at the times and angles given. For this, we will call the `feval()` function which takes a fit object and the values we wish to substitute into the expression specified in the fit object, and return the values predicted by the expression. Subtracting the predicted ln[A] and energy values from the measured values will then yield the residuals. Subsequently, we compute the standard deviation of the residuals and 3σ which we will add to, and subtract from the values `feval()` returned to generate the upper and lower boundaries around the fitted curve.

```
time = [0  1  2  3  4  5  6  7  8  9  10  11];
A = [1.568  1.124  0.858  0.747  0.475  0.459  0.339  0.313  0.168
    0.107  0.091  0.022];
ln_A = log(A);
fa = fit(time',ln_A','poly1'); % without fixing ln[A]0 (2-parameter fit)
ln_Ae = feval(fa,time); % evaluate expression fitted to data points at time points in t
r = ln_A - ln_Ae'; % compute residuals (ln_Ae column vector → transpose before subtracting)
m_r = mean(r); % mean of residuals (just to check it is zero)
s_r = std(r); % standard deviation of residuals
ln_Aucv = ln_Ae + 3*s_r; % upper critical values
ln_Alcv = ln_Ae - 3*s_r; % lower critical values
plot(fa,'-b',time,ln_A,'>m') % plot fitted expression and measured data points
hold on
plot(time,ln_Aucv,':r',time,ln_Alcv,':r') % plot 3-sigma boundaries
xlabel('time/min'); ylabel('ln[A]')
axis padded
legend('ln[A]','ln[A]_0 - kt','+3\sigma','-3\sigma')
title('2-parameter fit with critical values for identifying possible outliers')
```

Because the fitted expression, $\ln[A] = \ln[A]_0 - kt$, is linear, the points tracing out the 3σ boundaries around it will produce smooth curves.

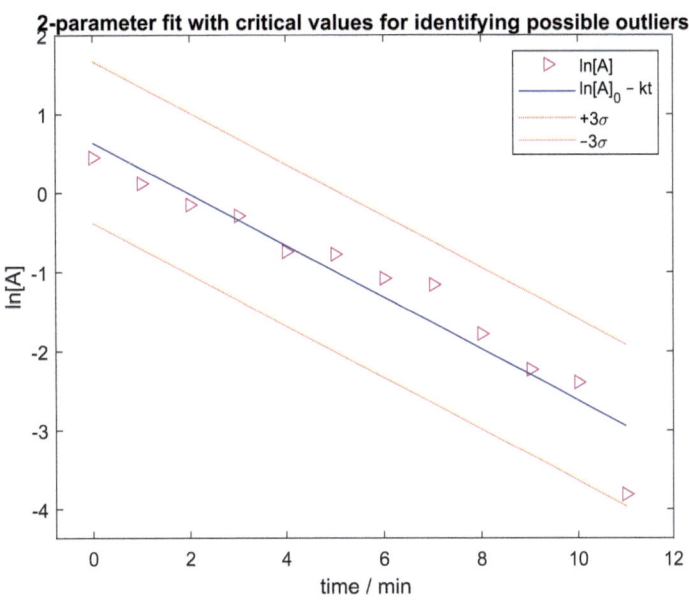

The 3σ boundaries will not be smooth curves for the double well potential because the fitted curve is not linear and the angles are sparsely and unevenly distributed along the horizontal axis. To be able to draw smooth boundaries, we will re-evaluate the fitted expression once the standard deviation has been computed, but this time by varying the dihedral angle in small equal steps across the domain. Drawing curves through the points at $\pm 3\sigma$ from the resulting energies will produce smooth boundaries.

```
ang=[139 149 155 160 165 169 175 180 185 194 200 204 210 219];
E=[-879 -2322 -2054 -2151 -2201 -2054 -1322 -1247 -1711 -2322 -2582 -2565 -2540 -1125];
f=fit(ang',E','poly4');
Ee=feval(f,ang); % evaluate expression fitted to energies
r=E - Ee'; % compute residuals (Ee column vector → transpose before subtracting)
m_r=mean(r); % mean of residuals (just to check it is zero)
s_r=std(r); % standard deviation of residuals
Eucv=Ee + 3*s_r; % smooth upper critical values
Elcv=Ee - 3*s_r; % smooth lower critical values

subplot(1,2,1) % rough 3-sigma boundaries (left)
plot(f,'-r',ang,E,'bd') % plot fitted expression and measured data points
hold on
plot(ang,Eucv,':k',ang,Elcv,':k') % plot 3-sigma boundaries
xlabel('dihedral angle/deg')
ylabel('E_{pot}/kJ mol^{-1}')
legend('data','Quartic fit','+3\sigma','-3\sigma','location','north')
axis padded

subplot(1,2,2) % smooth 3-sigma boundaries (right)
ang_fine=min(ang):max(ang); % vary angle in 1 degree increments
Eef=feval(f,ang_fine); % re-evaluate expression fitted to energies for angles in ang_fine
Eucv=Eef + 3*s_r; % smooth upper critical values
Elcv=Eef - 3*s_r; % smooth lower critical values
plot(f,'-r',ang,E,'bd')
hold on
plot(ang_fine,Eucv,':k',ang_fine,Elcv,':k')
xlabel('dihedral angle/deg')
ylabel('E_{pot}/kJ mol^{-1}')
legend('data','Quartic fit','+3\sigma','-3\sigma','location','north')
axis padded
```

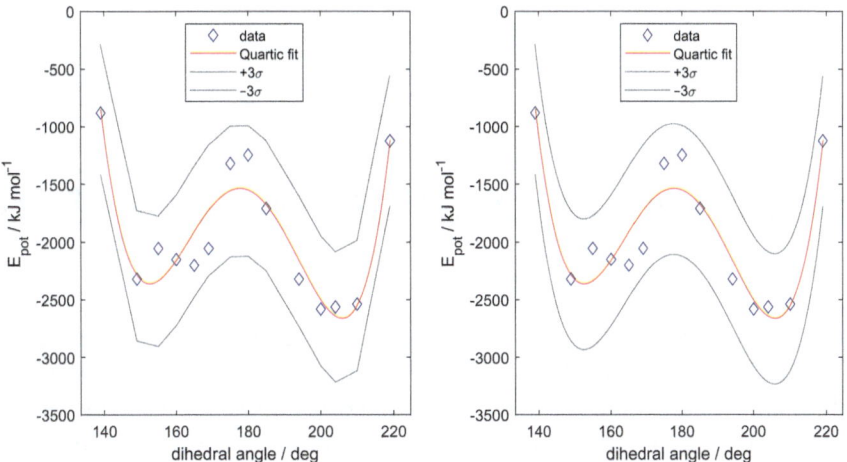

As the number of measurements grows, we are more likely to encounter values that despite being extremely far from the mean are not outliers. Therefore, flagging up possible true outliers requires adjusting the size of the envelope around the mean. For fewer than 10 measurements 2.5σ, and for over 100 measurements 3.5σ is more reasonable than 3σ.

6 Propagation of Uncertainties (Errors)

Having built an understanding of how uncertainties enter into measurements, we are now in the position to look at what happens to uncertainties in calculations. This question is important because the quantities we measure we usually need for calculations to find other quantities. For example, consider creating a calibration curve for determining the concentration of a substance. It could be the product of a reaction whose concentrations we would like to monitor throughout the course of the reaction to avoid shutting down the reaction while the compound is still forming. For the calibration curve, we prepare several solutions of the substance by carefully weighing out different amounts of the (commercially-sourced, high-purity) chemical and making them up to the same volume in volumetric flasks with the solvent used for the reaction. After measuring the absorbances of these solutions, we find that they do not line up perfectly on the absorbance plot, which results in some level of uncertainty in the position of our calibration line. Thus, the molar extinction coefficient of the product (ε), determined as the slope of the calibration curve, will be uncertain ($\varepsilon \pm \delta\varepsilon$), too. In turn, this will introduce uncertainty to the concentration of the product ($c \pm \delta c$) when we calculate it from the absorbances of the aliquots taken from the reaction mixture during the reaction. For all measurements being subject to uncertainty, the absorbances will also have some level of uncertainties ($A \pm \delta A$). The width of the cuvette we use for the absorbance measurements is not exact either ($\ell \pm \delta\ell$), even though it is too small to be seen with the naked eye. Accounting for all these uncertainties, the Beer–Lambert Law ($A = \varepsilon c \ell$) rearranged for concentration, $c = A\varepsilon^{-1}\ell^{-1}$, becomes

$$c \pm \delta c = \frac{A \pm \delta A}{(\varepsilon \pm \delta\varepsilon)(\ell \pm \delta\ell)}$$

To estimate (the true value of) the concentration of the product in the reaction mixture, we need to use this formula. We can usually disregard uncertainties that are small relative to others; in this case $\delta\ell$ is typically negligible compared to δA and $\delta\varepsilon$; yet we are still left with the latter two uncertainties to somehow pull through this expression to establish not only c, but δc as well.

A First Look at Coding in Chemistry: Solving Problems Using MATLAB
By Tamas Bansagi
© Tamas Bansagi 2025
Published by the Royal Society of Chemistry, www.rsc.org

We face the same problem when trying to calculate the concentration of H^+ in the reaction mixture from a pH measurement. Because pH measurements are subject to uncertainties, $[H^+]$ values must also somehow inherit the uncertainties in pH values. Now, we need to propagate the uncertainty in pH, $\delta(pH)$, through the expression linking $[H^+]$ to pH: $[H^+] = 10^{-pH}$, which becomes

$$[H^+] \pm \delta([H^+]) = 10^{-(pH \pm \delta(pH))}$$

when the calculation is carried out. In cases like these, there is always a mathematical expression linking the quantities we would like to determine ([Product], $[H^+]$) to the quantities we can measure (A, pH).

To begin with, we consider the simplest scenario: we are able to measure quantity x but we are interested in the value of quantity y, which can be calculated through a mathematical expression connecting x and y,

$$y = f(x)$$

The uncertainty in x (δx) will lead to uncertainty in y (δy); more specifically, the size of δy will depend on function f, the value of x and the size of δx. Figure 6.1 shows the graph of arbitrary expression f (solid black curve) connecting the independent variable x with the dependent variable y. We see that δy is determined not only by the size of δx, but also by the shape of the curve and the location on the horizontal axis, the value of x. In contrast, the value of y is determined by f and x only, δx has no effect on the value of y. Mathematically, when the values of x are uncertain (instead of $x_1, x_2, x_3, \ldots$ we have $x_1 \pm \delta x_1, x_2 \pm \delta x_2, x_3 \pm \delta x_3, \ldots$) the general expression $y = f(x)$ becomes

$$(y \pm \delta y) = f(x \pm \delta x)$$

To establish how δy can be determined from f, x and δx, we first look at the dashed lines each connecting an x value to its corresponding y value. The dotted lines on both sides of each dashed line trace out the confidence bounds associated with the uncertainty in the

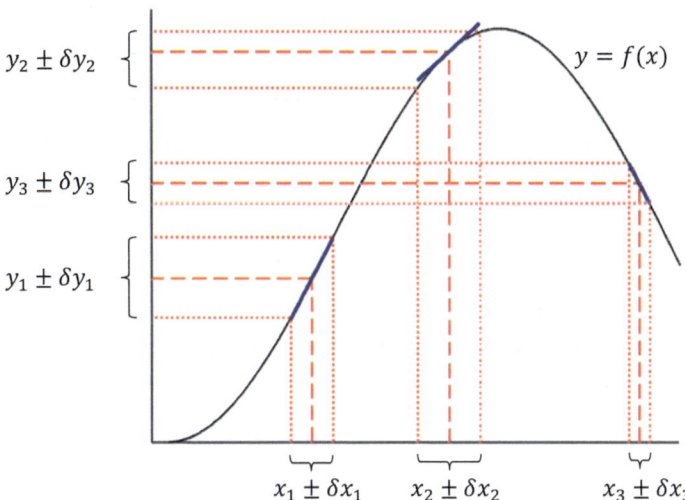

Figure 6.1 Propagation of uncertainties through the expression defining the black curve.

value of x translating into the corresponding confidence bounds around the respective value of y. This graphical representation can be directly followed to propagate uncertainties: substitute x_i, $(x_i - \delta x_i)$ and $(x_i + \delta x_i)$ into the expression of f, and you will readily obtain the corresponding y, $(y_i - \delta y_i)$ and $(y_i + \delta y_i)$ values. However, when y depends on multiple variables, their uncertainties can combine in many different ways, which may rapidly become difficult to trace as the number of independent variables increase. Instead of going through all possible combinations of uncertainties and picking the one resulting in the largest uncertainty, we look for a more robust general method.

If we draw the tangent line to f at x_i over the $x_i \pm \delta x_i$ interval (represented by the blue line segments), we notice that the end points of the tangent line segments could serve as good estimates for confidence bounds $y_i \pm \delta y_i$. The x coordinates of the end points of the blue line segments are $(x_i - \delta x_i)$ and $(x_i + \delta x_i)$. The y coordinates at these locations can be estimated from y_i (which we can calculate directly by substituting x_i into the $y = f(x)$ expression) and the gradient of f at x_i, $d(f(x_i))/dx$. To move forward to the end of the tangent line segment – like on any line – we need to multiply the slope of the tangent line, which is the gradient of f at x_i, by δx_i and add y_i to the result.[†] Thus, the end points of the tangent line segments can be given as

	end points of tangent line segments (–)	
	x coordinate	estimated y coordinate
forward	$x_i + \delta x_i$	$y_i + \dfrac{d}{dx}(f(x_i))\delta x_i$
backward	$x_i - \delta x_i$	$y_i - \dfrac{d}{dx}(f(x_i))\delta x_i$

	moving from backward end to forward end of a tangent line segment	
	change in x	estimated change in y
forward − backward	$2\delta x_i$	$2\dfrac{d}{dx}(f(x_i))\,\delta x_i$

Dividing both results in the last row by 2, we conclude that the change in y (δy_i) corresponding to δx_i can be estimated using the derivative of f as

$$\delta y_i \approx \frac{d}{dx}(f(x_i))\delta x_i$$

by multiplying the uncertainty in x by the value of the derivative of f at x_i. This sounds reasonably straightforward because we typically know $f(x)$ (for example $[H^+] = 10^{-pH}$; $y = 10^{-x}$) and the uncertainty in x (δx_i), which we can estimate from repeat/replicate

---

[†] The m slope of a line is equal to $\Delta y/\Delta x$ everywhere along the line ($m = \Delta y/\Delta x$). When the slope is multiplied by Δx, the corresponding Δy is obtained ($m\Delta x = \Delta y$). Therefore, starting from any (x, y) point on the line, the $(x + \Delta x, y + \Delta y)$ coordinates of a point in the forward direction can be given as $(x + \Delta x, y + m\Delta x)$. Here, the value of m for the tangent line at x_i is calculated by substituting x_i into the expression defining the df/dx derivative of f.

measurements of quantity x, or from the precision of the measuring device stated in its user's manual. Because uncertainties are positive, we should make sure that δy_i is always positive irrespective of the sign of the slope of $f(x)$. We will do that by taking the square of the derivative of f and then taking its square root as (while dropping i)

$$\delta y \approx \sqrt{\left(\frac{d}{dx}f\right)^2} \, \delta x$$

If y depends on multiple variables – let us say u, v, w, thus $y = f(u,v,w)$ – the formula above has to be amended[‡] to include all variables as

$$\delta y \approx \sqrt{\left(\frac{\partial}{\partial u}f\right)^2 (\delta u)^2 + \left(\frac{\partial}{\partial v}f\right)^2 (\delta v)^2 + \left(\frac{\partial}{\partial w}f\right)^2 (\delta w)^2}$$

This equation (prescribing to take the square root of the sum of the squared errors) may look complicated at first glance but the meaning of it is simple: If we want to calculate the uncertainty in a quantity (y) that depends on quantities we measured (u,v,w), we need to

1. calculate the uncertainties in the quantities we have measured: δu, δv, δw
2. obtain the partial derivatives $(\partial f/\partial u)$, $(\partial f/\partial v)$, and $(\partial f/\partial w)$ of expression f and evaluate them at the measured values of u, v and w (i.e. differentiate f with respect to u (while keeping v and w constant), with respect to v (while keeping u and w constant), and with respect to w (while keeping u and v constant), then substitute your measured values of u, v and w into the expressions of $(\partial f/\partial u)$, $(\partial f/\partial v)$ and $(\partial f/\partial w)$
3. substitute the values you obtained in steps 1 and 2 into the equation above and calculate the estimated value of uncertainty, δy
4. round δy to 1 significant figure and y to the same decimal place
5. quote y as $y \pm \delta y$ (followed by its units if applicable)

We can automate this process in MATLAB to save time and reduce the chances of making mistakes, which will be discussed in the last chapter. Now, however, let us look at first how we can propagate uncertainties by hand in some simple cases.

6.1 Addition and Subtraction

Assuming now that y depends on two quantities u and v only as $y = u + v$, then $\partial f/\partial u = 1$ and $\partial f/\partial v = 1$. Consequently,

$$\delta y \approx \sqrt{(\delta u)^2 + (\delta v)^2}$$

[‡] Summing the scaled squares of the uncertainties and taking the square root of the result is consistent with calculating distances in space. For example, when determining quantity y from two quantities u and v – i.e., through expression $y = f(u,v)$ – the uncertainty in y is calculated as $\delta y \approx \sqrt{[(\partial f/\partial u) \times \delta u]^2 + [(\partial f/\partial v) \times \delta v]^2}$, which is how we calculate the length of hypothenuse c from the lengths of the legs a and b of a right-angled triangle using Pythagoras' theorem: $c = \sqrt{a^2 + b^2}$. The legs can be taken to represent the two uncertainties scaled by their respective partial derivatives which translate to the δy overall uncertainty in y being equal to the hypothenuse.

If $y = u - v$, then $\partial f/\partial u = 1$ and $\partial f/\partial v = -1$ (with $(-1)^2 = 1$), which yields the same formula as above. Therefore, we now have a very useful formula for propagating uncertainties through expressions of the type $y = u \pm v$

$$\delta y \approx \sqrt{(\delta u)^2 + (\delta v)^2}$$

If there are more quantities added or subtracted in the expression linking the measured quantities to the quantity we are calculating the uncertainty for, we just need to extend this formula and keep adding more squared uncertainties before taking the square root of their sum.

6.2 Multiplication and Division

Assuming again that y depends on two quantities u and v only as $y = u \times v$, then $\partial f/\partial u = v$ and $\partial f/\partial v = u$. Consequently

$$\delta y \approx \sqrt{v^2 (\delta u)^2 + u^2 (\delta v)^2} \text{ or } (\delta y)^2 \approx v^2 (\delta u)^2 + u^2 (\delta v)^2$$

By using that $y^2 = u^2 v^2$, we can divide over by y^2 and obtain

$$\frac{(\delta y)^2}{y^2} \approx \frac{v^2 (\delta u)^2}{u^2 v^2} + \frac{u^2 (\delta v)^2}{u^2 v^2} = \frac{(\delta u)^2}{u^2} + \frac{(\delta v)^2}{v^2}$$

Rearrangement yields the useful formula for propagating uncertainties through expressions of the types $y = u \times v$ and $y = u/v$ (which we do not derive here)

$$\delta y \approx \sqrt{\left(\frac{\delta u}{u}\right)^2 + \left(\frac{\delta v}{v}\right)^2}\, y$$

If there are more quantities multiplied together (and/or there are divisions by multiple quantities) in the expression linking the measured quantities to the quantity we are calculating the uncertainty for, we just need to extend this formula.

6.3 Constants with Negligible Uncertainties

The expressions to propagate uncertainties through often contain constants and/or literature values with negligibly small uncertainties compared to the uncertainties of the quantities we determined. The error propagation formulae discussed above can be applied to these cases if we substitute in zero for the uncertainties of the constants and literature values. If the expression (or sub-expression) involves addition/subtraction, for example: $y = u + v + C$, with C having no or very little uncertainty relative to the error in u and v:

$$\delta y \approx \sqrt{(\delta u)^2 + (\delta v)^2 + 0^2} = \sqrt{(\delta u)^2 + (\delta v)^2}$$

thus, we can use the error propagation formula directly as before.

If the expression (or sub-expression) involves multiplication/division, for example: $y = uvC$ or $y = uv/C$, with C having no or very little uncertainty compared to the error in u and v:

$$\delta y \approx \sqrt{\left(\frac{\delta u}{u}\right)^2 + \left(\frac{\delta v}{v}\right)^2 + \left(\frac{0}{C}\right)^2}\, y = \sqrt{\left(\frac{\delta u}{u}\right)^2 + \left(\frac{\delta v}{v}\right)^2 + 0^2}\, y = \sqrt{\left(\frac{\delta u}{u}\right)^2 + \left(\frac{\delta v}{v}\right)^2}\, y$$

but we have to remember that now the formula for y does include C; therefore, δy will depend on C through y.

Exercise 6.1

How many moles of ideal gas evolved in a vessel with volume $V = 500.0 \pm 0.2$ dm^3 at temperature, $T = 292.4 \pm 0.5$ K, if the pressure of the gas is $p = 0.259 \pm 0.002$ MPa?

Answer:
First, convert to S.I. units. $V = 500.0 \pm 0.2$ dm^3 = 0.5000 ± 0.0002 m^3

$$n = \frac{pV}{RT} = \frac{0.259 \times 10^6\, \text{N m}^{-2} \times 0.5000\, \text{m}^3}{8.31415\, \text{J mol}^{-1}\, \text{K}^{-1} \times 292.4\, \text{K}}$$

```
>> n = 0.259e6 * 0.5/8.31415/292.4
n =
    53.2690
```

$$\delta n = \sqrt{\left(\frac{0.002}{0.259}\right)^2 + \left(\frac{0.2}{500}\right)^2 + \left(\frac{0.5}{292.4}\right)^2} \times 53.269$$

In this case, values given in the question can be used directly for error propagation. There is no need to use the values converted to S.I. units as this formula only involves relative uncertainties. However, when summing uncertainties, we must use converted uncertainties, because in that case absolute uncertainties are added together.

```
>> dn = sqrt((0.002/0.259)^2 + (0.2/500)^2 + (0.5/292.4)^2) * n
dn =
    0.4218
```

After rounding, we obtain $n = 53.3 \pm 0.5$ mol.
We could also think of this calculation as

$$n \pm \delta n = \left(\frac{pV}{T}\right) \times \left(\frac{1}{R}\right) \pm \delta\left(\frac{pV}{T}\right) \times \left(\frac{1}{R}\right) = \frac{pV}{RT} \pm \left(\sqrt{\left(\frac{\delta p}{p}\right)^2 + \left(\frac{\delta V}{V}\right)^2 + \left(\frac{\delta T}{T}\right)^2} \times \frac{pV}{T}\right) \times \left(\frac{1}{R}\right)$$

where we collect the quantities with uncertainties, perform the calculations involving only them, and then multiply both results by the sub-expression containing quantities without (or relatively small) uncertainties. This is useful to remember when we encounter expressions like $y = ab$ where only one of the quantities has uncertainty. If $\delta a \neq 0$, $\delta b = 0$, the expression becomes $y = (a \pm \delta a)b$ which should be evaluated as

$$y = ab \pm (\delta a)b$$

(and NOT $y = ab \pm \delta a$).

6.4 Propagating Uncertainties in MATLAB: *Addition and Subtraction*

An easy way to propagate uncertainties using MATLAB is *via* user-defined functions (see Section 2.3). Let us look at the first rule we derived.

```
function DY = pu_pm2(du,dv)
% function to propagate uncertainties for expressions: y=u±v
% uncertainty in y, dy, is computed as DY=√(du^2 + dv^2)
DY = sqrt(du^2 + dv^2);
end
```

Exercise 6.2

We need to transfer 7 cm³ of a solution into a flask. 7 ml pipettes are rare to find so we need to use a 5 ml and a 2 ml pipette which have manufacturer specified tolerances (written on the pipettes or listed on the manufacturer's website) ±0.03 ml for the 5 ml pipette and ±0.02 ml for the 2 ml pipette.

We take the tolerances written on glassware (shown above for assorted glassware) as uncertainties in the volumes delivered (pipettes, burettes) or contained (volumetric flasks).

What is the uncertainty in the volume transferred?

Answer:
(Next time you are using a piece of glassware, look for these tolerances and see how they compare across different types of glassware and volumes.)

Propagation of Uncertainties (Errors)

In the *Command Window*, we type

```
>> dv=pu_pm2(0.03,0.02)
dv =
     0.0361
```

Therefore we conclude, after rounding the uncertainty to 1 significant figure, that the combined volume delivered by the two pipettes is $V = 7.00 \pm 0.04$ cm^3. (You can try calculating this manually as well to see if you can obtain the same result.)

Exercise 6.3

Next time we need to transfer 7 cm^3 of a solution, we find only 5 and 1 ml pipettes in the lab (the latter with tolerance ± 0.015 ml). Now we use the 5 ml pipette once and the 1 ml pipette twice. What is the uncertainty in the volume transferred?

Can we still use our user-defined MATLAB function for three volumes?

Answer:

Yes, we can by nesting the functions, which means calling a function from inside a function. Here is how it is done:

```
>> dv=pu_pm2(pu_pm2(0.03,0.015),0.015)
dv =
     0.0367
```

This way, we first combine two uncertainties, the result of which we then combine with the third uncertainty. The total volume delivered by using the 5 ml pipette once and the 1 ml pipette twice is $V = 7.00 \pm 0.04$ cm^3. Even though the 1 ml pipette has a smaller tolerance than that of the 2 ml pipette, the rounded uncertainty remained the same (while the unrounded uncertainty even slightly increased, 0.0361 ml→0.0367 ml). Can you explain why?

Combining uncertainties from two (or three) sources is useful but how about being able to combine as many as we wish using just one simple user-defined MATLAB function? Now we make use of MATLAB's flexibility and high-level functions. We use the (.^) operator to square each element in array `du`, then the `sum()` function to add up the squares, before taking the square root.

```
function DY=pu_pm(du)
% function to propagate uncertainties for expressions: y=u1±u2±...
% uncertainty in y, dy, is computed as DY=√(Σdu^2)
DY=sqrt(sum(du.^2));
end
```
   NOTE: Nested operations are performed right to left.

> **Exercise 6.4**
>
> The 5 ml pipette has also gone missing, so we are left with only the 1 ml pipette. What is the uncertainty in the volume of the 7 ml solution transferred?
>
> **Answer:**
>
> ```
> >> pu_pm([0.015  0.015  0.015  0.015  0.015  0.015  0.015])
> dv =
>     0.0397
> ```
>
> The uncertainty in the total volume still rounds up to 0.04 cm^3, $V = 7.00 \pm 0.04$ cm^3; however, we notice that the uncertainty has further grown, even though the 1 ml pipette has the lowest tolerance. We conclude that *the overall uncertainty tends to increase as we combine more and more uncertainties.*

6.5 Propagating Uncertainties in MATLAB: *Multiplication and Division*

We are now developing a flexible user-defined function to do the hard work for us when we need to multiply or divide quantities with uncertainties. The operations we need MATLAB to perform are:

- divide each uncertainty by its corresponding measured quantity ($\delta u/u$, $\delta v/v$, ...)
- square each term calculated in the previous step (($\delta u/u)^2$, ($\delta v/v)^2$, ...)
- sum all the squares (($\delta u/u)^2 + (\delta v/v)^2 + ...$)
- take the square root of the sum of squares ($\sqrt{(\delta u/u)^2 + (\delta v/v)^2 + ...}$)
- multiply the result of the previous step by y which is obtained by substituting the measured quantities (u, v, ...) into the expression linking y to u, v, ... (which only contains multiplications and/or divisions)

This looks like a long list of tasks, but actually, thanks to MATLAB being a high-level programming language and our increasing knowledge of coding, these operations are easy to carry out. We even decide to endeavour into organising the output into a user-friendly output by using the `disp()` function. Inside the argument of `disp()`, we construct a text output defined within square brackets (Section 1.2) by turning the two values (Y: quantity we are calculating; DY: uncertainty in that quantity) into text characters using `num2str()` and by putting a "±" symbol in between them, with an added warning at the end that rounding and units are still needed before reporting the quantity.

Now we go back to the Beer–Lambert Law example we discussed at the start of this chapter and consider a real-life scenario where we need to calculate the molar extinction coefficient (ε) from the concentration, cuvette width and

```
function pu_md(u,du,Y)
% function to propagate uncertainties for expressions: y=u1 x/ u2 x/...
% uncertainty in y, dy, is computed as DY=√(Σ(du/u)^2) × Y
% Y must be supplied as it cannot be calculated from u
DY=sqrt(sum((du./u).^2))*Y;
disp([num2str(Y) '±' num2str(DY) ' **rounding & units**'])
end
```

absorbance all having uncertainties: $c = 0.50 \pm 0.07$ mol dm^{-3}, $l = 1.00 \pm 0.02$ cm and $A = 0.79 \pm 0.04$.

All we need to do is define the quantities and their uncertainties, and call the above function with the appropriate inputs.

```
>> c=0.50; dc=0.07; l=1.00; dl=0.02; A=0.79; dA=0.04;
>> pu_md([c l A],[dc dl dA],A/c/l)
1.58±0.23734 **rounding & units**
```

Notice that we did not even need to calculate the value of the molar extinction coefficient, $\varepsilon = A/(cl)$, separately as it was sufficient to simply pass the expression A/c/l as the last argument of pu_md(). With the necessary rounding and adding the units we quote the value of the molar extinction coefficient as:

$$\varepsilon = 1.6 \pm 0.3 \text{ dm}^3 \text{ mol}^{-1} \text{ cm}^{-1}$$

Exercise 6.5

Using a top pan balance you weight out 1.8 g of NaCl and dissolve it in 150 ml of water transferred into a beaker using a 50 ml pipette. What is the concentration of the NaCl solution $(c \pm \delta c)$ if the tolerance of the top pan balance is ± 0.1 g and the tolerance of the pipette is ± 0.05 ml, given that the molar mass of NaCl is 58.443 ± 0.001 g mol^{-1}?

Answer:

```
>> m=1.8; dm=0.1; M=58.443; dM=0.001; V=150/1e3; dV_pip=0.05/1e3;
% divide V and dV_pip by 1000 to convert their volumes from ml to dm^3
>> dV=pu_pm([dV_pip dV_pip dV_pip])
dV =
    8.6603e-05
>> pu_md([m M V],[dm dM dV],m/M/V)
0.20533±0.011408 **rounding & units**
```

After rounding, we obtain $c = 0.21 \pm 0.02$ mol/dm^3.

6.6 Propagating Uncertainties in MATLAB: *Functions*

Sometimes we need to use expressions to calculate quantities from measured quantities that contain some combinations of elementary functions like polynomials, exponential, logarithmic and/or trigonometric functions. In these cases, we should default to the general formula

$$\delta y \approx \sqrt{\left(\frac{\partial}{\partial u}f\right)^2 (\delta u)^2 + \left(\frac{\partial}{\partial v}f\right)^2 (\delta v)^2 + \left(\frac{\partial}{\partial w}f\right)^2 (\delta w)^2 + \cdots}$$

discussed earlier, which you often find written as

$$(\delta y)^2 \approx \sum_{i=1}^{n} \left[\left(\frac{\partial f}{\partial x_i}\right)^2 (\delta x_i)^2\right]$$

where $y = f(x_1, x_2, \ldots)$. For finding δy, work out the partial derivatives of f with respect to each variable $(\partial f/\partial x_i)$, then substitute into the resulting expressions and square the numbers you obtained. Multiply these squares by the squares of their corresponding uncertainties and sum of all resulting products. Finally, take the square root of the sum. If that seems too cumbersome, you can try finding the error propagation formula for the specific expression you have.

As we will see, MATLAB is able to compute partial derivatives of expressions, which makes it possible to construct a user-defined MATLAB function that can handle the propagation of uncertainties through any kind of expression we may ever encounter. At this point, we lack the knowledge to undertake this task, but we will return to this exciting problem in Chapter 12.

6.7 Appendix: Method of Least Squares for Linear Fitting

Propagating uncertainties plays a central role in data evaluation, not just in natural sciences, engineering and medicine but also far beyond these vast fields, for example in social sciences and finance. Here, we will look at how it can be deployed as part of the method of least squares to estimate the uncertainties in the slope and intercept of a line of best fit.

Our goal is to fit expression

$$y = a + bx$$

to a data set, where x denotes a quantity whose values (x_i) we set in an experiment and y is a quantity whose values (y_i) we determine in response to the values of x_i. Fitting parameters a and b are the intercept and slope of the line of best fit, respectively.

Finding the line of best fit requires assessing how good a fit is for a pair of a and b values. We would perhaps like to give priority to data points we have more confidence in and less priority to those that we are less sure about. One way of mathematically expressing the goodness of a fit is to calculate how far away the measured data points fall from the line fitted to them while weighing the individual distances proportionally to the uncertainties in the data points. The Method of Least Squares measures the goodness of fit through the quantity *chi-square*:

$$\chi^2 = \sum_{i=1}^{n} \frac{(y_i - \hat{y}_i)^2}{(\delta y_i)^2}$$

where $\hat{y}_i = a + bx_i$ ($i = 1, 2, \ldots, n$; with n being the total number of measurements while quantity x was varied) is the value estimated for a particular x_i using a pair of a and b values, and δy_i is the uncertainty in the corresponding y_i value we determined. Dividing the square of the difference between the observed and predicted values by $(\delta y_i)^2$ ensures that data points with smaller uncertainties are given more weight in χ^2. This formulation of assessing the goodness of a fit means that the line of best fit will have the smallest χ^2. If the uncertainties are the same we can replace all δy_i with δy. This is most often justified for data collected in undergraduate laboratories as each data point is typically determined only once, therefore we are unable to estimate the uncertainty (standard deviation) of each data point from multiple measurements. In these cases, we can use the uncertainty (δy) associated with the method (or measuring device) used to determine the data points, which therefore applies to the whole set of our data, in place of the individual uncertainties (δy_i). Another suitable estimate for δy is the standard deviation of the residuals[§]

$$s_e = \sqrt{\frac{\sum_{i=1}^{n}(y_i - \hat{y}_i)^2}{n-2}}$$

Assuming a uniform uncertainty across the determined y values, we can write (using $\sum_i$ instead of $\sum_{i=1}^{n}$ from now on)

$$\chi^2 = \sum_i \frac{(y_i - a - bx_i)^2}{(\delta y)^2}$$

$$= \frac{1}{(\delta y)^2} \sum_i (y_i^2 - 2y_i a - 2y_i b x_i + a^2 + 2abx_i + b^2 x_i^2)$$

where we collected $1/(\delta y)^2$ from all terms of the sum and brought it to the front. We know from calculus that at a minimum of a function its first derivative is zero. Thus, we will try to find the pair of a and b values that produces the best linear fit, the one with the smallest χ^2, by solving the equations

$$\frac{\partial \chi^2}{\partial a} = \frac{2}{(\delta y)^2} \sum_i (a + bx_i - y_i) = 0$$

$$\frac{\partial \chi^2}{\partial b} = \frac{2}{(\delta y)^2} \sum_i x_i(a + bx_i - y_i) = 0$$

[§] In line with the expressions for the population and sample standard deviations, σ and s_{n-1}, the expression for the standard deviation of residuals involves dividing by $(n-2)$ because calculating s_e first requires calculating two parameters (slope and intercept) which reduces the degrees of freedom by 2.

for a and b. Because $\sum_i a = a+a+\ldots = na$ and constants can be collected and moved before the Σ symbols, the equations – after separating the sums – become

$$na + b\sum_i x_i - \sum_i y_i = 0$$

$$a\sum_i x_i + b\sum_i x_i^2 - \sum_i x_i y_i = 0$$

from which the fitting parameters can be expressed as

$$a = \frac{\sum_i y_i}{n} - \frac{b\sum_i x_i}{n} = \bar{y} - b\bar{x}$$

$$b = \frac{\sum_i x_i y_i - \bar{y}\sum_i x_i}{\sum_i x_i^2 - \bar{x}\sum_i x_i}$$

(Above, we first rearranged for a while using the definition of mean ($\bar{x} = \sum_i x_i/n$ and $\bar{y} = \sum_i y_i/n$) and then substituted the expression for a into the second equation for finding b). The intercept and the slope of the line of best fit can be rearranged to

$$a = \frac{\sum_i y_i \sum_i x_i^2 - \sum_i x_i \sum_i x_i y_i}{n\sum_i x_i^2 - \left(\sum_i x_i\right)^2}$$

$$b = \frac{n\sum_i x_i y_i - \sum_i y_i \sum_i x_i}{n\sum_i x_i^2 - \left(\sum_i x_i\right)^2}$$

which means that a and b can be calculated directly, without having to determine b before we can calculate a.

In general, the uncertainty in a and b will depend on the uncertainties in the x and y values; therefore, to estimate δa and δb, we have to propagate all δx_i and δy_i uncertainties through the expressions above. As stated earlier, we are considering the uncertainties in the y values only, which is (believed to be) justified in undergraduate laboratories, for the uncertainty in a quantity set is assumed to be smaller than the uncertainty in a quantity determined. Applying the expression for propagating uncertainties yields

$$(\delta a)^2 \approx \sum_j \left[\left(\frac{\partial a}{\partial y_j}\right)^2 (\delta y_j)^2\right]$$

$$(\delta b)^2 \approx \sum_j \left[\left(\frac{\partial b}{\partial y_j} \right)^2 (\delta y_j)^2 \right]$$

where, for the sums involving the partial derivatives of the expressions for a and b, index j (with $j = 1,2,\ldots,n$) is used to separate this summation process from the sums already part of the formulae for finding a and b. Because we assume that the uncertainties are the same in all determined y values, $(\delta y_j)^2$ can be replaced with $(\delta y)^2$ which we can collect from each term of the sums and bring it in front of the Σ symbols.

$$(\delta a)^2 \approx (\delta y)^2 \sum_j \left(\frac{\partial a}{\partial y_j} \right)^2$$

$$(\delta b)^2 \approx (\delta y)^2 \sum_j \left(\frac{\partial b}{\partial y_j} \right)^2$$

Each partial derivative of a with respect to y_j is[¶]

$$\frac{\partial a}{\partial y_j} = \frac{\sum_i x_i^2 - x_j \sum_i x_i}{n \sum_i x_i^2 - \left(\sum_i x_i \right)^2}$$

where we used $\sum_j y_j \sum_i x_i^2 = (y_1 + y_2 + \ldots + y_j + \ldots y_n) \sum_i x_i^2 = y_1 \sum_i x_i^2 + y_2 \sum_i x_i^2 + \ldots + y_j \sum_i x_i^2 + \ldots y_n \sum_i x_i^2$, thus $\partial \left(\sum_j y_j \sum_i x_i^2 \right) / \partial y_j = 0 + 0 + \ldots + \sum_i x_i^2 + \ldots + 0 = \sum_i x_i^2 ; \sum_i x_i \sum_j x_j y_j$
$= (x_1 + x_2 + \ldots + x_i + \ldots x_n)(x_1 y_1 + x_2 y_2 + \ldots + x_j y_j + \ldots x_n y_n) = x_1 x_1 y_1 + x_1 x_2 y_2 + \ldots x_1 x_j y_j + \ldots$
$+ x_1 x_n y_n + x_2 x_1 y_1 + x_2 x_2 y_2 + \ldots x_2 x_j y_j + \ldots + x_2 x_n y_n + \ldots + x_i x_1 y_1 + x_i x_2 y_2 + \ldots x_i x_j y_j + \ldots + x_i x_n y_n + \ldots$
$+ x_n x_1 y_1 + x_n x_2 y_2 + \ldots x_n x_j y_j + \ldots + x_n x_n y_n$, hence $\partial \left(\sum_i x_i \sum_j x_j y_j \right) / \partial y_j = 0 + 0 + \ldots x_1 x_j + \ldots +$
$0 + 0 + 0 + \ldots x_2 x_j + \ldots + 0 + \ldots + 0 + 0 + \ldots x_i x_j + \ldots + 0 + \ldots + 0 + 0 + \ldots x_n x_j + \ldots + 0 = x_j (x_1 + x_2 + \ldots + x_i + \ldots x_n) = x_j \sum_i x_i$.

$$\left(\frac{\partial a}{\partial y_j} \right)^2 = \frac{\left(\sum_i x_i^2 \right)^2 - 2 x_j \sum_i x_i^2 \sum_i x_i + x_j^2 \left(\sum_i x_i \right)^2}{\left(n \sum_i x_i^2 - \left(\sum_i x_i \right)^2 \right)^2}$$

[¶] Because $n \sum_i x_i^2 - \left(\sum_i x_i \right)^2$ does not depend on y_j, the denominator can be taken outside the differentiation, thus we only need to differentiate the numerator.

Substituting this into the expression to propagate the uncertainties into a yields

$$(\delta a)^2 \approx (\delta y)^2 \sum_j \left(\frac{\partial a}{\partial y_j}\right)^2 = (\delta y)^2 \sum_j \frac{\left(\sum_i x_i^2\right)^2 - 2x_j \sum_i x_i^2 \sum_i x_i + x_j^2 \left(\sum_i x_i\right)^2}{\left(n \sum_i x_i^2 - \left(\sum_i x_i\right)^2\right)^2}$$

$$= (\delta y)^2 \frac{n\left(\sum_i x_i^2\right)^2 - 2\sum_i x_i^2 \left(\sum_i x_i\right)^2 + \sum_i x_i^2 \left(\sum_i x_i\right)^2}{\left(n \sum_i x_i^2 - \left(\sum_i x_i\right)^2\right)^2}$$

$$= (\delta y)^2 \frac{n\left(\sum_i x_i^2\right)^2 - \sum_i x_i^2 \left(\sum_i x_i\right)^2}{\left(n \sum_i x_i^2 - \left(\sum_i x_i\right)^2\right)^2}$$

$$= (\delta y)^2 \frac{\sum_i x_i^2 \left(n \sum_i x_i^2 - \left(\sum_i x_i\right)^2\right)}{\left(n \sum_i x_i^2 - \left(\sum_i x_i\right)^2\right)^2}$$

$$= (\delta y)^2 \frac{\sum_i x_i^2}{n \sum_i x_i^2 - \left(\sum_i x_i\right)^2}$$

where we used $\sum_i x_i = \sum_j x_j$, $\sum_i x_i^2 = \sum_j x_j^2$, $\sum_j \left(\sum_i x_i\right)^2 = n \left(\sum_i x_i\right)^2$, $\sum_j \left(\left(\sum_i x_i\right)^2\right)^2$

$= n\left(\left(\sum_i x_i\right)^2\right)^2$, $\sum_j \left(x_j \sum_i x_i^2 \sum_i x_i\right) = \sum_i x_i^2 \sum_i x_i \sum_j x_j = \sum_i x_i^2 \left(\sum_i x_i\right)^2$, $\sum_j \left(x_j^2 \left(\sum_i x_i\right)^2\right)$

$= \left(\sum_i x_i\right)^2 \sum_j x_j^2 = \left(\sum_i x_i\right)^2 \sum_i x_i^2$, and that the denominator could be collected from each term of the sum.

Similarly, each partial derivative of b with respect to y_j is

$$\frac{\partial b}{\partial y_j} = \frac{nx_j - \sum_i x_i}{n \sum_i x_i^2 - \left(\sum_i x_i\right)^2}$$

Propagation of Uncertainties (Errors)

where we used $\partial\left[n\sum_j(x_jy_j)\right]/\partial y_j = \partial[nx_1y_1 + nx_2y_2 + \ldots + nx_jy_j + \ldots + nx_ny_n]/\partial y_j = 0 + 0 + \ldots + nx_j + \ldots + 0 = nx_j$ and $\sum_j y_j \sum_i x_i = (y_1 + y_2 + \ldots + y_j + \ldots + y_n)(x_1 + x_2 + \ldots + x_n) = y_1x_1 + y_1x_2 + \ldots + y_1x_n + y_2x_1 + y_2x_2 + \ldots + y_2x_n + \ldots + y_jx_1 + y_jx_2 + \ldots + y_jx_n + \ldots + y_nx_1 + y_nx_2 + \ldots + y_nx_n$, therefore $\partial\left[\sum_j y_j \sum_i x_i\right]/\partial y_j = 0 + 0 + \ldots + 0 + 0 + 0 + \ldots 0 + \ldots + x_1 + x_2 + \ldots + x_n + 0 + 0 + \ldots + 0 = \sum_i x_i$.

$$\left(\frac{\partial b}{\partial y_j}\right)^2 = \frac{n^2 x_j^2 - 2nx_j \sum_i x_i + \left(\sum_i x_i\right)^2}{\left(n\sum_i x_i^2 - \left(\sum_i x_i\right)^2\right)^2}$$

Substituting this into the expression to propagate the uncertainties into b yields

$$(\delta b)^2 \approx (\delta y)^2 \sum_j \left(\frac{\partial b}{\partial y_j}\right)^2 = (\delta y)^2 \sum_j \frac{n^2 x_j^2 - 2nx_j \sum_i x_i + \left(\sum_i x_i\right)^2}{\left(n\sum_i x_i^2 - \left(\sum_i x_i\right)^2\right)^2}$$

$$= (\delta y)^2 \frac{n^2 \sum_i x_i^2 - 2n\left(\sum_i x_i\right)^2 + n\left(\sum_i x_i\right)^2}{\left(n\sum_i x_i^2 - \left(\sum_i x_i\right)^2\right)^2}$$

$$= (\delta y)^2 \frac{n^2 \sum_i x_i^2 - n\left(\sum_i x_i\right)^2}{\left(n\sum_i x_i^2 - \left(\sum_i x_i\right)^2\right)^2}$$

$$= (\delta y)^2 \frac{n\left(n\sum_i x_i^2 - \left(\sum_i x_i\right)^2\right)}{\left(n\sum_i x_i^2 - \left(\sum_i x_i\right)^2\right)^2}$$

$$= (\delta y)^2 \frac{n}{n\sum_i x_i^2 - \left(\sum_i x_i\right)^2}$$

where we used $\sum_j (n^2 x_j^2) = n^2 \sum_j x_j^2 = n^2 \sum_i x_i^2$, $\sum_j \left(x_j \sum_i x_i \right) = x_1 \sum_i x_i + x_2 \sum_i x_i + \ldots,$

$$= (x_1 + x_2 + \ldots + x_n) \sum_i x_i = \left(\sum_j x_j \right) \sum_i x_i = \left(\sum_i x_i \right) \sum_i x_i = \left(\sum_i x_i \right)^2.$$

In conclusion, propagating the δy uniform uncertainty in the y values (which is the standard deviation of the residuals when we have no information about the uncertainty in the data points) into the uncertainties of a and b yields

$$\delta a = \left(\frac{\sum_i x_i^2}{n \sum_i x_i^2 - \left(\sum_i x_i \right)^2} \right)^{1/2} \delta y$$

$$\delta b = \left(\frac{n}{n \sum_i x_i^2 - \left(\sum_i x_i \right)^2} \right)^{1/2} \delta y$$

Exercise 6.6

Using the method of least squares, fit a line to the data in Exercise 4.1. Calculate the slope and intercept, as well as their uncertainties at 95% confidence level for the line of best fit.

Answer:
In order to create a generic code, we will use x and y throughout instead of manganate concentration and absorbance, except in the beginning where we load the concentrations and the corresponding absorbances into arrays x and y, respectively. After that, we will calculate the various sums required for computing the slope and intercept. Once we have these, we will use them to calculate the estimated y values (ye) corresponding to the x values. Then, we compute the residuals from y and ye, and the standard deviation of the residuals which we will propagate to obtain the standard deviation of the slope and the intercept. In the next step, we will calculate the confidence intervals for the two fitting parameters within which their true values are 95% likely to be found. Finally, we collect, format and write out the results into the *Command Window*.

```
% Exercise 4.1 using Method of Least Squares
c = (1:10)*1e-4; % manganate ion concentrations
A = [0.2336 0.3396 0.5889 0.7654 0.9079
1.0838 1.2503 1.3832 1.5775 1.6971]; % Absorbances measured
x = c; % load concentrations into array x
y = A; % load Absorbances into array y
```

```
Sxy = sum(x.*y); % ∑(xi*yi)
Sx = sum(x); % ∑(xi)
Sy = sum(y); % ∑(yi)
Sx2 = sum(x.^2); % ∑(xi^2)
Sy2 = sum(y.^2); % ∑(yi^2)
n = length(x); % number of measurements (data points)
slope = (n*Sxy-Sx*Sy)/(n*Sx2-Sx^2);
intercept = mean(y) - slope*mean(x);
ye = slope*x + intercept; % evaluate the fitted linear expression for each value in array x, yei = f(xi)
r = y - ye; % compute residuals: (yi - yei)
se = sqrt(sum(r.^2)/(n-2)); % standard deviation of residuals (δy)
std_slope = se*sqrt(n/(n*Sx2-Sx^2)); % standard deviation of slope
std_intercept = se*sqrt(Sx2/(n*Sx2-Sx^2)); % standard deviation of intercept
CI = 0.95; % required confidence level
t = tinv(1-(1-CI)/2,n-2); % t-value for two-tailed Student's t-distribution
d_slope = t*std_slope; % uncertainty in slope
d_intercept = t*std_intercept; % uncertainty in intercept
disp('Linear Least Squares Fit:')
disp(['slope = ',num2str(slope),'±',num2str(d_slope),' (',num2str(slope-d_slope),', ',num2str(slope+d_slope),')'])
disp(['intercept = ',num2str(intercept),'±',num2str(d_intercept),' (',num2str(intercept-d_intercept),', ',num2str(intercept+d_intercept),')'])
```

```
Linear Least Squares Fit:
slope = 1662.9636±77.7983 (1585.1653, 1740.762)
intercept = 0.0681±0.048273 (0.019827, 0.11637)
```

which are the same as the slope and intercept calculated by the `fit()` function in Exercise 4.1. If the uncertainty in the absorbances is known, for example we previously determined the standard deviation of A for the spectrophotometer used in conjunction with our sample preparation method, we can use that instead of `se` in the code.

7 Vectors and Their Uses in Chemistry

When planning a reaction, an important thing to consider is the polarity of the reaction medium: which polar or non-polar solvent shall we use? Before separating the product from the reaction mixture the question about polarity arises again. During the development of a new drug molecule, attention must be paid to its polarity because that will decide whether the drug will be able to cross through cell membranes or not. The polarity of a molecule is a result of uneven spatial charge distribution within the molecule; some parts of the molecule are relatively more electron dense than others. Areas of the molecule where electrons are more likely to be found will have partial negative charges, whereas areas with lower possibilities of finding electrons will have partial positive charges, while the molecule remains overall electroneutral. Since it plays an important role in chemistry, it would be useful if we could not only visualise polarity but also quantitatively capture its key characteristics, which would enable us to handle it with the tools of mathematics. For example, how does partial charge separation (unevenly distributed electrons) along bonds contribute to the overall polarity of the molecule: do the imbalances in electron density present along bonds add up? Can they cancel out? To be able to answer these types of questions, we need to find a way to hold the information on the extent and spatial direction of partial charge separation. Thus, we would ideally want to use something that could quantitatively describe both of these aspects at the same time. Vectors are just what we need: the length could represent the magnitude, whilst the direction could indicate the spatial orientation of the partial charge separation. If we decide that the vector representing charge separation points from the negative $(-q)$ towards the positive charge $(+q)$ and its length is proportional to the extent of charge separation (how much $-q$ and $+q$ are different from zero), then the charge separation in methane could be represented as shown in Figure 7.1. The carbon has slightly higher electronegativity than hydrogen, resulting in electrons being more likely to be found near the centre of the CH_4, therefore the vectors point from the carbon to the hydrogen atoms. Because the bonds are identical, the vectors have the same length.

The length of the vectors depends on q and is not the same as the C–H bond length; however, they point along the bonds. Along each bond the charge separation is $-q \rightarrow +q$, thus the charge on the carbon atom is four times the charge on a hydrogen atom. (It is worthwhile noting that if we replaced one of the hydrogens with another atom to create

---

A First Look at Coding in Chemistry: Solving Problems Using MATLAB
By Tamas Bansagi
© Tamas Bansagi 2025
Published by the Royal Society of Chemistry, www.rsc.org

Vectors and Their Uses in Chemistry

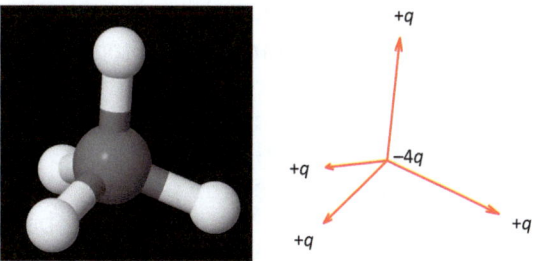

Figure 7.1 Vectors representing partial charge separations along the bonds of methane.

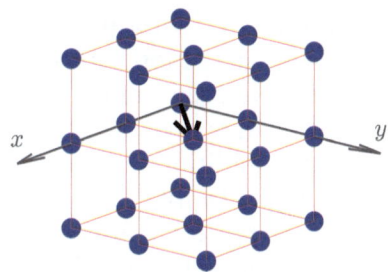

Figure 7.2 Displacement vector (→) representing moving from one of the atoms to another in the crystal lattice.

a charge separation of $-2q \rightarrow +2q$ along the bond connecting them, the charge on that atom and carbon would be $+2q$ and $-5q$, respectively. In that molecule the vector representing the $-2q \rightarrow +2q$ charge separation would be twice as long as the vectors representing the $-q \rightarrow +q$ charge separations along the three C–H bonds, irrespective of the bond lengths.)

How do polar bonds contribute to the overall polarity of a molecule? Does methane have two electric poles, one negatively and the other positively charged? Vectors enable us to answer these questions and much more. When we describe atoms and molecules in motion, the forces acting on atoms and molecules, the positions of electrons in an atom relative to the nucleus, or the strength and orientation of the electric field in the vicinity of a polar molecule, vectors make our lives a lot easier as they greatly simplify the treatment of these and numerous other problems in science and engineering.

7.1 Vector Algebra

Drawing a vector seems easy: it is just a straight arrow. On a flat sheet of paper – representing a plane – we simply first draw a line segment and then put an arrowhead at one end to indicate its direction. It seems equally straightforward to determine the length of the vector: align a ruler with it so that the zero on the ruler is at the starting point of the vector and simply read off the number at the division closest to the end point. However, when we attempt using this method to measure the distance travelled between two adjacent lattice points in a crystal, we would quickly run into difficulties (Figure 7.2). The sheet of paper taken to represent a plane cutting through a crystal lattice is too thick and we do not have a ruler with fine enough divisions for the task.

Instead of a direct measurement, we could work out the length of the displacement vector. For that we would need to have a frame of reference in which we view and solve the problem. A reference frame is constructed by specifying its origin, orientation and scale. The most convenient would be placing the origin at one of the lattice points – perhaps the one sitting at the centre of the crystal lattice – and orienting the two-dimensional frame of reference so that it aligns perfectly with the middle horizontal layer of the lattice, with the scale chosen to be the distance between the nearest lattice points in that layer measured along the edges of the lattice, which could be determined by X-ray crystallography. With a frame of reference now in place, we can apply vector algebra to solve the problem.

First we look at how vectors are defined in a two-dimensional frame of reference. A vector in 2D is conceptualised as a line section with a direction connecting the initial point with a terminal point as we have seen before. We can always draw a rectangle around a vector, making the vector the diagonal of the rectangle as shown in Figure 7.3, which can also be used to define the addition between vectors.

Imagine that vectors $\vec{i}$ and $\vec{j}$ are aligned with the edges of a sheet of paper or a crystal lattice and their sizes are chosen to represent the characteristic length scale of the reference frame (say, 1 cm on an A4 sheet, 1 Å (10^{-10} m) or even the distance between closest lattice points along an edge for a crystal lattice). They must be perpendicular to each other but their lengths can be different. Vectors $\vec{i}$ and $\vec{j}$ are called unit vectors and they can be used to define any vector in the plane of $\vec{i}$ and $\vec{j}$, like the vector in Figure 7.3 drawn diagonally ($\vec{v}$), which can be defined by the sides of the rectangle as

$$\vec{v} = \vec{i} + \vec{j}$$

If the vector is larger (or shorter), say, the sides of the rectangle drawn around it are $1.8\vec{i}$ and $3.2\vec{j}$ long as drawn in Figure 7.4, then it can be defined as

$$\vec{v} = a\vec{i} + b\vec{j}$$

where $a = 1.8$ and $b = 3.2$.

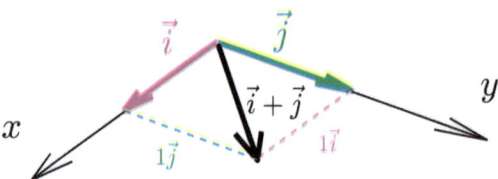

Figure 7.3 Defining a vector in 2D (resembling the one in Figure 7.2) using unit vectors.

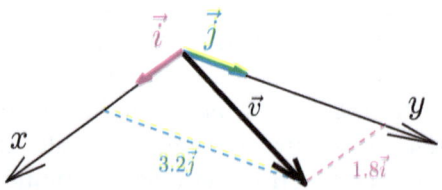

Figure 7.4 Defining a vector in 2D using unit vectors.

Vectors and Their Uses in Chemistry

a and b are called the components of $\vec{v}$, and they make it straightforward to calculate the length (modulus or Euclidean norm) of the vector by applying Pythagoras' theorem:

$$\vec{v} = \sqrt{a^2 + b^2}$$

In three dimensions, vectors can be similarly defined as $\vec{v} = a\vec{i} + b\vec{j} + c\vec{k}$, so that the vector drawn in Figure 7.5 is given as $\vec{v} = 3.5\vec{i} + 4\vec{j} + 5\vec{k}$. The length of $\vec{v} = \sqrt{a^2 + b^2 + c^2} = \sqrt{3.5^2 + 4^2 + 5^2} \approx 7.3$. MATLAB's `norm()` or `vecnorm()` functions provide an easy way to compute the modulus of a vector.

```
>> v=[3.5  4  5]; % define vector by assigning its components to a 1D array
>> modulus_v=norm(v)
modulus_v=
    7.2973
```

The advantage of defining vectors this way is that the addition and subtraction of vectors – which may otherwise seem complicated, especially in 3D – are reduced to elementary algebra tasks: adding or subtracting respective vector components. For vectors $\vec{a}$ and $\vec{b}$ given as

$$\vec{a} = a_1\vec{i} + a_2\vec{j} + a_3\vec{k}$$

$$\vec{b} = b_1\vec{i} + b_2\vec{j} + b_3\vec{k}$$

the vectors resulting from summing or subtracting them are:

$$\vec{a} \pm \vec{b} = (a_1 \pm b_1)\vec{i} + (a_2 \pm b_2)\vec{j} + (a_3 \pm b_3)\vec{k}$$

For example, the sum of vectors $\vec{a} = 1\vec{i} + 3\vec{j} + 2\vec{k}$ and $\vec{b} = 4\vec{i} + 2\vec{j} + 3\vec{k}$ shown in Figure 7.6 is $(\vec{a} + \vec{b}) = 5\vec{i} + 5\vec{j} + 5\vec{k}$.

The *scalar* or *dot product* of vectors, $\vec{a} \cdot \vec{b}$, is defined as

$$c = \vec{a} \cdot \vec{b} = a_1 b_1 + a_2 b_2 + a_3 b_3 = |\vec{a}||\vec{b}|\cos\theta$$

where θ is the angle between $\vec{a}$ and $\vec{b}$. The dot product enables us to calculate bond angles in molecules using simple algebra (which can be performed quickly and easily

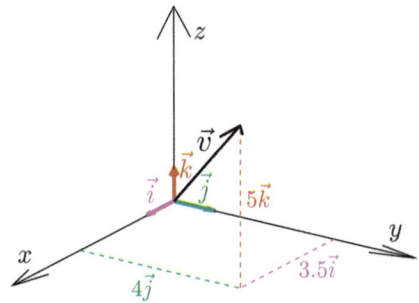

Figure 7.5 Defining a vector in 3D using unit vectors.

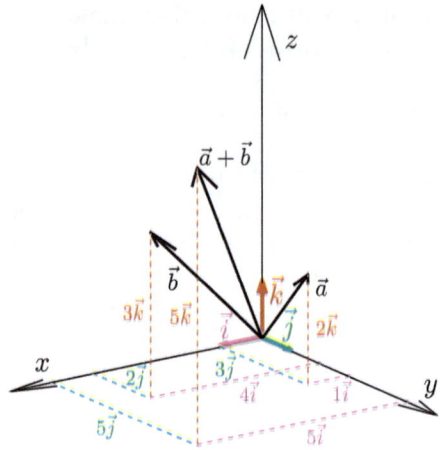

Figure 7.6 Addition of vectors.

on computers). If we have the coordinates of the atoms (o) in a molecule, the bond angle between three atoms in a non-cyclic arrangement (o←$\vec{a}$—o—$\vec{b}$→o) can be computed as

$$\theta = \cos^{-1}\left(\frac{\vec{a}\cdot\vec{b}}{|\vec{a}||\vec{b}|}\right) = \cos^{-1}\left(\frac{a_1 b_1 + a_2 b_2 + a_3 b_3}{\sqrt{a_1^2 + a_2^2 + a_3^2}\sqrt{b_1^2 + b_2^2 + b_3^2}}\right)$$

instead of performing lengthy trigonometric calculations.

Exercise 7.1

The spatial coordinates of the atoms in methane can be given (in Å) as

Atom	X	Y	Z
C	0.000	0.000	0.000
H	0.000	0.000	1.089
H	1.026	0.000	−0.363
H	−0.513	−0.889	−0.363
H	−0.513	0.889	−0.363

Using MATLAB, calculate the H–C–H bond angle in methane. Use the coordinates of the first and second hydrogen atoms as the components of the vectors pointing from carbon to the hydrogen atoms. (i) First, use the expression above, (ii) then the `norm()` function and the element-wise multiplication operator (.*). The inverse of cosine in degrees is calculated by the `acosd()` function.

Answer:

```
>> at_crd=[0.000  0.000  0.000; 0.000  0.000  1.089; 1.026  0.000 −0.363;
−0.513  −0.889  −0.363;  −0.513  0.889  −0.363];    % assign atomic
coordinates
>> a=at_crd(2,:);  % define C→H(1) vector
>> b=at_crd(3,:);  % define C→H(2) vector
```

```
>> ang_i=acosd((a(1)*b(1)+a(2)*b(2)+a(3)*b(3))/(sqrt(a(1)^2+a(2)^2
+a(3)^2)*sqrt(b(1)^2+b(2)^2+b(3)^2)))   % compute bond angle using
formula directly
>> ang_ii=acosd(sum(a.*b)/(norm(a)*norm(b)))  % compute bond angle using
collapsed formula
ang_i=
   109.4838
ang_ii=
   109.4838
```

The cross product of vectors, $\vec{a} \times \vec{b}$, is defined as

$$\vec{c} = \vec{a} \times \vec{b} = \vec{a}\vec{b}\sin\theta$$

with the resulting vector being perpendicular to the plane containing $\vec{a}$ and $\vec{b}$. The direction of the resulting vector depends on how the angle is measured between $\vec{a}$ and $\vec{b}$. If the angle is measured counterclockwise from $\vec{a}$ to $\vec{b}$, $\vec{c}$ will point upward; whereas if the angle is measured clockwise from $\vec{a}$ to $\vec{b}$, $\vec{c}$ will point downward. This means that $\vec{a} \times \vec{b} = -\left(\vec{b} \times \vec{a}\right)$. (Please remember that the scalar or dot product of vectors is a scalar (a number), whereas the cross product of vectors is a vector.) We will return to the cross product and its applications in Exercise 7.3 and Exercise 7.4.

7.2 Drawing Vectors in MATLAB

MATLAB is fully equipped to handle and visualise vectors. The `quiver()` and `quiver3()` functions are easy to use to draw 2D and 3D vectors. To draw a 2D vector we need to specify the coordinates of the starting point of the vector and the displacements along the x and y axes to the end point. MATLAB usually prefers to automatically scale vectors. The auto scale functionality is often very useful when we have many overlapping vectors and it would be too difficult to see which is which. Now, however, we would only like to draw a single vector, so we need to insist that MATLAB draw our vector scaled exactly to our specification (Figure 7.7), which we can do by writing `'AutoScale','off'` following the vector components within `quiver()` and `quiver3()`.

```
xo=-1; yo=-0.5; % x and y coordinates of the origin of the vector
xd=1.5; yd=2; % displacements along x and y axes to end point
quiver(xo,yo,xd,yd,'r','AutoScale','off') % draw vector; 'r': red line
axis square % draw square diagram
axis([-2 2 -2 2]) % ranges of x and y axes on the diagram
grid on % add grid lines to aid the eye
xlabel x % shorthand for labelling x axis
ylabel y % shorthand for labelling y axis
```

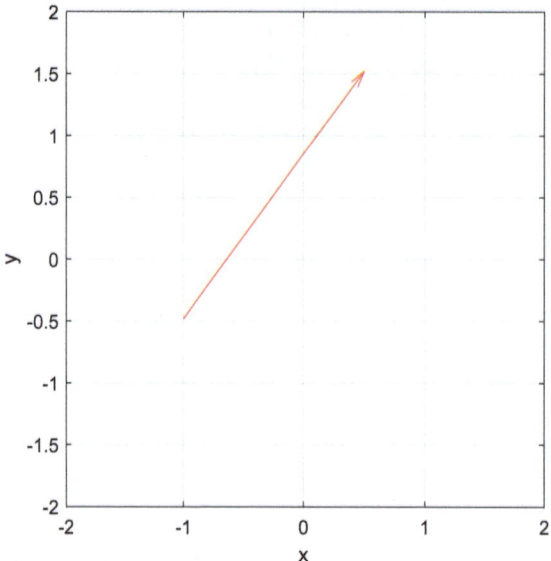

Figure 7.7 Vector drawn by MATLAB's `quiver()` function.

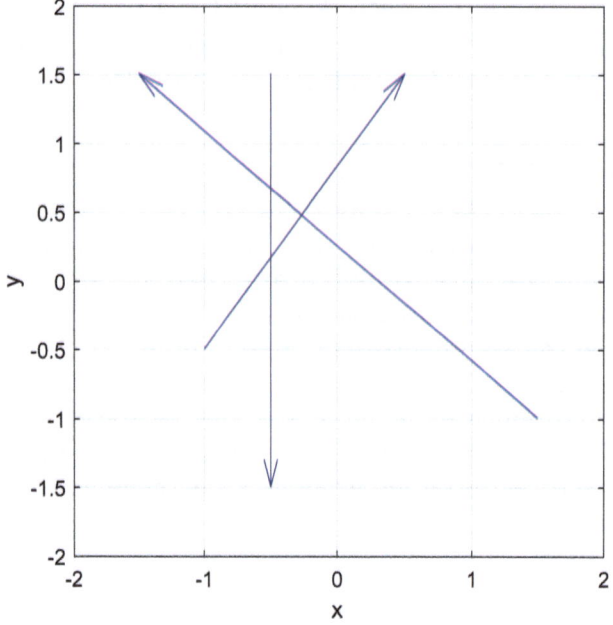

Figure 7.8 Passing arrays to `quiver()` generates multiple vectors in one step.

MATLAB supports drawing large amounts of vectors all at once. All we need to do is to expand xo, yo, xd and yd to store starting points and displacements for multiple vectors. Let us draw three vectors first (Figure 7.8).

```
xo = [-1   1.5 -0.5]; % x coordinates of origins
yo = [-0.5 -1   1.5]; % y coordinates of origins
xd = [1.5 -3   0];    % displacements along x axis to end points
```

```
yd = [2  2.5 −3]; % displacements along y axis to end points
quiver(xo,yo,xd,yd,'b','AutoScale','off') % draw vectors; 'b': blue lines
axis square % draw square diagram
axis([−2  2 −2  2]) % ranges of x and y axes on the diagram
grid on % add grid lines to aid the eye
xlabel x % shorthand for labelling the x axis
ylabel y % shorthand for labelling the y axis
```

MATLAB also makes it simple to draw vectors in 3D. For that use the `quiver3()` function as shown below (Figure 7.9).

```
xo = [0   1.5 −1.5]; % x coordinates of origin
yo = [0   0   1.5]; % y coordinates of origin
zo = [−1.5  0  0.5]; % z coordinates of origin
xd = [0 −3  3]; % displacements along x axis to end points
yd = [0  0 −3]; % displacements along y axis to end points
zd = [3  0 −1]; % displacements along z axis to end points
quiver3(xo,yo,zo,xd,yd,zd,'m','AutoScale','off') % draw vectors;
axis square % draw cubic diagram
axis([−2  2 −2  2 −2  2]) % ranges of x y z axes on the diagram
grid on % add grid lines to aid the eye
xlabel x % shorthand for labelling the x axis
ylabel y % shorthand for labelling the y axis
zlabel z % shorthand for labelling the z axis
```

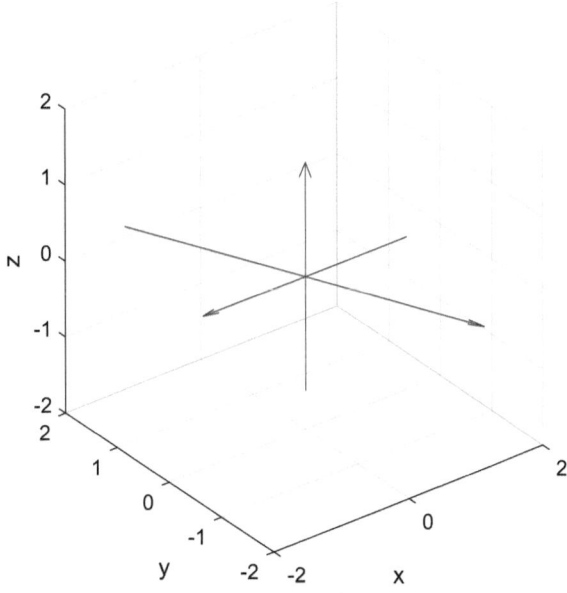

Figure 7.9 Vectors drawn by MATLAB's `quiver3()` function.

7.3 Vector Operations in MATLAB: *Addition and Subtraction*

Broadly speaking, a vector is a set of numbers in a row or column. Three such numbers can be taken as the *x,y,z* coordinates (or components) of the end point of a three-dimensional vector starting from the origin. In MATLAB, we can define two vectors simply as

```
>> v1=[2  1   4];
>> v2=[1 -2 -3];
```

If we want to quickly plot them we can just use:

```
>> quiver3(0,0,0,v1(1),v1(2),v1(3),'m','AutoScale','off'); hold on;
>> quiver3(0,0,0,v2(1),v2(2),v2(3),'g','AutoScale','off'); hold on;
>> xlabel x; ylabel y; zlabel z;
```

Adding and subtracting vectors are done by simply (be advised that the minus sign may not copy over correctly so you might need to retype the latter two commands)

```
>> v3 = v1 + v2
>> v4 = v2 + v1
>> v5 = v1 - v2
>> v6 = v2 - v1
```

The *x,y,z* components of the resulting 4 vectors are written into the *Command Window*. Let us also plot these vectors **v3**, **v4**, **v5** and **v6**, with **v4** drawn in a dashed yellow line so that the overlap with **v3** is easy to see (Figure 7.10).

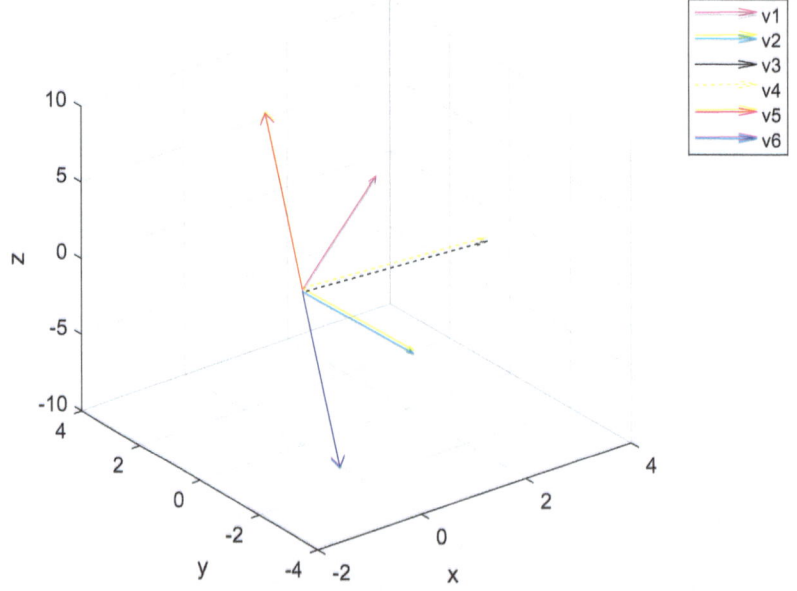

Figure 7.10 Vectors v1 and v2, and the vectors resulting from the additions and subtractions of v3 = v1 + v2; v4 = v2 + v1; v5 = v1 − v2; v6 = v2 − v1. Grab diagram on the screen and rotate around.

Vectors and Their Uses in Chemistry 121

```
>> quiver3(0,0,0,v3(1),v3(2),v3(3),'k','AutoScale','off'); hold on;
>> quiver3(0,0,0,v4(1),v4(2),v4(3),'--y','AutoScale','off'); hold on;
>> quiver3(0,0,0,v5(1),v5(2),v5(3),'r','AutoScale','off'); hold on;
>> quiver3(0,0,0,v6(1),v6(2),v6(3),'b','AutoScale','off'); hold on;
>> legend('v1','v2','v3','v4','v5','v6')
```

As expected, v5≠v6 because during subtraction the order of vectors matters. (Subtraction is a *noncommutative* operation.)

Exercise 7.2

Show using vector addition in MATLAB that despite the partial charge separation along the bonds, methane does not possess two electric poles (*i.e.* a methane molecule is not an electric dipole).

Answer:

Because an equal amount of charge separation occurs along each bond, we can use the atomic coordinates in Exercise 7.1 to represent the spatial directions of all charge separation in the molecule.

```
>> at_crd=[0.000  0.000  0.000;  0.000  0.000  1.089;  1.026  0.000
-0.363;  -0.513 -0.889 -0.363;  -0.513  0.889 -0.363];  % assign
atomic coordinates
>> vect_of_sum_of_charge_sep=at_crd(2,:)+at_crd(3,:)+at_crd(4,:)+
at_crd(5,:) % sum the vectors representing the directions of charge
separation along bonds
vect_of_sum_of_charge_sep=
     0       0       0
```

The vector resulting from summing the individual vectors representing charge separation along the bonds has zero components. This means that the overall charge separation in methane is given by a vector of zero length; consequently, methane does not possess two overall electric poles; *i.e.*, methane molecules are not electric dipoles.

A molecule not being an electrical dipole, does not mean that there is no charge separation in it. Methane still has an electric field around it that other molecules in its close vicinity feel the effect of. *Methane*, given its spatial distribution of charge separation, appears to the outside microscopic world as an *electric quadrupole*.

7.4 Vector Operations in MATLAB: Dot (Scalar) Product and Cross Product

We assign new components for vectors **v1** and **v2**, before computing their two different vector products.

```
>> v1=[-2  3  0.5];
>> v2=[-1.5 -2  1];
```

```
>> v7=dot(v1,v2)
>> v8=dot(v2,v1)
>> v9=cross(v1,v2)
>> v10=cross(v2,v1)
```

The x, y, z components of the resulting vectors are written into the *Command Window*: **v7** and **v8** are scalars; whereas, **v9** and **v10** are vectors shown in Figure 7.11. (Note that vectors **v9** and **v10** are pointing in opposite directions. That is because the cross product is an anticommutative operation, which means that the modulus of **v9** and **v10** are the same ($|v9| = |v10|$), despite **v9** = −**v10**.)

```
>> quiver3(0,0,0,v1(1),v1(2),v1(3),'m','AutoScale','off'); hold on;
>> quiver3(0,0,0,v2(1),v2(2),v2(3),'g','AutoScale','off'); hold on;
>> quiver3(0,0,0,v9(1),v9(2),v9(3),'k','AutoScale','off'); hold on;
>> quiver3(0,0,0,v10(1),v10(2),v10(3),'y','AutoScale','off'); hold on;
>> legend('v1','v2','v9','v10')
>> xlabel x; ylabel y; zlabel z;
>> axis equal % make axis scales the same
>> axis padded % add a bit of space around objects displayed
```

If you are having difficulty seeing the vectors, type

```
>> view([10,50])
```

or

```
>> view(10,50)
```

into the Command Window, which should rotate the 3D diagram (see above) so that you can see better that **v9** and **v10** are both perpendicular to the plane of **v1** and **v2**.

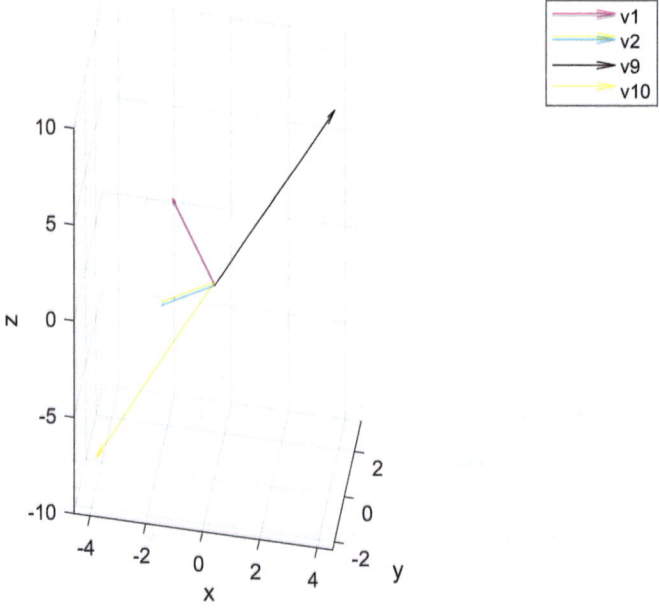

Figure 7.11 Vectors v1 and v2, and the vectors resulting from their cross products v9 = v1 × v2; v10 = v2 × v1. Grab diagram on the screen and rotate around.

Vectors and Their Uses in Chemistry

The cross product of vectors follows the so-called *right-hand rule*, which means that if you align your index finger with **v1** and your middle finger with **v2**, their cross product, **v9**, will align with your thumb. The direction of the cross product is determined by how the angle is measured between **v1** and **v2**. If the angle is measured counterclockwise from **v1** and **v2**, the cross product, **v9**, will point upward; whereas if the angle is measured clockwise from **v1** and **v2** (*i.e.* if we swap the two vectors in the formula), the cross product, **v10**, will point downward.

Exercise 7.3

In mass spectrometers, charged particles (fragments of molecules) fly in rarefied space under the influence of a magnetic field. The force, ***F***, acting on a moving charged particle (measured in newtons, N) due to a magnetic field depends on the charge, q, of the particle (measured in coulombs, C), the velocity, ***v***, of the particle, and the magnetic induction, ***B***, (often called the "magnetic field", measured in tesla, T):

$$\boldsymbol{F} = q\boldsymbol{v} \times \boldsymbol{B}$$

Find the force on an electron flying along the x axis left to right in a magnetic field with a speed of 2.5×10^5 m s^{-1} in a magnetic field of 3.0 T orienting along the z axis (pointing from North (+) to South (−)). The value of the charge of an electron is $q = -1.6 \times 10^{-19}$ C. (If the force is perpendicular to the velocity, it will cause the trajectory of the particle to curve, instead of changing the speed of the particle.)

Answer:

```
clear all
format shortG
v=[2.5e5,0,0]; % m/s; define velocity vector
B=[0,0,-3]; % T; define magnetic field vector
q=-1.60e-19; % C; define charge of electron (scalar)
F=q*cross(v,B) % compute Force vector
v_sc=v/1e5; % scale v for better displaying
F_sc=F/1e-13; % scale F for better displaying
% resultant direction of motion of the electron
R=v+F
R_sc=R./[1e5 1e-13 1]; % scale v_res for better displaying
quiver3(0,0,0,v_sc(1),v_sc(2),v_sc(3),'m','AutoScale','off'); hold
on; % draw velocity vector of the electron
quiver3(0,0,0,B(1),B(2),B(3),'g','AutoScale','off'); hold on; % draw
vector representing the magnetic field acting on the electron
quiver3(0,0,0,F_sc(1),F_sc(2),F_sc(3),'k','AutoScale','off'); hold
on; % draw the (scaled) force acting on the electron
quiver3(0,0,0,R_sc(1),R_sc(2),R_sc(3),'b','AutoScale','off'); hold
on; % draw the (scaled) resultant vector representing the direction of
motion of the electron under the influence of the magnetic field applied
plot3(0,0,0,'r.','markersize',20) % draw red dot representing the
electron
legend('v','B','F','R','e^-') % label diagram
```

```
xlabel('x / 10^5 m'); ylabel('y / 10^{-13} m'); zlabel('z / m'); % label
axes including the scaling applied
axis padded
view(40,15) % set view point of diagram
```

F =

    0   −1.2e−13   0

which means that a force of $|F| = 1.2 \times 10^{-13}$ N will act on the electron directed along the negative y axis.

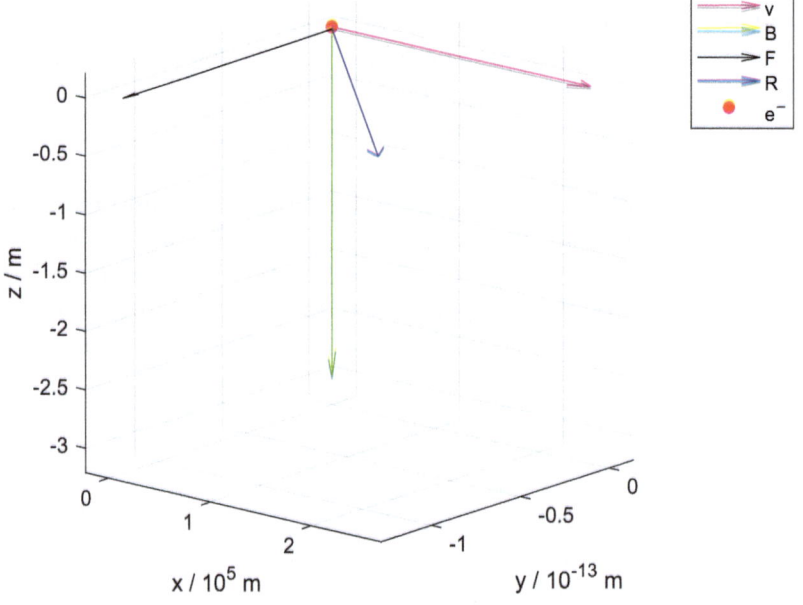

As a result of the force (black arrow) due to the magnetic field (green arrow), the path of the electron will deviate from its current trajectory (magenta vector, moving from left to right), towards negative y values; *i.e.* the trajectory of the electron will bend towards us in the v-F plane (indicated by the blue arrow).

In what direction would the path of an alpha particle (He^{2+}) bend under the same conditions?

7.5 Vector Operations in MATLAB: *Calculating Volumes*

Vector operations are useful in calculating volumes of unit cells of crystal lattices (Figure 7.12). For unit cell edge lengths represented by vectors $\vec{a}$, $\vec{b}$, and $\vec{c}$, and internal angles α, β, and γ the volume of the unit cell can be calculated simply as

$$\left(\vec{a} \times \vec{b}\right) \cdot \vec{c}$$

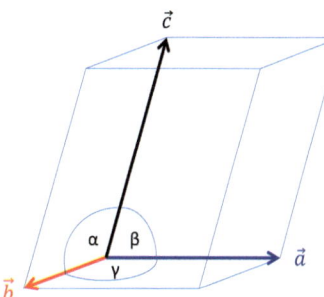

Figure 7.12 Unit cell of a crystal lattice and the vectors defining it.

If we know the coordinates of the end points of the vectors the calculation becomes very simple using vector operations in MATLAB.

Exercise 7.4

What is the volume (in Å^3) of a unit cell if the vectors representing its edges are given as vectors $\vec{a}$, = (5.2 Å, 0 Å, 0 Å); $\vec{b}$, = (1.5 Å, 4.3 Å, 0 Å); $\vec{c}$ = (0.5 Å, 0.4 Å, 6.1 Å)? What are the angles (in degrees) between the edges of the unit cell?

Answer:

```
a=[5.2  0   0];
b=[1.5  4.3 0];
c=[0.5  0.4 6.1];
vol=dot(cross(a,b),c) % calculate volume
% vol_2=dot(cross(b,c),a) % same as vol
% vol_3=dot(cross(c,a),b) % same as vol => volume is the same
irrespective of the order of vectors
ang_a_b=acosd(sum(a.*b)/(norm(a)*norm(b))) % angle between a and b
ang_a_c=acosd(sum(a.*c)/(norm(a)*norm(c))) % angle between a and c
ang_b_c=acosd(sum(b.*c)/(norm(b)*norm(c))) % angle between b and c
quiver3(0,0,0,a(1),a(2),a(3),'b','AutoScale','off'); hold on;
quiver3(0,0,0,b(1),b(2),b(3),'r','AutoScale','off'); hold on;
quiver3(0,0,0,c(1),c(2),c(3),'k','AutoScale','off'); hold on;
legend('a','b','c')
axis equal % make axis scales the same
axis padded % add a bit of space around objects displayed
xlabel('x / Å'); ylabel('y / Å'); zlabel('z / Å');
view([30,30])
```

vol =
     136.4

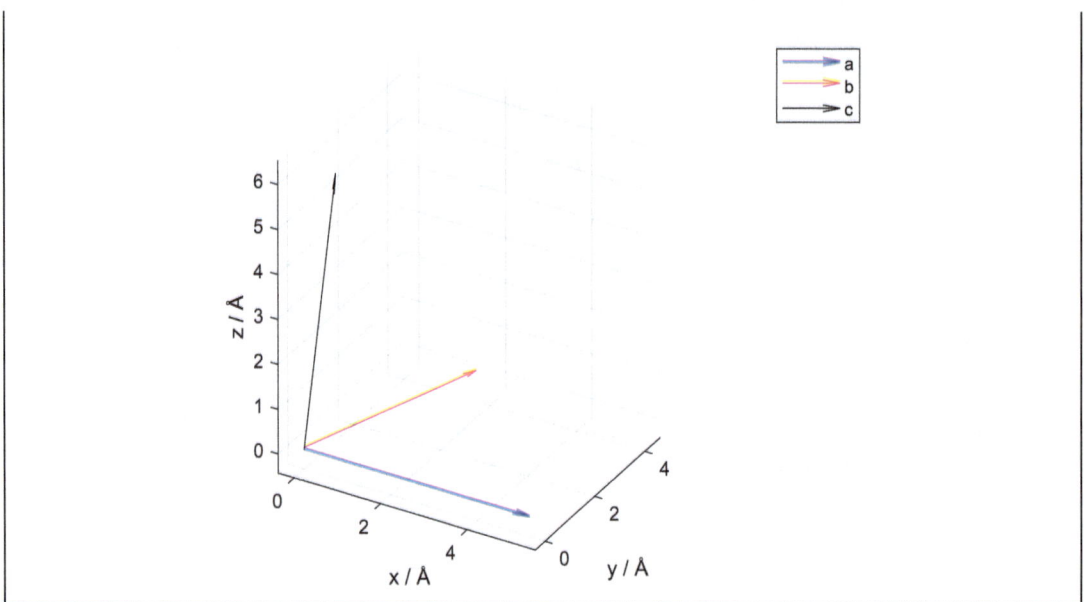

Exercise 7.5

Construct a model of a system composed of a singly positive ion, $Q=+e$, and a dipole where partial charges $+q=+\frac{1}{2}e$ and $-q=-\frac{1}{2}e$ are separated by 2 Å. The three point-charges are perfectly aligned with the $+q$ end of the dipole pointing toward the ion ($\theta=0°$). The midpoint of the dipole is 6 Å from the ion.

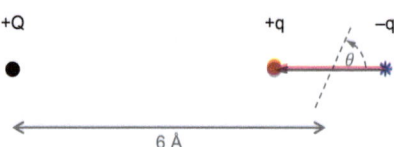

a. Create a visual representation of the system that shows the rotation of the dipole in two equal steps from its initial position ($\theta=0°$) until its side faces the ion (θ: $0° \rightarrow 45° \rightarrow 90°$) and calculate the potential energies of the initial and the two subsequent orientations using Coulomb's Law.
b. Starting from the initial position, rotate the dipole in a full circle in 30° steps, calculate the ion–dipole potential energies (U_{ID}) of all orientations using Coulomb's Law and establish which trigonometric function describes the relationship between rotation angle θ and U_{ID}.
c. Using a dipole orientation with non-zero U_{ID}, investigate the effect of distance and determine n in the expression $U_{ID} \propto r^n$.

Answer:

```
ke=1; % Coulomb constant 1ke=1 x 8.988×10^9 N m^2 C^{-2}
Q=1; % charge of ion, 1e=1 x 1.602×10^{-19} C
```

```
q=0.5; % partial charges of dipole, 0.5e=0.5 x 1.602176634×10^{-19} C
r=6; % distance between ion and midpoint of dipole in Å (1Å=100 pm)
r_dpl=[r 0]; % x,y coordinates of midpoint of dipole
th_d=[0  45  90]'; % rotation angles in degrees with respect to y=0
line
th=deg2rad(th_d); % convert rotation angles to radians
m=length(th); % number of rotation angles generated
rQ=repmat([0  0],m,1); % create rows of the same x,y coordinates of ion
L=2; % size of dipole in in Å (1Å=100 pm)
r_pq=[r_dpl(1)-L/2*cos(th) r_dpl(2)-L/2*sin(th)]; % x,y
coordinates of +q
r_mq=[r_dpl(1)+L/2*cos(th) r_dpl(2)+L/2*sin(th)]; % x,y
coordinates of -q
rQ_pq=vecnorm(r_pq - rQ,2,2); % distance between Q and +q
rQ_mq=vecnorm(r_mq - rQ,2,2); % distance between Q and -q
U=ke*Q*q./rQ_pq+ke*Q*(-q)./rQ_mq; % compute ion-
dipole potential energy
% Magnitude of U in J: 8.988e9*(1.602e-19)^2/1e-10=2.3067e-18
subplot(2,2,[1,2]) % Diagram showing model of ion-dipole system
plot(rQ(1),rQ(2),'ko','MarkerFaceColor','k');
hold on
plot(r_pq(1:m),r_pq(m+1:end),'ro','MarkerFaceColor','r');
hold on
plot(r_mq(1:m),r_mq(m+1:end),'b*');
hold on
dip_vec(:,1)=r_pq(1:m)-r_mq(1:m); % x coordinates of +q
dip_vec(:,2)=r_pq(m+1:end)-r_mq(m+1:end); % y coordinates of +q
% draw dipole vectors from [x y] of (-q) to [x y] of (+q)
quiver(r_mq(1:m,1),r_mq(1:m,2),dip_vec(:,1),dip_vec
(:,2),'AutoScale','off','MaxHeadSize',0.4)
% generate text locations for displaying rotational angle theta
text_poz=[r_dpl(1)+1.5*L/2*cos(th) r_dpl(2)+1.5*L/2*sin(th)];
% generate text equivalent to numerical rotational angles
ang_text=[num2str(th_d),repmat('^{\circ}',m,1)];
% add rotational angles to the diagram
text(text_poz(1:m),text_poz(m+1:end),ang_text,'fontsize',8,
'horizontalalignment','center','verticalalignment','middle')
legend({'+Q' '+q' '-q'},'location','northeast')
axis equal
axis([-1  11 -2  2])
xlabel ('x / Å'); ylabel ('y / Å')
subplot(2,2,3) % diagram showing the Q to (-q) and (+q) distances
plot(th_d,rQ_mq,'-bv',th_d,rQ_pq,'-r^')
xlabel('\theta / deg'); ylabel('|r_{q-Q}| / Å')
legend({'-q' '+q'},'location','west')
subplot(2,2,4) % diagram showing ion-dipole energies as dipole rotates
plot(th_d,U,'-gs')
xlabel('\theta / deg'); ylabel('U x (k_ee^2/Å) / J')
```

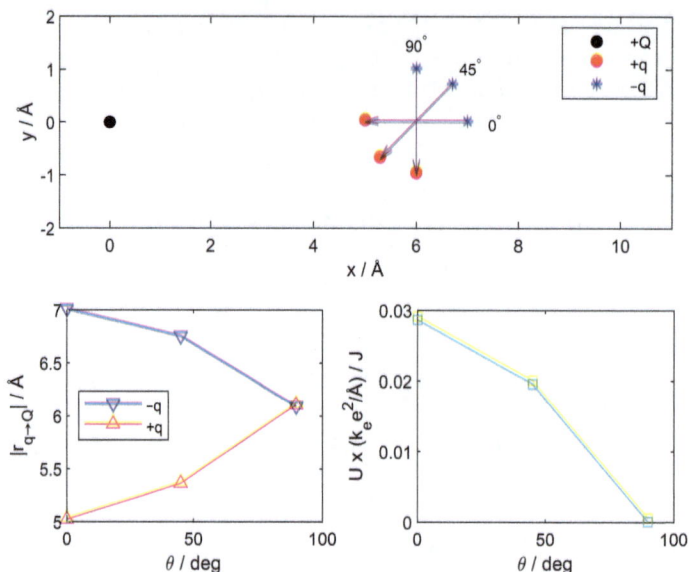

With the graphics and calculations tested, we continue our investigation by fully rotating the dipole. This will require (after clearing all variables by running `clear all`) replacing `th_d=[0 45 90]';` with `th_d=(0:30:360)';`. To erase the set of characters rendering the last angle (360°) to avoid writing over 0°, we fill the last row of array `ang_text` with white space characters as `ang_text(end,:)=' '`, which we insert into the code just under the line generating `ang_text`.

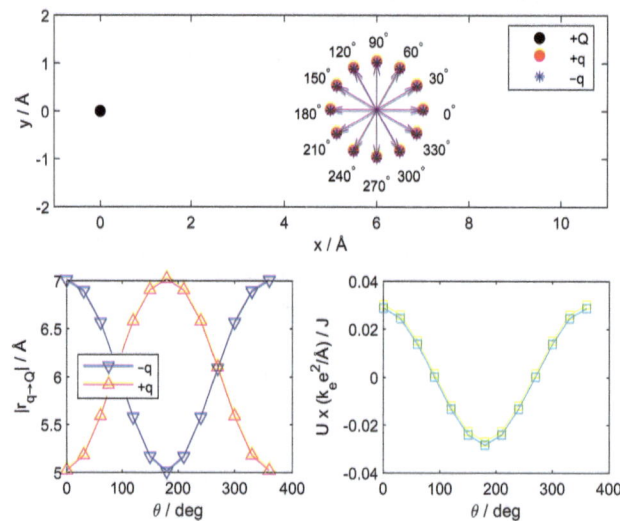

By inspecting the bottom right subplot (green curve: symmetric shape, maxima at 0° and 360°, minimum at 180°, 0 J at 90° and 270°), we conclude that the ion–dipole energy (U_{ID}) changes with $\cos\theta$, thus

$$U_{ID} \propto -\cos\theta$$

Now we investigate the role of distance. We set the angle to 0 as this will lead to the largest non-zero interaction energy and change r gradually from 3 Å to 9 Å, while computing U. We will determine n as the slope of the line fitted to the ln U vs. ln r data points.

```
th=0; % angle giving largest non-zero U (0 deg=0 rad)
r=(3:0.5:9)'; % distances: 3Å...9Å
m=length(r); % number of distances generated
rQ=repmat([0 0],m,1); % create rows of the same x,y coordinates of ion
r_dpl=[r zeros(m,1)]; % x,y coordinates of midpoint of dipole
r_pq=[r_dpl(:,1)-L/2*cos(th) r_dpl(:,2)-L/2*sin(th)]; % x,y coords of +q
r_mq=[r_dpl(:,1)+L/2*cos(th) r_dpl(:,2)+L/2*sin(th)]; % x,y coords of -q
rQ_pq=vecnorm(r_pq - rQ,2,2); % distance between Q and +q
rQ_mq=vecnorm(r_mq - rQ,2,2); % distance between Q and -q
U=ke*Q*q./rQ_pq+ke*Q*(-q)./rQ_mq; % compute ion-dipole potential energy
% Magnitude of U in J: 8.988e9*(1.602e-19)^2/1e-10=2.3067e-18
f=fit(log(r),log(U),'poly1')
subplot(1,2,1)
plot(r,U,'-gs')
xlabel('r / Å'); ylabel('U x (k_ee^2/Å) / J')
subplot(1,2,2)
plot(f,'-m',log(r),log(U),'-g^')
xlabel('ln(r / Å)'); ylabel('ln(U x (k_ee^2/Å) / J)')
legend('ln(U)','ln(U) =n ln(r) +const')
axis padded
```

```
Linear model Poly1:
f(x) =p1*x+p2
Coefficients (with 95% confidence bounds):
  p1=-2.086   (-2.105, -2.068)
  p2=0.1906   (0.1576, 0.2236)
```

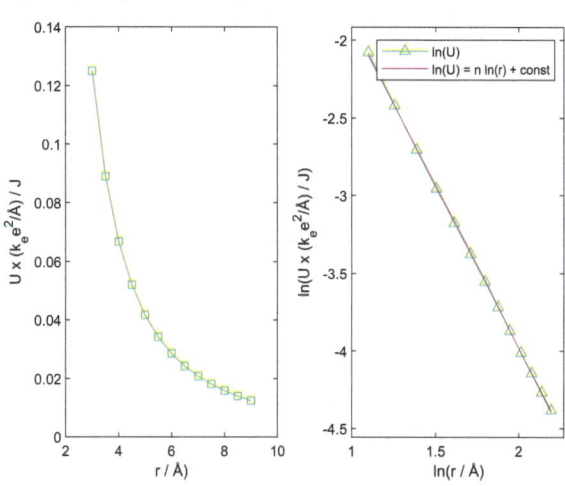

> From `p1 = -2.086`, we conclude that $n = -2$ and therefore
>
> $$U_{\text{ID}} \propto -1/r^2$$
>
> which means that the interaction energy of an ion with a dipole is proportional to the inverse square of the distance between them. Combining our results yields the relationship
>
> $$U_{\text{ID}} \propto -\frac{\cos\theta}{r^2}$$

7.6 Appendix: Vector Fields (Force Fields)

Now that we can create large numbers of vectors with very little effort in MATLAB, we can apply this to visualising fields. Electric and magnetic fields are very important in chemistry, for example, when we are trying to understand the electrostatic interactions between ions and molecules through space. Being able to visualise the electric field around ions and molecules can greatly help us imagine and understand forces acting at the molecular level. We use the `meshgrid()` function to generate the coordinates of the grid points resolving space around the point charges representing atoms in molecules with polar bonds. In each grid point we calculate the direction and relative strength of the electric field to create the impression of the electric field surrounding the molecule (Figure 7.13). (The local field at a point in space around a molecule is the sum of the fields generated by the point charges representing partially charged atoms in the molecule. The local fields in each grid point are calculated using Coulomb's Law, before being vectorially summed up to produce the local electric field a positive point charge would experience at that point in space.)

The script below draws the electric field (magnitude and orientation of $\vec{E}$: pointing from + charges to − charges) on a 2D mesh around a HCl molecule. (Vectors represent forces acting on a positive point charge placed at the grid points. This charge would travel, while continuously accelerating, along trajectories resolved by subsequent vectors.)

Modify the script below to visualise the electric field around a CO_2 molecule (Figure 7.14).

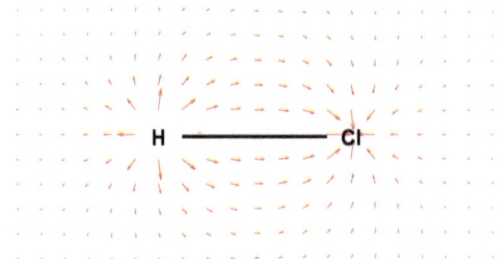

Figure 7.13 Vector field illustrating the direction and strength of the electric field around a hydrogen chloride molecule if the atoms were point charges.

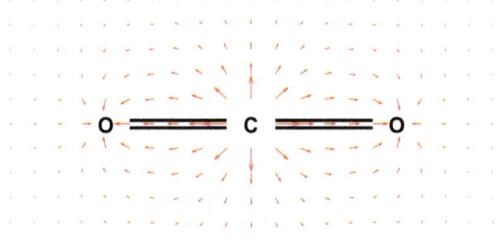

Figure 7.14 Vector field representing the direction and strength of the electric field surrounding a carbon dioxide molecule if the atoms were point charges.

These representations show what the electric field would look like around and in between the atoms if they were point charges existing in the classical world. We must view them as rough approximations as the microscopic world can be accurately described only by quantum mechanics and not classical field equations.

```
x=0:1:20; y=0:1:10; % set space coordinates on plane
x1=6; y1=5; q1=1; % set location of charge 1
x2=14; y2=5; q2=-1; % set location of charge 2
[xp1,yp1]= meshgrid(x-x1,y-y1); % generate mesh points relative to charge 1
[xp2,yp2]= meshgrid(x-x2,y-y2); % generate mesh points relative to charge 2
rs1=xp1.^2+yp1.^2; % calculate distances^2 (r1^2) from charge 1
rs2=xp2.^2+yp2.^2; % calculate distances^2 (r2^2) from charge 2
quiver(x,y,q1*xp1./rs1+q2*xp2./rs2,q1*yp1./rs1+q2*yp2./rs2,0.75,'r');
% Components of local electric field: Ex=∑qx/r^2; Ey=∑qy/r^2
hold on
plot([x1+1 x2-1],[y1 y2],'-k','linewidth',2) % draw line to connect atoms
text(x1,y1,'H','HorizontalAlignment','center','FontSize',14,'FontWeight',
'bold') % write H on the diagram
text(x2,y2,'Cl','HorizontalAlignment','center','FontSize',14,'FontWeight',
'bold') % write Cl on the diagram
axis equal
axis off
```

8 Matrices in Chemistry

The more we learn about chemistry the more complex the molecules we encounter tend to become. Usually, the difficulty in keeping track of the positions of atoms and how they are connected rapidly increases with molecular mass. Imagine that you work for a company specialising in synthetizing new drug molecules for medical testing. As part of your job, you are tasked with setting up a system for storing the structures of all test molecules. You quickly realise that you need to build a searchable digital archive and the first question you face is how to best capture molecular structure? After some thinking it becomes clear that a table containing the three-dimensional – x, y, z – coordinates of each atom in a molecule together with another table containing which atoms are connected *via* bonds and the bond order would nicely do the job. For a quick test of your chemical information storage system you choose the molecule: 1,3-dichloroallene (Tables 8.1 and 8.2).

To generate a rudimentary image of the 3D structure of the molecule, we can use `scatter3()` for the atoms and `plot3()` for the bonds. Circles drawn at the locations specified in Table 8.1 could give the impression of atoms, while lines connecting them could represent the bonds. The sizes and colours of circles could be set to reflect the relative atomic diameters and match the colours used in molecule building kits and visualisation software.

Figure 8.1 shows that the structure of 1,3-dichloroallene can be captured in just two simple tables. It also demonstrates that complex information, like molecular structure, can be neatly stored in simple two-dimensional arrays.

Because chirality has vital importance in drug discovery, you wonder if it is possible to use Table 8.1 directly to establish whether this molecule is chiral. As chirality results from no mirror symmetry, you could test for chirality by flipping all signs in each column and redrawing the molecule. This is because the central carbon atom is at the (0,0,0) location, so flipping the signs of all numbers in the column for the x-coordinates of the atoms would have the effect of projecting the atoms across the yz plane. Similarly, multiplying all numbers in the columns for the y-coordinates and z-coordinates of the atoms by (−1) would result in all atoms projected across the xz and xy planes,

---

A First Look at Coding in Chemistry: Solving Problems Using MATLAB
By Tamas Bansagi
© Tamas Bansagi 2025
Published by the Royal Society of Chemistry, www.rsc.org

Table 8.1 Atoms and their coordinates in 1,3-dichloroallene in pm.

Atom	x coordinate	y coordinate	z coordinate
C	−131	0	0
C	0	0	0
C	131	0	0
Cl	−217	−150	0
Cl	217	0	150
H	−185	94	0
H	185	0	−94

Table 8.2 Bonds between atoms (numbers in columns 1 and 2 represent the atoms given in rows in Table 8.1).

Bond from atom...	To atom...	Bond order
1	2	2
2	3	2
1	4	1
1	6	1
3	5	1
3	7	1

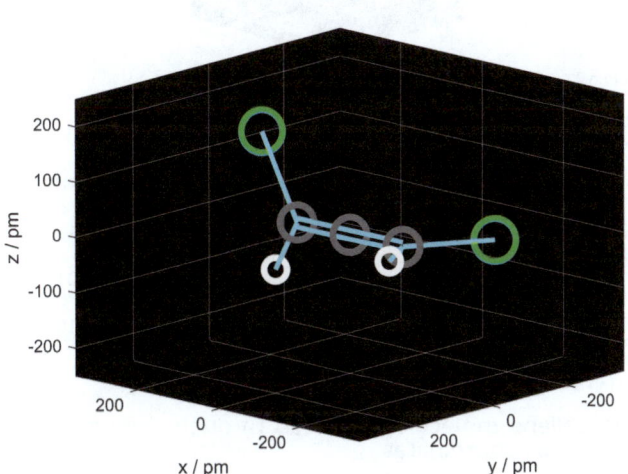

Figure 8.1 Rudimentary 3D molecular structure of 1,3-dichloroallene drawn using a short, user-defined function `simp_mol_draw()` given in Section 8.7.

respectively. The rules of these transformations, the mirror projections they generate and the corresponding atomic coordinates are given in Figure 8.2.

You conclude that the molecule (despite having no stereogenic centres) is chiral because none of the projections are identical to the initial molecule, which is also obvious from the tabulated coordinates next to the 3D structures. More importantly, you have discovered that representing molecular structures using tabulated atomic coordinates is not only a concise way of storing chemical information but it also allows

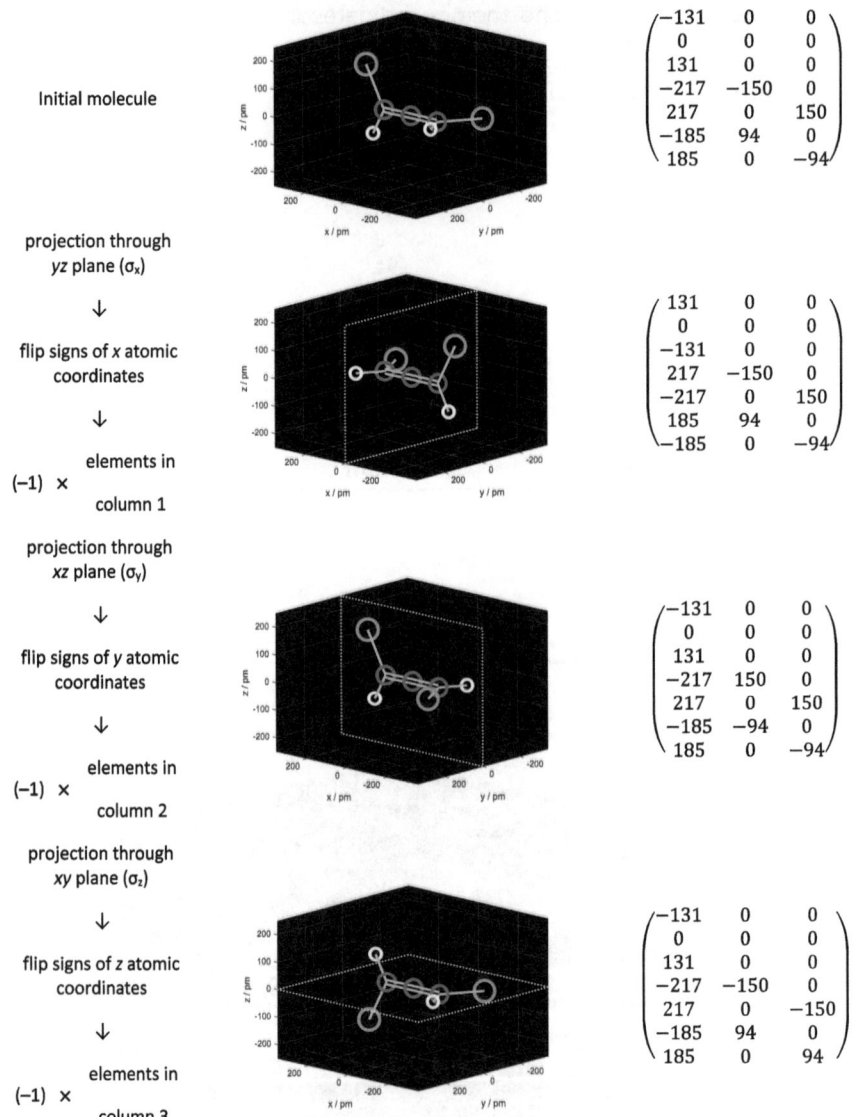

Figure 8.2 1,3-Dichloroallene molecule projected through the *yz*, *xz* and *xy* planes and the resulting atomic coordinates.

us to investigate the symmetry and chirality of molecules without even looking at their structures.

The power of condensing complex information into a set of tabulated numbers and the fact that tangible physical meaning can be assigned to simple rules transforming a set of tabulated numbers into another were recognised a long time ago. In fact, manipulating arrays of numbers proved so powerful in a wide range of problems that it was developed into an important field of mathematics.

Two-dimensional arrays of elements or objects, most often numbers, are called matrices. As we have seen, the appeal of using matrices is that these tabulated arrays lend themselves naturally to systematic manipulation following some rules which are easy to implement on computers. Matrices are written in the form below, with the elements of a matrix arranged in rows and columns.

$$\mathbf{A} = \begin{pmatrix} 1 & 2 & 3 \\ 4 & -5 & 6 \\ 7 & 8 & 9 \\ -2 & 6 & 1 \end{pmatrix}$$

This matrix has elements in 4 rows and 3 columns, which makes $\mathbf{A}$ a 4×3 matrix. We refer to the elements of $\mathbf{A}$ by their row and column numbers, so we write a_{32} and a_{41} when we are referring to 8 and -2, respectively. Matrices can represent both the entities we are transforming (like molecular structures) and the transformations themselves (like projection through a plane). In the concise language of matrices, the task "project the structure of the molecule through the *xy* plane" is simply written as

$$\mathbf{V} = \mathbf{U}\boldsymbol{\sigma}_z$$

where reflection through the *xy* mirror plane, $\boldsymbol{\sigma}_z$, and the atomic coordinates before and after the transformation – $\mathbf{U}$ and $\mathbf{V}$, respectively – are all matrices. $\mathbf{U}\boldsymbol{\sigma}_z$ means multiplying the matrix holding the atomic coordinates for the atoms making up a molecule of interest ($\mathbf{U}$) by the matrix representing the symmetry operation reflection in the *xy* mirror plane ($\boldsymbol{\sigma}_z$). If the molecule has reflection symmetry in the *xy* mirror plane, the 3D structures of the molecule before and after reflection will be indistinguishable.

8.1 Matrix Algebra

- Matrices can be *added* and *subtracted*. For matrices $\mathbf{A}$ and $\mathbf{B}$ of the same size (*i.e.* both having the same number of rows and the same number of columns), $\mathbf{C} = \mathbf{A} \pm \mathbf{B}$ is defined as

$$\begin{pmatrix} a_{11} \pm b_{11} & a_{12} \pm b_{12} & \cdots & a_{1n} \pm b_{1n} \\ a_{21} \pm b_{21} & a_{22} \pm b_{22} & \cdots & a_{2n} \pm b_{2n} \\ \vdots & \vdots & \ddots & \vdots \\ a_{m1} \pm b_{m1} & a_{m2} \pm b_{m2} & \cdots & a_{mn} \pm b_{mn} \end{pmatrix} = \begin{pmatrix} a_{11} & a_{12} & \cdots & a_{1n} \\ a_{21} & a_{22} & \cdots & a_{2n} \\ \vdots & \vdots & \ddots & \vdots \\ a_{m1} & a_{m2} & \cdots & a_{mn} \end{pmatrix} \pm \begin{pmatrix} b_{11} & b_{12} & \cdots & b_{1n} \\ b_{21} & b_{22} & \cdots & b_{2n} \\ \vdots & \vdots & \ddots & \vdots \\ b_{m1} & b_{m2} & \cdots & b_{mn} \end{pmatrix}$$

- The product of a scalar c and a matrix $\mathbf{A}$ is defined as

$$\begin{pmatrix} c \times a_{11} & c \times a_{12} & \cdots & c \times a_{1n} \\ c \times a_{21} & c \times a_{22} & \cdots & c \times a_{2n} \\ \vdots & \vdots & \ddots & \vdots \\ c \times a_{m1} & c \times a_{m2} & \cdots & c \times a_{mn} \end{pmatrix} = c \times \begin{pmatrix} a_{11} & a_{12} & \cdots & a_{1n} \\ a_{21} & a_{22} & \cdots & a_{2n} \\ \vdots & \vdots & \ddots & \vdots \\ a_{m1} & a_{m2} & \cdots & a_{mn} \end{pmatrix}$$

- The *product* of two matrices $\mathbf{C}=\mathbf{AB}$ is defined *only* if the number of rows in matrix **B** is the same as the number of columns in matrix **A**:

$$\begin{array}{cccccc} \text{Matrices} & \mathbf{C} & = & \mathbf{A} & \times & \mathbf{B} \\ \text{dimensions:} & m \text{ by } k & & m \text{ by } n & & n \text{ by } k \\ \textit{rows by columns} & & & & & \end{array}$$

where the elements of **C** are defined as

$$c_{ij} = \sum_{k=1}^{n} a_{ik} b_{kj}$$

so, for example,

$$\begin{pmatrix} -2 & 3 \\ 1 & 5 \end{pmatrix} \times \begin{pmatrix} 1 & 0 \\ 0 & -1 \end{pmatrix} = \begin{pmatrix} (-2 \times 1)+(3 \times 0) & (-2 \times 0)+(3 \times (-1)) \\ (1 \times 1)+(5 \times 0) & (1 \times 0)+(5 \times (-1)) \end{pmatrix} = \begin{pmatrix} -2 & -3 \\ 1 & -5 \end{pmatrix}$$

Notice how multiplying the first matrix by the second has resulted in flipping only specific signs in the first matrix.
- Is it possible to *divide* matrices? Because division is the inverse operation of multiplication, we could approach this question as whether it is possible to define matrices such that matrix **A** multiplied by $\mathbf{A}^{-1}$ would give 1, just as numbers would. As we have seen, multiplying matrices result in matrices, thus $\mathbf{AA}^{-1}$ (or $\mathbf{A}^{-1}\mathbf{A}$) cannot yield 1; however, its matrix equivalent, the so-called identity matrix (a square matrix with ones on the main diagonal and zeros elsewhere)

$$\mathbf{I} = \begin{pmatrix} 1 & 0 & \cdots & 0 & 0 \\ 0 & 1 & \cdots & 0 & 0 \\ 0 & 0 & \ddots & 0 & 0 \\ 0 & 0 & \cdots & 1 & 0 \\ 0 & 0 & \cdots & 0 & 1 \end{pmatrix}$$

can be achieved as a result of $\mathbf{AA}^{-1}$ (or $\mathbf{A}^{-1}\mathbf{A}$). Matrix $\mathbf{A}^{-1}$ is called the inverse of **A** so that

$$\mathbf{AA}^{-1} = \mathbf{A}^{-1}\mathbf{A} = \mathbf{I}$$

Note that *only square matrices* (having the same number of rows and columns) *can have an inverse* if it exists. Not every square matrix has an inverse; those that have no inverse are called *singular matrices*.

Just like with vectors, the rules of matrix algebra (addition, subtraction, multiplication) are straightforward; however, it is easy to make mistakes during these lengthy and repetitive calculations. Therefore, we can save time and effort by using computers to perform matrix operations so that we can focus on how matrices can be used in chemistry. Matrix operations are particularly easy to do with MATLAB as it was first developed specifically with that aim in mind (hence its name MATRIX LABORATORY) in the 1970s.

8.2 Matrices in MATLAB

The above matrix, **A**, is defined in MATLAB as

```
A=[1  2  3; 4  -5  6; 7  8  9; -2  6  1]
```

For which MATLAB returns

```
A=
     1     2     3
     4    -5     6
     7     8     9
    -2     6     1
```

For accessing specific elements of a matrix, use the same syntax as for arrays: use the row and column indices (as shown below for inquiring the element of A in row 4 and column 1)

```
A(4,1)
ans=
    -2
```

If we need an entire row or column we use the (:) operator as

```
A(2,:)
ans=
     4    -5     6
```

We can use the (:) operator more subtly to specify a particular domain within A if necessary:

```
A(3:4,2:3)
ans=
     8     9
     6     1
```

8.3 Matrix Algebra in MATLAB: *Addition* and *Subtraction*

Let us define another matrix first (which must have the same dimensions, 4 × 3, as A)

```
B=[3  1  -5; 2  -1  -6; -7  8  9; -6  4  2];
```

then calculate

>> C = A + B			>> C = A − B			>> C = B − A		
C =			C =			C =		
4	3	−2	−2	1	8	2	−1	−8
6	−6	0	2	−4	12	−2	4	−12
0	16	18	14	0	0	−14	0	0
−8	10	3	4	2	−1	−4	−2	1

8.4 Matrix Algebra in MATLAB: *Multiplication*

Before we go ahead with asking MATLAB to compute C = A * B, we need to inspect the dimensions of the matrices we wish to multiply together to check that they are compatible for multiplication. Matrix multiplication requires that the number of rows in the second matrix, B, is the same as the number of columns in the first matrix, A. This means we need to redefine B first to have 3 rows. The number of columns we can chose freely; however, we need to bear in mind that the dimensions of product matrix C will be determined by our choice.

>> B = [3 2 −7 −6; 1 −1 8 4; −5 −6 9 2];				>> B = [3 2; 1 −1; −5 −6];	
C = A * B				C = A * B	
C =				C =	
−10	−18	36	8	−10	−18
−23	−23	−14	−32	−23	−23
−16	−48	96	8	−16	−48
−5	−16	71	38	−5	−16

If we set B as a 3 × 4 matrix, C will be a 4 × 4 matrix; if B is a 3 × 2 matrix, C will be a 4 × 2 matrix.

Now we can look back at the scenario at the beginning of this chapter and test – using matrices and MATLAB – if 1,3-dichloroallene has reflection symmetry in the *xy* plane. We define the atomic matrix (U), the matrix representing reflection about the *xy* plane (sz), then multiply them as U*sz:

```
>> U = [−131  0   0;  0  0  0;  131  0  0;  −217  −150  0;  217  0  −150;
−185  94  0;  185  0  94]  % atomic matrix of 1,3-dichloroallene

U =
    −131      0      0
       0      0      0
     131      0      0
    −217   −150      0
     217      0   −150
    −185     94      0
     185      0     94
```

```
>> sz = [1  0  0; 0  1  0; 0  0  -1]  % matrix for reflection through the xy
plane

sz =
     1   0    0
     0   1    0
     0   0   -1

>> V = U * sz  % perform symmetry operation

V =
   -131      0      0
      0      0      0
    131      0      0
   -217   -150      0
    217      0    150
   -185     94      0
    185      0    -94
```

Matrices **U** and **V** are different; in the third column (storing the z coordinates of the atoms) the signs have been flipped by this symmetry operation. Because the magnitudes of the changes in the z coordinates are different for the different types of atoms, this transformation has generated a molecule distinguishable from the original molecule, which can also be seen by comparing the top and bottom structures in Figure 8.2. Consequently, 1,3-dichloroallene has no reflection symmetry in the xy plane.

When we have large molecules containing many atoms, it is difficult to see by looking at the atomic matrices if the molecule has been left unchanged or not by a symmetry operation. In this particular case, an easy way to catch that the magnitudes of the coordinate changes are different is by subtracting one atomic matrix from the other and adding up the differences in each set of coordinates (*i.e.* each matrix elements) in each column. Using the 1,3-dichloroallene atomic matrices before (**U**) and after reflection in the xy plane (**V**) we compute

```
>> sum(U - V)
```

which returns

```
ans =
     0     0    -112
```

The first number is the sum of all changes in the x coordinates of atoms; the second number is the sum of all changes in the y coordinates of atoms and the third number is the sum of all changes in the z coordinates of atoms during the symmetry operation. Here, these three numbers capture that during reflection through the xy plane there is no change in the x and y coordinates of the atoms in 1,3-dichloroallene, and that the displacements of atoms in the z direction do not cancel out. The displacement of one atom cancelling out the displacement of another atom is often indicative of the two

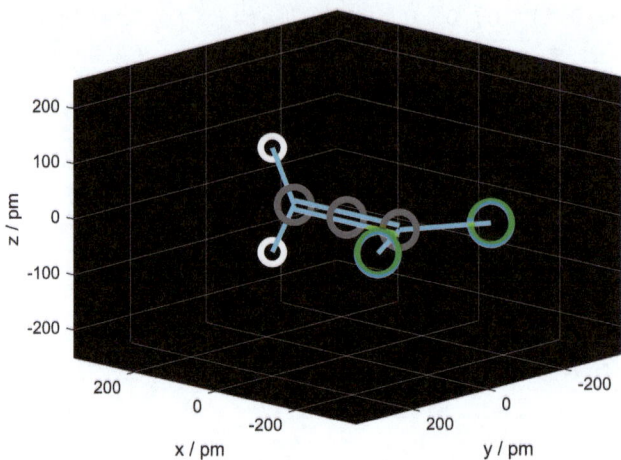

Figure 8.3 Simple 3D molecular structure of 1,1-dichloroallene drawn using a short, user-defined function simp_mol_draw() given in Section 8.7.

swapping places during a symmetry operation, which results in a molecule indistinguishable from the original molecule. Because this is not the case for 1,3-dichloroallene during reflection in the *xy* plane – shown by the non-zero sum of displacements – we conclude that 1,3-dichloroallene has no reflection symmetry in the *xy* plane.

We will now look at the effect of the same symmetry operation on 1,1-dichloroallene whose structure is shown in Figure 8.3.

Exercise 8.1

Using matrices and MATLAB, establish whether 1,1-dichloroallene ($Cl_2C=C=CH_2$) has reflection symmetry in the *xy* plane.

(Hint: Start by formulating the atomic coordinate matrix of 1,1-dichloroallene by altering the atomic coordinate matrix of 1,3-dichloroallene such that both H atoms are at one end, and both Cl atoms are at the other end of the molecule. Think about how many ways this could be achieved, given that the H and Cl atoms are now in perpendicular planes.)

Answer:
```
>> U=[-131 0 0; 0 0 0; 131 0 0; -217 -150 0; -217 150 0;
185 0 94; 185 0 -94]

U =
    -131       0       0
       0       0       0
     131       0       0
    -217    -150       0
    -217     150       0
     185       0      94
     185       0     -94
```

```
V = U * sz
V =
    -131       0       0
       0       0       0
     131       0       0
    -217    -150       0
    -217     150       0
     185       0     -94
     185       0      94

sum(U - V)

ans =
     0     0     0
```

In this representation, the Cl atoms are in the *xy* plane; hence, their *z* coordinates remain unchanged. The displacement of one H atom is cancelled out by the displacement of the other H atom (94 pm→−94 pm and −94 pm→94 pm) during the summation (yielding 0 0 0) indicating that the two H atoms simply swap places during the symmetry operation, *i.e.* leaving the molecule looking the same. As the resulting molecule is indistinguishable from the original molecule (because the sums of displacements are zero, despite the changes in the atomic matrix), reflection in the *xy* plane leaves 1,1-dichloroallene unchanged; therefore, 1,1-dichloroallene has reflection symmetry in the *xy* plane. (Consequently, 1,1-dichloroallene is not chiral.)

Back to the properties of matrix multiplication, we shall now look at what happens when we swap the order of multiplying two matrices. We define two square matrices of the same size (**A** and **B**) so that they can be multiplied as **AB** and **BA**.

```
A = [1  -1   2; 3  -4   3; 6   5  -5];
B = [1   2   3; 4   5   6; 7   8   9];
```

```
>> C = A * B              >> C = B * A
C =                       C =
    11    13    15            25     6    -7
     8    10    12            55     6    -7
    -9    -3     3            85     6    -7
```

Because **AB** ≠ **BA** we conclude that *matrix multiplication is non-commutative*. (There are matrices for which **AB** = **BA**, but in general this is not the case, therefore we should never assume that **AB** = **BA**.)

However, *matrix multiplication is associative*, which means that **A(BC)** = **(AB)C**. Using the above matrices **A** and **B** along with matrix **C** on the right:

```
>> A * (B*C)                    >> (A*B) * C
ans =                           ans =
    2265    234   -273              2265    234   -273
    1770    180   -210              1770    180   -210
    -135    -54     63              -135    -54     63
```

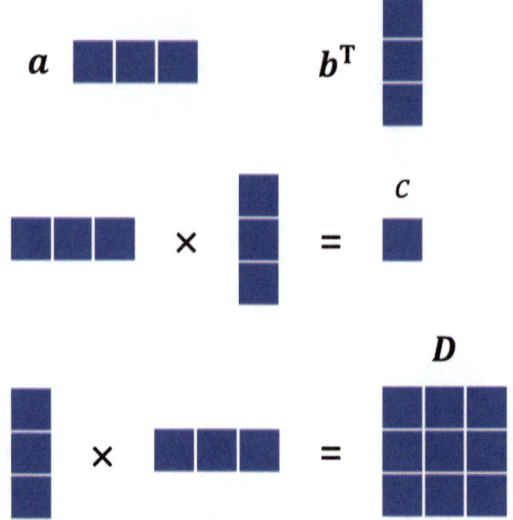

Figure 8.4 Depending on their order, multiplying a row and a column vector of the same number of elements yields either a scalar or a square matrix. b^T denotes the transpose of vector b which has the same dimensions as a.

Matrices can be multiplied and divided by constants; or constants can be added to, or subtracted from matrices, which are all simple to do in MATLAB, for instance:

```
C = A * 2.5
C =
     2.5000    -2.5000     5.0000
     7.5000   -10.0000     7.5000
    15.0000    12.5000   -12.5000
```

The rules of matrix multiplication can be applied to vectors which are essentially one-dimensional matrices. For a row and a column vector of the same number of elements, multiplication results in either a number (scalar) or a square matrix depending on the order of the vectors (Figure 8.4).

For example,

```
>> a = 1:3
a =
     1     2     3
>> b = 4:6
b =
     4     5     6
>> bT = b'
bT =
     4
     5
     6
```

where bT is the transpose of b.

```
>> c=a*bT
c =
    32
>> D=bT*a
D =
    4    8   12
    5   10   15
    6   12   18
```

8.5 Matrix Algebra in MATLAB: *Matrix Inversion*

For calculating the inverse of a matrix, use the `inv()` function. (Remember that only square matrices can be inverted, if they are not singular.) Let us define a matrix now (**A**) and use MATLAB to compute its inverse A^{-1} which we will call matrix **B** here, because we cannot name a matrix A^{-1} in MATLAB. In other words, the task is to find matrix **B** (playing the role of A^{-1}) for which

$$AB = I$$

```
A = [7 2 1; 0 3 -1; -3 4 -2];
B = inv(A)
B =
   -2.0000    8.0000   -5.0000
    3.0000  -11.0000    7.0000
    9.0000  -34.0000   21.0000
```

Let us test now whether **B** is indeed the inverse of **A**. Multiply **A** by **B** and see if the result (**C**) is the identity matrix (**I**).

```
>> C = A * B                          >> C = B * A
C =                                   C =
    1.0000         0   -0.0000            1.0000   -0.0000         0
         0    1.0000         0                 0    1.0000         0
         0         0    1.0000           -0.0000    0.0000    1.0000
```

(Don't worry about the "−0.0000" elements of C. Due to imperfect rounding during calculations some results are not spot on 0 but instead tiny numbers (positive or negative) which can be treated as 0 for any practical purpose.)

Finally, it is worthwhile mentioning that when we multiply a matrix (**A**) with the identity matrix (**I**) we obtain the same matrix (**A**) as a result:

$$AI = IA = A$$

which you can test easily in MATLAB.

8.6 Applications of Matrices

8.6.1 Solving Systems of Linear Equations

We often encounter problems that can be formulated as sets of linear equations, even if it is hard to see straight away. (It takes a bit of getting used to spotting opportunities for applying maths and programming to solve problems you have learned to solve differently. It is well worth the effort though, because by using these tools you will be able to solve more difficult problems quicker.)

We start by solving the system of linear equations

$$2x_1 + x_2 = 7$$
$$-3x_1 + 7x_2 = 32$$

using matrices. The aim is to determine the values of x_1 and x_2. (Of course, you could subtract one equation from the other, or you could rearrange one equation for one of the variables and then substitute the resulting expression into the other equations to determine the values of the two unknowns. But what if you had more than two equations? It would be very useful to have a quick way of solving systems of linear equations of any size.) We can look at this problem from the perspective of matrix algebra and rewrite the above set of linear equations using matrix representation. We turn the coefficients in the equations into a square matrix of coefficients, $\mathbf{A}$; and the unknowns and the right hand sides of the equations into column matrices as

$$\mathbf{A} = \begin{pmatrix} 2 & 1 \\ -3 & 7 \end{pmatrix}, \quad \mathbf{x} = \begin{pmatrix} x_1 \\ x_2 \end{pmatrix}, \quad \mathbf{B} = \begin{pmatrix} 7 \\ 32 \end{pmatrix}$$

then, using the rules of matrix multiplication, we rewrite the set of linear equations as a matrix equation

$$\mathbf{Ax} = \mathbf{B}$$

which can be solved for x by multiplying both sides by the inverse of $\mathbf{A}$, $\mathbf{A}^{-1}$ as

$$\mathbf{A}^{-1}\mathbf{Ax} = \mathbf{A}^{-1}\mathbf{B}$$

Because $\mathbf{A}^{-1}\mathbf{A} = \mathbf{I}$, the equation becomes

$$\mathbf{Ix} = \mathbf{A}^{-1}\mathbf{B}$$

where $\mathbf{Ix} = \mathbf{x}$, therefore

$$\mathbf{x} = \mathbf{A}^{-1}\mathbf{B}$$

In MATLAB, the entire process of solving sets of linear equations can be completed in a few lines; here

```
A=[2 1; -3 7];
B=[7; 32];
x=inv(A)*B

x =
    1
    5
```

8.6.2 Balancing Reaction Equations

Let us take balancing reaction equations as our next chemistry example. We have done these problems so many times, yet we have probably not thought of them as problems ideally suited for using matrices. The combustion of dodecane is described by the unbalanced reaction equation below.

$$C_{12}H_{26} + O_2 \rightarrow CO_2 + H_2O$$

If we would like MATLAB to balance the equation for us by using matrices, we need to reformulate the problem into a system of linear equations which can then be readily solved using matrix operations. What we are looking to find is the values of x_1, x_2, x_3 and x_4 to balance the reaction equation

$$x_1 C_{12}H_{26} + x_2 O_2 \rightarrow x_3 CO_2 + x_4 H_2O$$

We know that the amount of matter must be conserved during any reaction; therefore, we devise a table to keep track of the atoms through the reaction, requiring that their numbers remain unchanged. We have three kinds of atoms: C, H, and O in four different molecules: $C_{12}H_{26}$, O_2, CO_2, and H_2O. In a table, we tabulate the number of atoms in each molecule multiplied by the unknown stoichiometric coefficient of the molecule. For example, there are $12x_1$ carbon atoms and $26x_1$ hydrogen atoms in $x_1 C_{12}H_{26}$ molecules on the left-hand side of the equation; therefore, there must be the same number of them on the right-hand side, which dictates that $12x_1 = 1x_3$ and $26x_1 = 2x_4$.

	$C_{12}H_{26}$	O_2	CO_2	H_2O
C	$12x_1$	$0x_2$	$1x_3$	$0x_4$
H	$26x_1$	$0x_2$	$0x_3$	$2x_4$
O	$0x_1$	$2x_2$	$2x_3$	$1x_4$

$$12x_1 + 0x_2 = 1x_3 + 0x_4 \qquad 12x_1 + 0x_2 - 1x_3 - 0x_4 = 0$$
$$26x_1 + 0x_2 = 0x_3 + 2x_4 \Rightarrow 26x_1 + 0x_2 - 0x_3 - 2x_4 = 0$$
$$0x_1 + 2x_2 = 2x_3 + 1x_4 \qquad 0x_1 + 2x_2 - 2x_3 - 1x_4 = 0$$

(Subtracting the left-hand sides of the equations from the right-hand sides would have been equally correct.) We realise that we have three equations and four variables. Luckily, this can be remedied by adding an auxiliary equation:

$$x_1 = 1$$

which can be interpreted as a guess for the stoichiometric coefficient of dodecane. (Since dodecane is composed of the most amounts of atoms, we expect its stoichiometric coefficient to be the smallest, so we assume that $x_1 = 1$.) It does not matter which variable we set to what value, because the auxiliary equation will not affect the ratio between the compounds. With the addition of the auxiliary equation ($x_1 = 1$, which yields the fourth equation $1x_1 + 0x_2 - 0x_3 - 0x_4 = 1$, which also makes **A** a square matrix), we can rewrite our four linear equations into the matrix equation

$$A x = B$$

where

$$A = \begin{pmatrix} 12 & 0 & -1 & 0 \\ 26 & 0 & 0 & -2 \\ 0 & 2 & -2 & -1 \\ 1 & 0 & 0 & 0 \end{pmatrix}, \quad x = \begin{pmatrix} x_1 \\ x_2 \\ x_3 \\ x_4 \end{pmatrix}, \quad B = \begin{pmatrix} 0 \\ 0 \\ 0 \\ 1 \end{pmatrix}$$

Instead of writing all the atom balance equations and solving the resulting set of simultaneous equations, it is much more convenient to just create matrix **A** directly based on the entries in the table and to simply solve the above matrix equation by multiplying both sides by the inverse of **A**, A^{-1}, as

$$x = A^{-1} B$$

As before, the whole process in MATLAB can be completed by writing only three lines of script!

```
A = [12 0 -1 0; 26 0 0 -2; 0 2 -2 1; 1 0 0 0];
B = [0; 0; 0; 1];
x = inv(A)*B

x =
    1.0000
    5.5000
   12.0000
   13.0000
```

The solution reflects the correct ratio between compounds; however, the stoichiometric coefficient of O_2 is not a whole number. This is because we have fixed the stoichiometric coefficient of dodecane at 1 (by imposing $x_1 = 1$), which puts constraints on the values of the rest of the stoichiometric coefficients. To make all stoichiometric coefficients whole numbers, we multiply x by 2 (or impose $x_1 = 2$ via B = [0; 0; 0; 2]), to obtain the balanced reaction equation

$$2 C_{12}H_{26} + 11 O_2 \rightarrow 24 CO_2 + 26 H_2O$$

Exercise 8.2

Find the stoichiometric coefficients for the combustion of pentadecanol, heptadecanoic acid (margaric acid) and dipropyl sulphide using matrices

$$CH_3(CH_2)_{13} CH_2OH + O_2 \rightarrow CO_2 + H_2O$$

$$CH_3(CH_2)_{15} COOH + O_2 \rightarrow CO_2 + H_2O$$

$$(C_3H_7)_2 S + O_2 \rightarrow CO_2 + SO_2 + H_2O$$

Answer:

$$2CH_3(CH_2)_{13}CH_2OH + 45O_2 \rightarrow 30CO_2 + 32H_2O$$

$$2CH_3(CH_2)_{15}COOH + 49O_2 \rightarrow 34CO_2 + 34H_2O$$

$$2(C_3H_7)_2 S + 21O_2 \rightarrow 12CO_2 + 2SO_2 + 14H_2O$$

Redox equations are usually a bit more complicated to balance, but luckily, they can also be done by using this method. When a chemical process involves charged species, the charge balance provides an additional linear equation which we may include among the set of linear equations, if necessary, to find the stoichiometric coefficients. For example, balancing the redox equation

$$MnO_4^- + Fe^{2+} + H^+ \rightarrow Mn^{2+} + Fe^{3+} + H_2O$$

requires finding six stoichiometric coefficients, but there are only four elements involved in the process (Mn, Fe, H and O); therefore, even with an auxiliary equation added we would only have five equations and six unknowns. Now, however, the charge balance

$$-1x_1 + 2x_2 + 1x_3 = +2x_4 + 3x_5 + 0x_6 \Rightarrow -1x_1 + 2x_2 + 1x_3 - 2x_4 - 3x_5 + 0x_6 = 0$$

enables us to increase the total number of equations to six and formulate the matrix equation

$$Ax = B$$

where

$$A = \begin{pmatrix} 1 & 0 & 0 & -1 & 0 & 0 \\ 0 & 1 & 0 & 0 & -1 & 0 \\ 0 & 0 & 1 & 0 & 0 & -2 \\ 4 & 0 & 0 & 0 & 0 & -1 \\ -1 & 2 & 1 & -2 & -3 & 0 \\ 1 & 0 & 0 & 0 & 0 & 0 \end{pmatrix}, \quad x = \begin{pmatrix} x_1 \\ x_2 \\ x_3 \\ x_4 \\ x_5 \\ x_6 \end{pmatrix}, \quad B = \begin{pmatrix} 0 \\ 0 \\ 0 \\ 0 \\ 0 \\ 1 \end{pmatrix}$$

with x_1, x_2, x_3, x_4, x_5, and x_6 denoting the stoichiometric coefficients of MnO_4^-, Fe^{2+}, H^+, Mn^{2+}, Fe^{3+} and H_2O, respectively; which we solve using MATLAB as

```
A = [1  0  0  -1  0  0; 0  1  0  0  -1  0; 0  0  1  0  0  -2; 4  0  0  0  0
     -1; -1  2  1  -2  -3  0; 1  0  0  0  0  0];
B = [0; 0; 0; 0; 0; 1];
x = inv(A)*B

x =
    1
    5
    8
    1
    5
    4
```

yielding
$$MnO_4^- + 5Fe^{2+} + 8H^+ \rightarrow Mn^{2+} + 5Fe^{3+} + 4H_2O$$

It is worthwhile mentioning that this method cannot be used if we have fewer equations (including the auxiliary equation) than elements involved in the chemical process and when **A** is a singular matrix (thus, it has no inverse).

Exercise 8.3

Find the stoichiometric coefficients for the redox reactions below using matrices.

$$Na_2S_2O_4 + NaOH \rightarrow Na_2SO_3 + Na_2S + H_2O$$
$$HNO_2 + H^+ + Cr_2O_7^{2-} \rightarrow NO_3^- + Cr^{3+} + H_2O$$
$$KMnO_4 + C_2H_5OH + H^+ \rightarrow Mn^{2+} + CH_3COOH + K^+ + H_2O$$
$$KMnO_4 + C_2H_5OH + H^+ \rightarrow Mn^{2+} + C_2H_4O + K^+ + H_2O$$

Answer:

$$3\,Na_2S_2O_4 + 6\,NaOH \rightarrow 5\,Na_2SO_3 + Na_2S + 3\,H_2O$$
$$3\,HNO_2 + 5\,H^+ + Cr_2O_7^{2-} \rightarrow 3\,NO_3^- + 2\,Cr^{3+} + 4H_2O$$
$$2\,KMnO_4 + 5\,C_2H_5OH + 6\,H^+ \rightarrow 2\,Mn^{2+} + 5\,C_2H_4O + 2\,K^+ + 8\,H_2O$$
$$4\,KMnO_4 + 5\,C_2H_5OH + 12\,H^+ \rightarrow 4\,Mn^{2+} + 5\,CH_3COOH + 4\,K^+ + 11\,H_2O$$

8.6.3 Representing Reaction Mechanisms (Reaction Networks)

During most chemical transformations many reactions take place simultaneously. Likewise, chemical species present in complex biological and environmental systems react in multiple concurrent reactions. Organising compounds and their reactions into matrices, called *stoichiometric matrices*, allows us to analyse these systems (also called reactions networks) using computers. The stoichiometric matrix of a reaction system/mechanism/network is constructed by placing the stoichiometric coefficients of each chemical in a separate row in the order of reactions, putting zeros where the chemical does not participate in a reaction. Signs are assigned based on whether a compound is a reactant or a product in a reaction; its stoichiometric coefficient will be negative if it is a reactant, and positive if it is a product. For example, the process

$$A \rightarrow B \rightarrow C$$

can be translated into a stoichiometric matrix

$$N = \begin{pmatrix} -1 & 0 \\ 1 & -1 \\ 0 & 1 \end{pmatrix} \begin{matrix} A \\ B \\ C \end{matrix} \quad \begin{matrix} R1 & R2 \end{matrix}$$

a single entity concisely representing all simultaneous chemical processes in the mechanism. Compound A is removed in reaction 1 (R1), therefore $N_{1,1} = -1$. It does not participate in reaction 2 (R2), thus $N_{1,2} = 0$. Compound B is produced in R1, hence $N_{2,1} = 1$, while it is removed in R2, and consequently $N_{2,2} = -1$.

Like matrices used for representing molecular structure and geometric transformations, stoichiometric matrices also have great practical uses. One of these is the succinct formulation of the steps of computing how quickly the concentrations of the compounds in a chemical network change. In our example system, the rates of changes of [A], [B] and [C] (d[A]/dt, d[B]/dt, d[C]/dt, respectively) can be viewed as the three elements of the vector (d[c]/dt) pointing from the set of concentration values at an instant, $c(t)$, to the next set of concentrations an instant later, $c(t+dt)$. With [A] = c_1, [B] = c_2, and [C] = c_3 determined (or calculated) at a point in time, we define vector $c = (c_1, c_2, c_3)$ which we can be used to compute the two reactions rates $r_1 = k_1$[A] and $r_2 = k_2$[B], where k_1 and k_2 are the rate constants of R1 and R2, respectively, as vector $r = (k_1 c_1, k_2 c_2)$. All rates of change of concentrations can be then computed *via* a single matrix multiplication as

$$\frac{d\boldsymbol{c}}{dt} = \boldsymbol{N}\,\boldsymbol{r}^T$$

If we have a more complex chemical network, the steps of the calculations remain the same, we will just have larger matrices to multiply and possibly more varied expressions in r. Large reaction networks such as metabolic pathways, atmospheric models and complex chemical mechanisms can have hundreds of reactions and chemical species. Instead of having to define one by one the net rate of formation equation for each species from the rate laws of the reaction steps the species is involved in, they are generated *via* matrix multiplication from the reaction rate laws conveniently given only once in r. This not only speeds up the workflow but also reduces errors. Once the $\boldsymbol{Nr}^T$ matrix is generated, it can be used to compute the net rates of changes of the species, which in turn are used to approximate their concentration–time profiles during reactions.

Exercise 8.4

Create the stoichiometric matrix for the reaction mechanism below and compute the rate of change of the chemical species using the given concentrations and rate constants.

$$2A \underset{k_{1r}}{\overset{k_1}{\rightleftharpoons}} B$$

$$B + C \xrightarrow{k_2} D$$

[A] = 0.502 M, [B] = 0.029 M, [C] = 0.102 M, [D] = 0.394 M, $k_1 = 1.8 \times 10^{-1}$ M^{-1}s^{-1}, $k_{1r} = 5.7 \times 10^{-2}$ s^{-1}, $k_2 = 7.3 \times 10^{2}$ M^{-1}s^{-1}.

Answer:

```
% 2A <=> B  r1=k1*[A]^2; r1r=k1r*[B]
% B+C -> D  r2=k2*[B]*[C]
%
% R1    R1r    R2
% ─────────────────
% -2    2      0    | A
% 1     -1     -1   | B
% 0     0      -1   | C
% 0     0      1    | D
k1=1.8e-1; k1r=5.7e-2; k2=7.3e2; % rate constants in 1/Ms 1/s 1/Ms
c=[0.502  0.029  0.102  0.394]; % conc. vector: [A], [B], [C], [D]
in M
r=[k1*c(1)^2 k1r*c(2) k2*c(2)*c(3)]; % rate expressions in a vector
N=[-2 2 0; 1 -1 -1; 0 0 -1; 0 0 1]; % stoichiometric matrix
dcdt=N*r'  % d[A]/dt; d[B]/dt; d[C]/dt; d[D]/dt in M/s
```

```
dcdt =
    -0.0874
    -2.1156
    -2.1593
     2.1593
```

8.7 Molecular Symmetry and Transformation Matrices

Molecular symmetry is important in medicine, solid state chemistry and spectroscopy. Not only the chirality of molecules but, for example, the presence of certain vibrational and rotational transitions in the spectrum of a molecule can be predicted based on the symmetry of the molecule.

As we have seen, matrices enable us to perform geometric transformations on molecules, like reflection and rotation, without having to draw or build molecules using molecular model kits. Computing the new positions of atoms in space following a symmetry operation can be easily done by matrix multiplication. We just need (i) the atomic matrix of a molecule and (ii) the symmetry operation resolved as a matrix (called the transformation matrix). The resulting atomic matrix will represent the new x', y', z' coordinates of the atoms in the geometrically-transformed molecule. Atomic matrices can be downloaded from databases (or generated using computational chemistry software). Matrices for some 3D symmetry operations are listed in Table 8.3.

We can obtain atomic coordinates from databases, like ChemSpider (chemspider.com), PubChem (pubchem.ncbi.nlm.nih.gov) and the National Institute of Standards and Technology (NIST) database (webbook.nist.gov/chemistry/name-ser/) or from computational chemistry software. If we choose to download the atomic coordinates, we simply need to search for the molecule of interest and download its structure data file (SDF or MOL) which is a text file containing all necessary data on the arrangement of the atoms and the bonds between them within the molecule.

Matrices in Chemistry

Table 8.3 Some transformation matrices representing symmetry operations often used in chemistry.

Reflection about the xz-plane, σ_{xz}

$$\begin{pmatrix} 1 & 0 & 0 \\ 0 & -1 & 0 \\ 0 & 0 & 1 \end{pmatrix}$$

Reflection about the yz-plane, σ_{yz}

$$\begin{pmatrix} -1 & 0 & 0 \\ 0 & 1 & 0 \\ 0 & 0 & 1 \end{pmatrix}$$

Reflection about the xy-plane, σ_{yz}

$$\begin{pmatrix} 1 & 0 & 0 \\ 0 & 1 & 0 \\ 0 & 0 & -1 \end{pmatrix}$$

Inversion, i

$$\begin{pmatrix} -1 & 0 & 0 \\ 0 & -1 & 0 \\ 0 & 0 & -1 \end{pmatrix}$$

Reflection across a vertical plane cutting the xy plane through the line given by the equation $y = mx$

$$\frac{1}{1+m^2}\begin{pmatrix} 1-m^2 & 2m & 0 \\ 2m & m^2-1 & 0 \\ 0 & 0 & 1+m^2 \end{pmatrix}$$

Rotation about the x-axis, $R_x(\theta)$

$$\begin{pmatrix} 1 & 0 & 0 \\ 0 & \cos\theta & -\sin\theta \\ 0 & \sin\theta & \cos\theta \end{pmatrix}$$

Rotation about the y-axis, $R_y(\theta)$

$$\begin{pmatrix} \cos\theta & 0 & \sin\theta \\ 0 & 1 & 0 \\ -\sin\theta & 0 & \cos\theta \end{pmatrix}$$

Rotation about the z-axis, $R_z(\theta)$

$$\begin{pmatrix} \cos\theta & -\sin\theta & 0 \\ \sin\theta & \cos\theta & 0 \\ 0 & 0 & 1 \end{pmatrix}$$

These files have multiple fields, of which we are only focusing on the atomic and bond blocks.

First we consider a small and highly symmetric molecule: sulphur hexafluoride (SF_6). The atomic and bond blocks for SF_6 look like as follows, with the x-y-z atomic co-ordinates and the connections between atoms highlighted in bold.

⋮

```
   0.0000    0.0000    0.0000 S  0 0 0 0 0 0 0 0 0 0 0 0
   1.6410    0.0000    0.0000 F  0 0 0 0 0 0 0 0 0 0 0 0
  -1.6410    0.0000    0.0000 F  0 0 0 0 0 0 0 0 0 0 0 0
   0.0000    1.6410    0.0000 F  0 0 0 0 0 0 0 0 0 0 0 0
   0.0000   -1.6410    0.0000 F  0 0 0 0 0 0 0 0 0 0 0 0
   0.0000    0.0000    1.6410 F  0 0 0 0 0 0 0 0 0 0 0 0
   0.0000    0.0000   -1.6410 F  0 0 0 0 0 0 0 0 0 0 0 0
  1  2  1  0  0  0  0
  1  3  1  0  0  0  0
  1  4  1  0  0  0  0
  1  5  1  0  0  0  0
  1  6  1  0  0  0  0
  1  7  1  0  0  0  0
```

⋮

By looking at the atomic block, we see that the sulphur atom is at the origin, as its coordinates are (0,0,0), and that all fluorine atoms are located 1.64 Å (164 pm) from it. The rudimentary 3D structure of the molecule seen in Figure 8.5 is generated by the following MATLAB function `simp_mol_draw()`.

```
function simp_mol_draw(ac,b,d,c,ax,sf,va)
% ac  atom block
% b   bond block
% d   atomic radii
% c   colour codes for atoms
% ax  axis limits
% sf  scaling factor for atoms
% va  viewing angle
% (only draws single bonds, amend to create double bonds for Figure 8.1,
8.2, 8.3)
ac=round(ac*100); % switch from Å to pm (10^{-12} m) while rounding to 1pm
fig=figure; % create figure object
scatter3(ac(:,1),ac(:,2),ac(:,3),d.*sf,c,'LineWidth',4); % draw atoms
hold on
% draw bonds (single bonds only)
linkx=reshape([ac(b(1:end,1),1)  ac(b(1:end,2),1)  NaN(length(b),1)]',
1,[]);
linky=reshape([ac(b(1:end,1),2)  ac(b(1:end,2),2)  NaN(length(b),1)]',
1,[]);
linkz=reshape([ac(b(1:end,1),3)  ac(b(1:end,2),3)  NaN(length(b),1)]',
1,[]);
plot3(linkx,linky,linkz,'-c','LineWidth',3)
xlabel('x/pm'); ylabel('y/pm'); zlabel('z/pm');
grid on
box on
axis equal
axis(ax)
view(va)
set(gca,'color','k')
set(gca,'GridColor','w','GridAlpha',0.4)
set(gcf,'color','w')
fig.Renderer='painters'; % set renderer for smooth display
end
```

Open a new script in MATLAB, copy and paste the above code into the script window. Save the script as "simp_mol_draw.m". In the next step we assign the arrays; first the atomic and bond blocks (by copying the highlighted numbers from the above molecular structure data file)

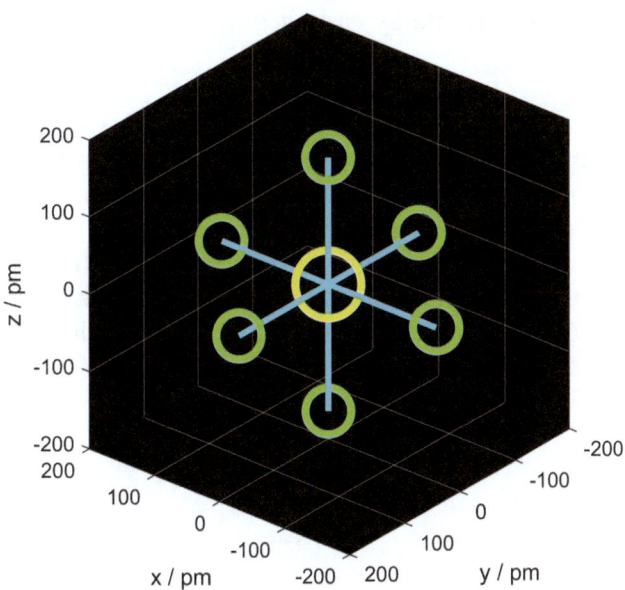

Figure 8.5 Crude 3D molecular structure of sulphur hexafluoride.[†]

```
>> AM = [0.0000  0.0000  0.0000; 1.6410  0.0000  0.0000; -1.6410  0.0000
0.0000; 0.0000  1.6410  0.0000; 0.0000  -1.6410  0.0000; 0.0000  0.0000
1.6410; 0.0000  0.0000  -1.6410];
>> B = [1 2 1; 1 3 1; 1 4 1; 1 5 1; 1 6 1; 1 7 1];
```

followed by the diameters of the atoms (which you can find in textbooks or online) in the order of atoms in the atomic block:

```
>> AD = [104 72 72 72 72 72 72]';
```

The colours of the atoms are assigned using their RGB colour values (a set of three numbers each between 0 and 255 for each atom) available online (https://sciencenotes.org/wp-content/uploads/2019/07/CPK-Jmol.png) divided by 255, as the intensities of colour components in MATLAB must be between 0 and 1:

```
>> C = [255  255  48; 144  224  80; 144  224  80; 144  224  80; 144  224  80;
144  224  80; 144  224  80]/255;
```

Then, we set the axis limits (to define the space containing the rudimentary molecular structure) ensuring that no atoms will be outside the 3D domain defined to contain the molecule.

```
>> S = [-200  200  -200  200  -200  200];
```

[†] In Octave, markers may be scaled differently. If the circles representing atoms appear too large, reduce the scaling factor; for example, run `simp_mol_draw(AM,B,AD,C,S,2,[50 30])`.

Finally, we display the molecule by calling `simp_mol_draw()` with the arrays defined above:

```
>> simp_mol_draw(AM,B,AD,C,S,10,[50 30])
```

which should result in a figure identical to the one shown in Figure 8.5. The last two input arguments of `simp_mol_draw()` are the scaling factor for the size of atoms as they appear on the screen and the viewing angle (horizontal angle about the z axis with respect to the y axis and the elevation angle with respect to the xy plane).

Now that we have a rudimentary molecule drawing MATLAB function, we can continue exploring molecular symmetry. Let us rotate the molecule by 45 ($\pi/4$), 60 ($\pi/3$) and 90 ($\pi/2$) degrees about the z axis.

```
>> th_45=pi/4
th_45 =
    0.7854
>> th_60=pi/3
th_60 =
    1.0472
>> th_90=pi/2
th_90 =
    1.5708
```

Generate the transformation matrices (R_z) for 45, 60 and 90-degree rotations about the z axis (bottom right matrix in the table above)

```
>> Rz_45=[cos(th_45) -sin(th_45) 0; sin(th_45) cos(th_45) 0; 0 0 1]
Rz_45 =
    0.7071   -0.7071        0
    0.7071    0.7071        0
         0         0   1.0000
>> Rz_60=[cos(th_60) -sin(th_60) 0; sin(th_60) cos(th_60) 0; 0 0 1]
Rz_60 =
    0.5000   -0.8660        0
    0.8660    0.5000        0
         0         0   1.0000
>> Rz_90=[cos(th_90) -sin(th_90) 0; sin(th_90) cos(th_90) 0; 0 0 1]
Rz_90 =
    0.0000   -1.0000        0
    1.0000    0.0000        0
         0         0   1.0000
```

then perform the symmetry operations on atomic matrix AM and output the resulting atomic matrices into the corresponding AM_R45, AM_R60, AM_R90 arrays:

```
>> AM_R45=AM * Rz_45;
>> AM_R60=AM * Rz_60;
>> AM_R60=AM * Rz_90;
```

Finally, we display the results as

```
>> simp_mol_draw(AM_R45,B,AD,C,S,10,[50 30])
>> simp_mol_draw(AM_R60,B,AD,C,S,10,[50 30])
>> simp_mol_draw(AM_R90,B,AD,C,S,10,[50 30])
```

which are organised in Table 8.4 for easy comparison. Atomic coordinates have been rounded to 1 pm, for example: AM = round(AM*100).

By looking at the 3D plots and the atomic matrices, we conclude that neither a 45 nor a 60 degree rotation of SF_6 about the z-axis leaves the molecule unchanged.

Deciding whether a molecule has been unchanged by a symmetry operation *via* visual inspection still requires displaying the molecule or outputting its atomic coordinates before and after the transformation, followed by careful inspection, which may be quite difficult, especially if the molecule is large and/or complex. Since we are already using matrices and computers to perform symmetry operations on molecules it would be very efficient if we could compare the transformed molecule against the initial molecule computationally. (Subtracting the post-transformation atomic matrix from the pre-transformation atomic matrix does not work in this case because displacements cancel out for all rotational angles.)

For this task we can use the ismember() function which is designed to determine if elements of an array are present in another array as well. Comparing a molecule with its transformed version requires going through the atomic matrices of the two molecules row by row and checking if each row of atomic coordinates for the initial molecule appears (somewhere) in the atomic matrix of the transformed molecule. Finding all the sets of atomic coordinates means that the atoms of the two molecules appear at the same locations in space; as if they have not moved during the symmetry operation, even if their coordinates have shifted to a different row within the atomic matrix. For example, deciding whether all spatial positions of atoms in AM are present in AM_R45 (within a 1 pm range) can be done using the following command:

```
>> AM_vs_AM_R45 = ismember(round(AM*100),round(AM_R45*100), 'rows')'
```

Components of atomic position vectors in Å are multiplied by 100 to switch to pm and then rounded to provide ismember() with integers to compare row-wise. The result, which is a column vector, is then transposed using operator (') so that its components will be written out in a row for easier comparison. The output generated is

```
AM_vs_AM_R45 =

  1×7 logical array
   1  0  0  0  0  1  1
```

where 1 means: *yes*, all three atomic coordinates within that row of AM were found (somewhere) in AM_R45. Similarly, 0 represents: *no*, not all or none of the three atomic coordinates within that row of AM were found (anywhere) in AM_R45. Going from left to right through the elements of AM_vs_AM_R45 while comparing each corresponding row of AM top to bottom with the rows of AM_R45 in Table 8.4, we can confirm that the logical array AM_vs_AM_R45 correctly represents the result of finding matching rows. For example, the first element of AM_vs_AM_R45 is 1, which means that the first row of AM (the atomic coordinates of sulphur before the rotation

Table 8.4 Rotation of SF$_6$ about the z-axis

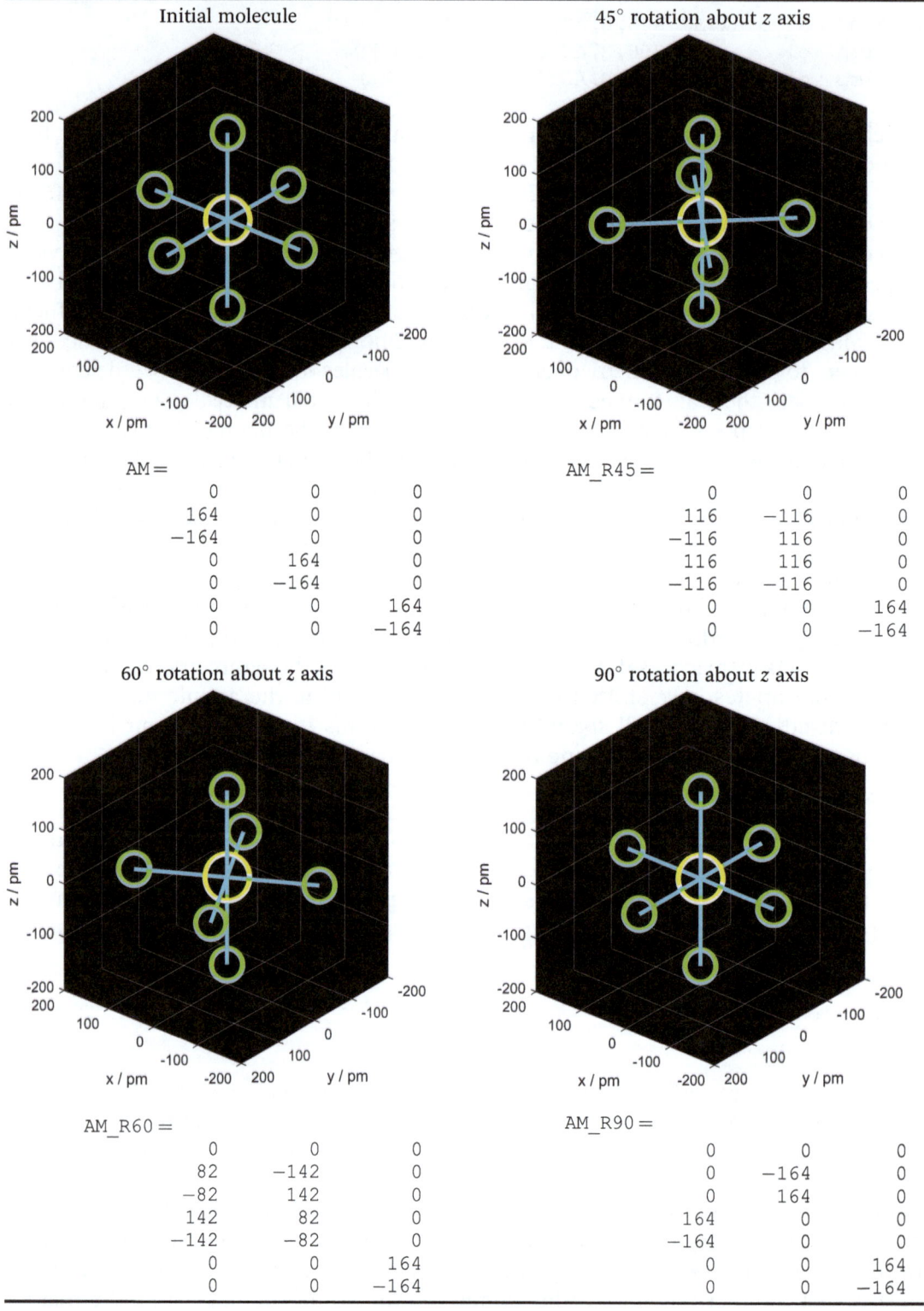

by 45° about the z-axis) are found in AM_R45 (the atomic coordinates of sulphur after rotation); whereas the second element of AM_vs_AM_R45 is 0, which represents that the second row of AM (the atomic coordinates of the right most fluorine atom in the top left figure in Table 8.4) are *not* found in AM_R45, indicating that no fluorine atom moved into the [164, 0, 0] position during 45° rotation about the z-axis. In turn, this also means that 45° rotation about the z-axis does not leave SF_6 unchanged. For 60° and 90° rotations,

```
>> AM_vs_AM_R60 = ismember(round(AM*100),round(AM_R60*100),'rows')'
AM_vs_AM_R60 =
  1×7 logical array
   1   0   0   0   0   1   1
>> AM_vs_AM_R90 = ismember(round(AM*100),round(AM_R90*100),'rows')'
AM_vs_AM_R90 =
  1×7 logical array
   1   1   1   1   1   1   1
```

the output arrays show that 60° rotation about the z-axis leaves only sulphur, the top and bottom fluorine atoms unchanged, and the rest changed; while 90° rotation about the z-axis leaves all the atoms (the entire SF_6 molecule) unchanged.

The MATLAB function `all()` is ideal for testing if all atoms in a molecule have remained unchanged during a symmetry operation. It returns 1 (representing *"true"*) if all the elements of a vector are nonzero, which on the above outputs of `ismember()` would mean "unchanged by rotation". On the other hand, 0 (representing *"false"*) returned would correspond to "changed by rotation".

```
>> SF6_unchanged_by_R_45 = all(AM_vs_AM_R45)
SF6_unchanged_by_R_45 =
  logical
   0
>> SF6_unchanged_by_R_60 = all(AM_vs_AM_R60)
SF6_unchanged_by_R_60 =
  logical
   0
>> SF6_unchanged_by_R_90 = all(AM_vs_AM_R90)
SF6_unchanged_by_R_90 =
  logical
   1
```

which conveniently condenses the response to the statement "this symmetry operation leaves the molecule unchanged" into a single digit: 1 *"true"* or 0 *"false"*.

Following the example of SF_6, where bonds conveniently align with the axes of the Cartesian coordinate system, we return to methane to look at a molecule with less obvious symmetry. Methane can be oriented (as in Exercise 7.1, see the atomic coordinates) so that the atoms are in three vertical planes that cut through the *xy* plane at

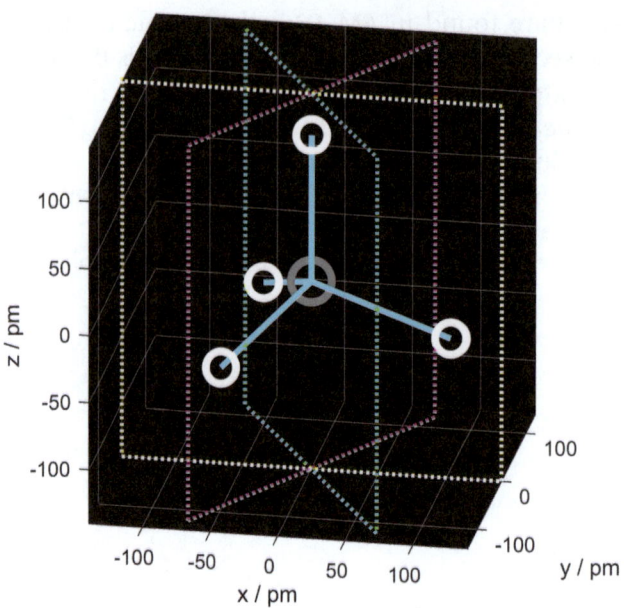

Figure 8.6 Rudimentary 3D molecular structure of methane and its mirror planes.

the origin along lines given by the equation $y=mx$, where $m=0=\tan(0°)$ for the yellow, $m=-1.7329=\tan(120°)$ for the green and $m=1.7329=\tan(240°)$ for the magenta plane (Figure 8.6). The values for m can be obtained from the atomic coordinates of the hydrogens in these planes located outside the z axis.

H atom	Vertical plane		
	Yellow	Green	Magenta
x/pm	1.026	−0.513	−0.513
y/pm	0	0.889	−0.889
$m=y/x$	0	−1.7329	1.7329

```
>> m=[0/1.026  0.889/(-0.513)   (-0.889)/(-0.513)]

m =
   0   -1.7329   1.7329

>> m=tand([0 120 240])

m =
   0   -1.7321   1.7321
```

Transformation matrices for reflections across these vertical planes are generated as

```
>> s_0=1/(1+m(1)^2)*[1-m(1)^2 2*m(1) 0; 2*m(1) m(1)^2-1 0; 0 0 1+m(1)^2]
s_0 =
   1   0   0
   0   0   0
   0   0   1
```

```
>> s_120=1/(1+m(2)^2)*[1-m(2)^2 2*m(2) 0; 2*m(2) m(2)^2-1 0; 0 0 1+m(2)^2]
s_120 =
    -0.5000   -0.8660         0
    -0.8660    0.7500         0
          0         0    1.0000

>> s_240=1/(1+m(3)^2)*[1-m(3)^2 2*m(3) 0; 2*m(3) m(3)^2-1 0; 0 0 1+m(3)^2]
s_240 =
    -0.5000    0.8660         0
     0.8660    0.7500         0
          0         0    1.0000
```

Using the atomic coordinates given in Exercise 7.1 (having loaded them into array AM), we can now perform the reflection operations on methane through the three mirror planes by multiplying the atomic matrix of methane by the transformation matrices s_0, s_120 and s_240 generated above.

```
>> AM_re_0=AM * s_0;    % reflect methane across yellow plane
>> AM_re_120=AM * s_120; % reflect methane across green plane
>> AM_re_240=AM * s_240; % reflect methane across magenta plane
```

Finally, we display the results.

```
>> B=[1 2 1; 1 3 1; 1 4 1; 1 5 1]; % bond block for methane
>> ad=[140; 80; 80; 80; 80]; % atomic diameter
>> ac=[144 144 144; 255 255 255; 255 255 255; 255 255 255; 255 255 255]/255; % color codes for the atoms
>> ax=[-120 120 -120 120 -120 120]; % axes limits
>> sf=3; % scale factor for displaying atoms
>> v=[10 20]; % 3D view point
>> simp_mol_draw(AM,B,ad,ac,ax,sf,v) % draw original methane molecule
>> simp_mol_draw(AM_re_0,B,ad,ac,ax,sf,v) % draw methane molecule reflected across yellow plane
>> simp_mol_draw(AM_re_120,B,ad,ac,ax,sf,v) % draw methane molecule reflected across green plane
>> simp_mol_draw(AM_re_240,B,ad,ac,ax,sf,v) % draw methane molecule reflected across magenta plane
```

All being well, the resulting four figures should look identical visually confirming that, indeed, reflection through these planes leaves the molecule unchanged. We can also use MATLAB to judge whether these symmetry operations have left the molecule unchanged.

```
>> CH4_unchanged_by_s_0=all(ismember(round(AM*100),round(AM_re_0*100),'rows')) % r. a. yellow plane
CH4_unchanged_by_s_0 =
```

```
  logical
    1
>> CH4_unchanged_by_s_120=all(ismember(round(AM*100),
round(AM_re_120*100),'rows')) % r. a. green plane
CH4_unchanged_by_s_0=
  logical
    1
>> CH4_unchanged_by_s_240=all(ismember(round(AM*100),
round(AM_re_240*100),'rows')) % r. a. magenta plane
CH4_unchanged_by_s_0=
  logical
    1
```

Indeed, these outputs agree with our visual inspection. These, however, required no human interaction (beyond coding up how to carry out symmetry operations and compare the molecules before and after a transformation), which opens up the possibility of fully automating the process of investigating molecular symmetry.

> **Exercise 8.5**
>
> *Inversion* is a type of reflection that takes place across a point. Projecting each atom of a molecule with inversion symmetry through the molecule's centre of inversion (a point that does not have to overlap with any atom) and across to the other side of the molecule leaves the molecule unchanged.
>
> Using the transformation matrix for inversion, given in Table 8.3, determine if the chair conformation of 1,4 dimethyl cyclohexane has inversion symmetry.
>
>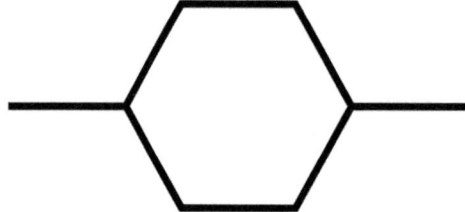
>
> **Answer:**
>
> 1,4 dimethyl cyclohexane (C_8H_{16}) is a relatively simple molecule (as far as organic molecules are concerned), yet because it is made up of 24 atoms it would take a while to build it using a molecular model kit for establishing whether it has inversion symmetry. Building both the *cis* and *trans* conformations might be too much even for an avid organic chemist. Luckily, we can download their molecular structure files and use matrix multiplication and MATLAB to answer the question without touching a molecular model kit.
>
> We go to PubChem, search for the molecule, download the SDF files and load the atomic blocks into arrays AM_C (*cis*) and AM_T (*trans*). Their bond blocks are the same, thus we only need one array (B) for storing the information on the connections between the atoms.

```
>> AM_C= [1.4481    0.0001    0.3345;  -1.4478   -0.0001   -0.3344;  -0.7429
-1.2625    0.1776;   0.7430   -1.2625   -0.1770;   -0.7431    1.2626    0.1775;
 0.7429    1.2628   -0.1769;  2.9192   -0.0002   -0.0763;  -2.9193   -0.0002
 0.0750;   1.4043    0.0001    1.4319;  -1.4032   -0.0001   -1.4318;  -0.8535
-1.3371    1.2670;  -1.2158   -2.1543   -0.2508;  -1.2161    2.1542   -0.2512;
-0.8539    1.3373    1.2669;   1.2160   -2.1541    0.2517;   0.8536   -1.3374
-1.2664;   0.8537    1.3379   -1.2664;   1.2158    2.1544    0.2520;   3.4315
-0.8851    0.3157;   3.0291   -0.0002   -1.1660;   3.4317    0.8846    0.3157;
-3.0302   -0.0001    1.1647;  -3.4315    0.8846   -0.3173;  -3.4313   -0.8851
-0.3173];
>> AM_T= [1.4774    0.0001    0.2825;  -1.4773   -0.0001   -0.2824;  -0.7425
 1.2652    0.1822;   0.7426    1.2651   -0.1824;   -0.7424   -1.2652    0.1821;
 0.7428   -1.2651   -0.1823;   1.6968    0.0000    1.7983;  -1.6972    0.0000
-1.7980;   2.4729    0.0001   -0.1809;  -2.4725   -0.0002    0.1812;  -1.2210
 2.1561   -0.2422;  -0.8450    1.3549    1.2707;   -0.8450   -1.3550    1.2706;
-1.2208   -2.1561   -0.2424;   0.8450    1.3546   -1.2709;   1.2211    2.1562
 0.2417;   1.2213   -2.1560    0.2421;   0.8453   -1.3547   -1.2708;   2.2671
-0.8837    2.1040;    2.2670    0.8837    2.1040;   0.7563   -0.0001    2.3565;
-2.2676    0.8837   -2.1035;  -0.7568    0.0000   -2.3566;  -2.2675   -0.8837
-2.1037];
>> B= [1    4    1;  1    6    1;  1    7    1;  1    9    1;  2    3    1;  2    5    1;  2    8    1;
 2   10    1;  3    4    1;  3   11    1;  3   12    1;  4   15    1;  4   16    1;  5    6    1;  5   13    1;
 5   14    1;  6   17    1;  6   18    1;  7   19    1;  7   20    1;  7   21    1;  8   22    1;
 8   23    1;  8   24    1];
```

Before we can continue, we have to make sure that the molecule's supposed inversion centre is located at the origin. Some structural data files come with the atomic block centred on the origin while others do not. Because it is difficult to decide just by looking at the tabulated atomic coordinates, it is best practice to always perform this coordinate adjustment before investigating symmetry. For example, the means of the x, y, z coordinates stored in AM_C and AM_T are computed using the `mean()` function which returns the average of each atomic position component column as a row vector:

```
>> centre_AM_C=mean(AM_C)
centre_AM_C=
  1.0e-04 *
    0.1250    -0.1667    -0.6667
>> centre_AM_T=mean(AM_T)
centre_AM_T=
  1.0e-05 *
    0.4567    -0.8333    -0.6971
```

which can be regarded as $\bar{x}=0, \bar{y}=0, \bar{z}=0$ for our purposes on the basis that they are on the order of 10^{-5} Å $\ll$ Bohr radius ($a_0 = 0.53$ Å, the radius of the lowest energy electron orbit in the hydrogen atom according to the Bohr model of atoms).

For the sake of future reference, however, we can still try further centring the atomic coordinates on the origin by subtracting each mean from the corresponding components of all atomic coordinates:

```
>> AM_C = AM_C − mean(AM_C);
>> AM_T = AM_T − mean(AM_T);
```

For brevity, we have loaded the adjusted values for AM_C and AM_T back into to AM_C and AM_T, respectively. The resulting adjusted centres are (performed on the newly assigned AM_C and AM_T arrays):

```
>> adj_centre_AM_C = mean(AM_C)
adj_centre_AM_C =
   1.0e−15 *
     0.1110    0.0185    0

>> adj_centre_AM_T = mean(AM_T)
adj_centre_AM_T =
   1.0e−16 *
     0.2313    0.0925    0.1850
```

which show seemingly considerable improvement. However, these values, despite being many orders of magnitude closer to zero, bring no practically appreciable advantage to our analysis.

What is more concerning is that when we look at the components of the atomic positions in AM_C and AM_T we see that some components would not perfectly invert to each other like before. For example, when carefully looking at AM_C we spot amongst the last few sets of components 0.8846 Å and −0.8851 Å. These and similar small discrepancies in atomic positions tend to be found more frequently as molecules become larger. Because the computational cost of establishing molecular structure increases with size, the coordinates of atoms in larger molecules are often not refined to the same degree as the positions of atoms in smaller molecules. Atoms being slightly off in a molecular structure file are indiscernible when a molecule is displayed; they, on the other hand, can affect the analysis of molecular symmetry as we will see below.

At this point, if you prefer a visual aid you can draw the molecules using `simp_mol_draw()`, and with the commands

```
>> ad = [repmat(140,1,8) repmat(80,1,16)]'; % assign atomic diameters
```
concisely: 140 repeated 8 times followed by 80 repeated 16 times

```
>>   ac = [repmat([144 144 144],8,1);  repmat([255 255 255],16,1)]/
255; % assign colour codes concisely the same way as above
>> ax = [−400 400 −400 400 −300 300];
>> sf = 1.5; % assign scaling factor for circles representing the atoms (we
```
have 24 atoms so we would like to draw them small to avoid overlaps)

```
>> v = [−167 17]; % assign view point
```

```
>> simp_mol_draw(AM_C,B,ad,ac,ax,sf,v) % draw cis molecule; use AM_T for
trans
```

you should obtain these molecular structures (*cis*: left, *trans*: right).

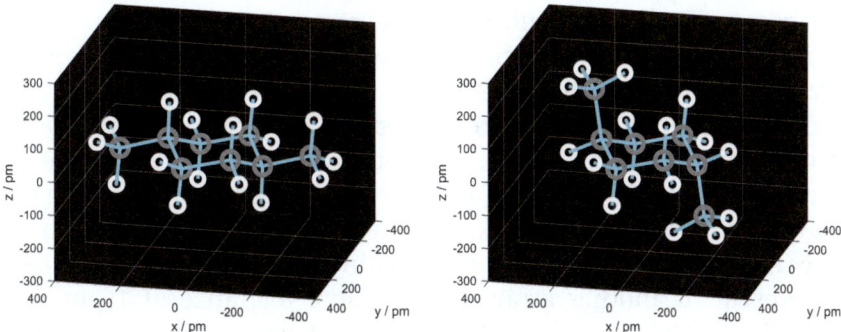

Obtaining 0,0,0 (or very close to 0,0,0 as seen above) for the components of the position of the adjusted centre of a molecule is usually an indication that the molecule has inversion symmetry; although not always, so we have to be careful. Think of methane, for example, where the central carbon atom sits at [0,0,0], yet inversion across that point results in a methane molecule not overlapping with – *i.e.* distinguishable from – the original molecule.

We perform the inversion operation on our 1,4 dimethyl cyclohexane molecules by first creating the transformation matrix, i, (see Table 8.3) and subsequently inverting the molecules using matrix multiplication:

```
>> i=[-1 0 0; 0 -1 0; 0 0 -1]; % transformation matrix for
inversion
>> AM_C_INV=AM_C * i; % invert cis-diMe cyclohexane
>> AM_T_INV=AM_T * i; % invert trans-diMe cyclohexane
```

At this point, you could draw the resulting molecules (whose atomic coordinates are stored in AM_C_INV and AM_T_INV) to visually establish if they are indistinguishable from the original molecules or not. We use MATLAB instead to make the comparisons.

```
>> cis_1_4_diMe_cyclohexane_unchanged_by_i=all(ismember(round(AM_C*100),
round(AM_C_INV*100), 'rows')) % round coordinates to 1 pm and compare
cis_1_4_diMe_cyclohexane_unchanged_by_i =
    logical
     0
>> trans_1_4_diMe_cyclohexane_unchanged_by_i =all(ismember(round
(AM_T*100),round(AM_T_INV*100), 'rows')) % round coordinates to 1 pm and
compare
trans_1_4_diMe_cyclohexane_unchanged_by_i =
    logical
     1
```

which suggests (wrongly) that *cis*-1,4 dimethyl cyclohexane has no inversion symmetry; whereas *trans*-1,4 dimethyl cyclohexane has. We now recall that some atomic positions are slightly off in molecular structure files. Using the above example, during inversion (flipping their signs) the components in AM_C 0.8846 Å and −0.8851 Å become −0.8846 Å and 0.8851 Å, respectively. After rounding to 1 pm (resulting in 88 pm, −89 pm, −88 pm, and 89 pm), MATLAB checks whether 88 pm is equal to 89 pm and whether −89 pm is equal to −88 pm, which both rightly evaluate to "false", *i.e.* 0, as they are not equal, even though they are within the precision of atomic positions. Thus, we should rather make comparisons within a 1 pm tolerance instead of rounding to 1 pm before comparing. MATLAB function ismembertol() is ideal for the task. It can take a user-defined, absolute tolerance value for comparing numbers while evaluating whether they (or a sequence of them) are present in another array. The absolute tolerance of 1 pm is passed to ismembertol() as "0.01,'DataScale',1" (1 pm = 0.01 Å) and the row-wise comparison is specified by the "'ByRows',true" name–value pair.

```
>>    cis_1_4_diMe_cyclohexane_unchanged_by_i=all(ismembertol(AM_C,
AM_C_INV,0.01,'DataScale',1,'ByRows',true))
cis_1_4_diMe_cyclohexane_unchanged_by_i=
    logical
     1
>>    trans_1_4_diMe_cyclohexane_unchanged_by_i=all(ismembertol(AM_T,
AM_T_INV,0.01,'DataScale',1,'ByRows',true))
trans_1_4_diMe_cyclohexane_unchanged_by_i=
    logical
     1
```

This result, obtained without using a molecular model kit, confirms that *cis*- and *trans*-1,4 dimethyl cyclohexane have inversion symmetry.

The analysis above can be extended to any molecule and symmetry operation, and turned into a user-defined MATLAB function which would then simply take the atomic coordinates and the matrix of the required symmetry, and output whether the molecule has that symmetry or not. With that function, it would be easy to set up a workflow to automatically classify molecules in terms of their symmetry for further analysis, for example deciding if they are electric dipoles or whether certain spectroscopic transitions are expected to be present in their spectra.

9 One-line Functions and Kinetic Modelling

By now, we have written multiple user-defined MATLAB functions and seen their usefulness, especially in performing complex tasks like displaying environmental data on interactive maps or propagating uncertainties. These functions are ideally suited for problems that would otherwise require scripts to handle. Often, however, we encounter tasks that we think would be best handled by user-defined functions, yet they do not seem complex enough to justify writing user-defined functions to solve them. For these types of problems we can use user-defined *one-line functions*.

Imagine that we are synthesising a library of fragrant esters found in nature and comparing the yields of the reactions under identical conditions. We react carboxylic acids with alcohols to form esters according to the scheme below (Scheme 9.1).

If the reactants are not added in a 1:1 molar ratio, the product mixture at full conversion will contain the leftovers of the reactants added in excess. If the initial reaction mixtures contain more alcohol than carboxylic acid, the reactions stop when the alcohols, the limiting reactants (LR), run out. Under such conditions, 100% yield is achieved when the amount of a product formed (n_P) has reached the initial amount of the limiting reactant ($n_{LR,0}$), which can be calculated by dividing the mass of LR added to the reaction mixture ($m_{LR,0}$) by its molar mass (M_{LR})

$$n_P = n_{LR,0} = \frac{m_{LR,0}}{M_{LR}}$$

therefore, 100% yield corresponds to $n_P/n_{LR,0} = 1$. During the course of reactions $0 \leq n_P/n_{LR,0} \leq 1$, which we can use to define *percent yield* as

$$\text{percent yield} = \frac{n_P}{n_{LR,0}} \times 100\%$$

to conveniently show the extent of the reaction: at the start ($n_P = 0$), percent yield = 0%; at the end ($n_P = n_{LR,0}$), percent yield = 100%. Because $n_P = m_P/M_P$, we can

---

A First Look at Coding in Chemistry: Solving Problems Using MATLAB
By Tamas Bansagi
© Tamas Bansagi 2025
Published by the Royal Society of Chemistry, www.rsc.org

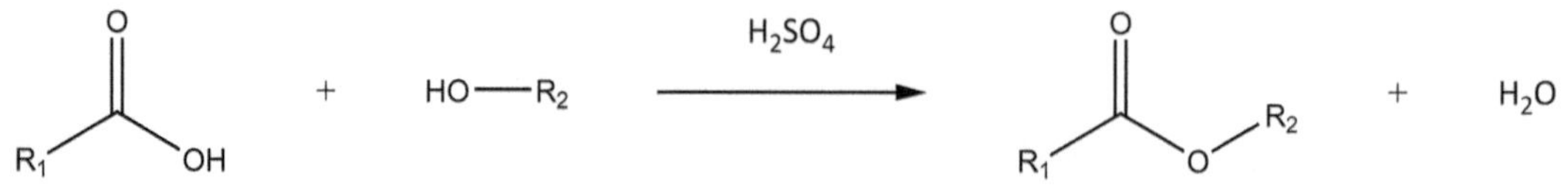

Scheme 9.1

calculate the *percent yield* directly from the masses of the limiting reactant and product as

$$\text{percent yield} = \frac{m_P M_{LR}}{m_{LR,0} M_P} \times 100\%$$

You may think this formula is too simple for turning it into a user-defined function, yet at the same time you perhaps feel that it would take up too much space in the *Command Window* if you kept using it over and over again. In situations like this, we can use MATLAB's *one-line functions*.

One-line functions are quick and easy to create in the *Command Window* and can be called afterwards just like any other function until we close MATLAB or delete them. As the name suggests, they can be defined concisely in a single command, without having to open a separate script window for creating and saving them.

We turn the above formula into a one-line function we name `percent_yield()`. It will take four numeric input arguments separated by commas in parentheses in the order of m_P, M_P, $m_{LR,0}$, M_{LR} to compute the *percent yield* according to the formula specified. Defining a one-line function has a strict syntax which starts with the name of the function object followed by assigning a one-line function to the object in the overall format of

name = @(input arguments) expression

Creating `percent_yield()` would, therefore, look like this:

```
>> percent_yield=@(m_P,M_P,m_LR0,M_LR) m_P*M_LR/m_LR0/M_P*100
percent_yield=
  function_handle with value:
    @(m_P,M_P,m_LR0,M_LR)m_P*M_LR/m_LR0/M_P*100
```

With our user-defined, one-line function `percent_yield()` created, we can quickly calculate the yields for our ester syntheses. Once we have separated and weighed the product of each reaction, we just need to feed m_{ester} to `percent_yield()` along with the corresponding M_{ester}, $m_{\text{alcohol},0}$ and M_{alcohol} values to fill the last column of the table in a lab notebook, as shown in Figure 9.1.

For example, if you weighed out 10.038 g of pentanol (and more than 70.092 g of butyric acid to ensure that pentanol is the limiting reactant), and obtained 13.885 g of product, the percent yield can be computed as simply as

```
>> pentyl_butyrate_yield=percent_yield(13.885,158.240,10.038,88.148)
pentyl_butyrate_yield=
    77.0539
```

You would calculate the rest of the percent yields the same way by calling `percent_yield()` each time with a set of input arguments specific to a particular ester synthesis.

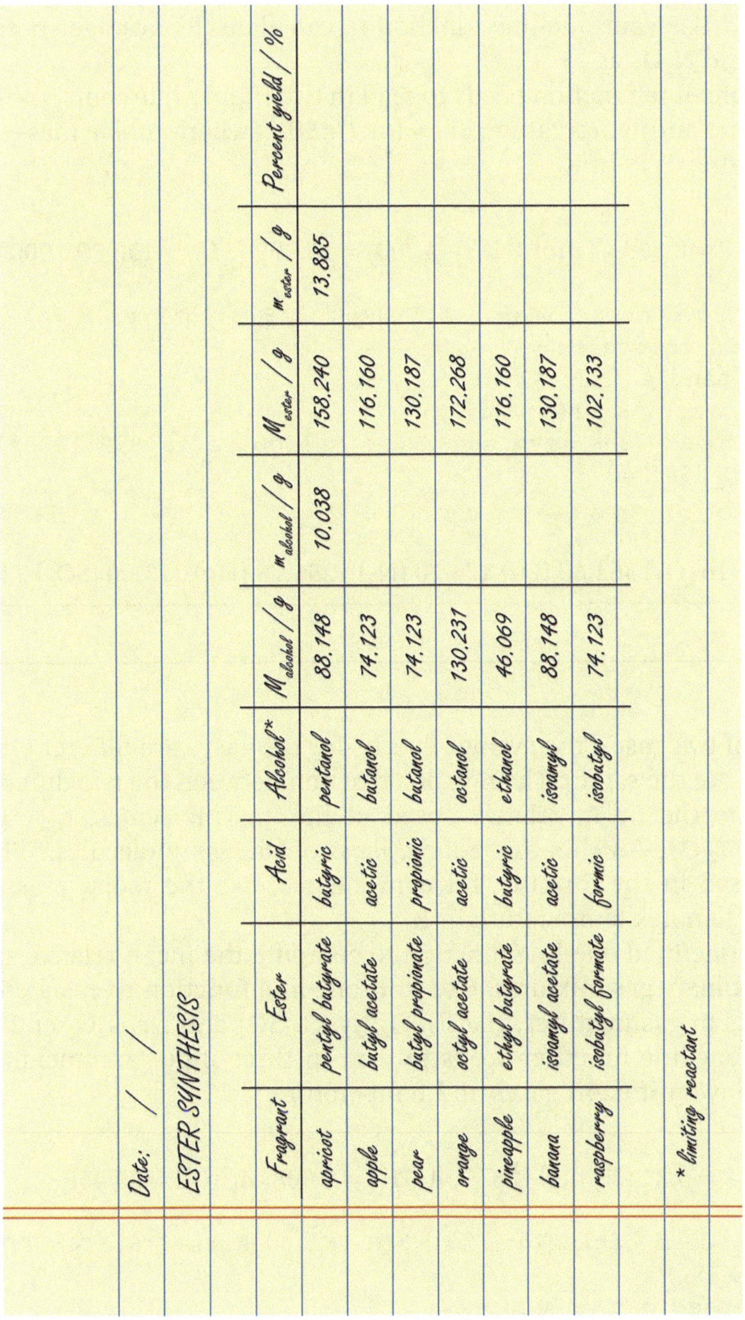

ESTER SYNTHESIS

Date: / /

Fragrant	Ester	Acid	Alcohol*	$M_{alcohol}$ / g	$m_{alcohol}$ / g	M_{ester} / g	m_{ester} / g	Percent yield / %
apricot	pentyl butyrate	butyric	pentanol	88.148	10.038	158.240	13.885	
apple	butyl acetate	acetic	butanol	74.123		116.160		
pear	butyl propionate	propionic	butanol	74.123		130.187		
orange	octyl acetate	acetic	octanol	130.231		172.268		
pineapple	ethyl butyrate	butyric	ethanol	46.069		116.160		
banana	isoamyl acetate	acetic	isoamyl	88.148		130.187		
raspberry	isobutyl formate	formic	isobutyl	74.123		102.133		

* limiting reactant

Figure 9.1 Excerpt of a lab notebook.

Exercise 9.1

The speed of gas molecules effects the rate of gas phase reactions.

Write a user-defined one-line function that calculates the average speed of gas molecules, $\langle v \rangle$. Use your one-line function to calculate the average speed of O_2, O_3, N_2, He, SO_2 and N_2O_5 at 25 °C.

Hint: your one-line function needs to read in two input arguments, the molar mass of the molecule and temperature; $\langle v \rangle = (8RT/\pi M)^{1/2}$ where molar mass, M, is in kg and temperature, T, is in K.

Answer:
$T = 298.15$ K, $R = 8.3145$ J mol^{-1} K^{-1}, $A_r°(O) = 15.999$ ($M_{O_2} = 15.999/1000*2$)

```
>> avg_speed_gas_molecules = @(M,T) sqrt(8*8.3145*T/pi/M)
avg_speed_gas_molecules =
    function_handle with value:
       @(M,T)sqrt(8*8.3145*T/pi/M)
>> O2_avg_speed_298K = avg_speed_gas_molecules(15.999/1000*2, 298.15)
O2_avg_speed_298 K =
    444.1648
```

$\langle v \rangle$/m s^{-1}: 444.16 (O_2) 362.66 (O_3) 474.70 (N_2) 1255.78 (He) 313.90 (SO_2) 241.75 (N_2O_5)

Exercise 9.2

In a mixture of two gases, the average speeds for the gases are different (unless their molar masses are the same). The rate of a reaction between the two different gases is proportional to the mean relative speed of the gas molecules, $\bar{c}_{rel} = (8RT/\pi\mu)^{1/2}$ where $\mu = M_1 M_2/(M_1 + M_2)$ is the *reduced mass* of the gas molecules. When the gas constant is used in the formula, for convenience, use the molar masses in kg to calculate μ. T denotes temperature in K.

Write a user-defined one-line function to compute the mean relative speed of gas molecules in binary gas mixtures. Use your one-line function to calculate the mean relative speeds in gas mixtures: H_2/Cl_2, O_2/N_2, O_3/SO_2 and Ne/N_2O_5 at 25 °C.

Hint: your one-line function needs to read in three input arguments, the molar masses of the two different gases and temperature.

Answer:
$T = 298.15$ K, $R = 8.3145$ J mol^{-1} K^{-1}, $A_r°(H) = 1.008$, $A_r°(Cl) = 35.450$

```
>> mean_rel_speed_gas_molecules = @(M1,M2,T) sqrt(8*8.3145*T/pi/(M1*M2/(M1+M2)))
mean_rel_speed_gas_molecules =
    function_handle with value:
       @(M1,M2,T)sqrt(8*8.3145*T/pi/(M1*M2/(M1+M2)))
>> mean_rel_speed_gas_molecules(1.008/1000*2, 35.450/1000*2, 298.15)
```

```
ans =
   1.7945e+03
```

$\bar{c}_{rel}$/m s^{-1}: 1794.5 (H_2/Cl_2) 650.09 (O_2/N_2) 479.64 (O_3/SO_2) 1278.8 (Ne/N_2O_5)

A one-line function can contain multiple expressions in square brackets separated by semicolons or colons (or white spaces). A one-line function with multiple expressions loads the outputs into an array with appropriate dimensions. Here, we define a one-line function to compute and output separately the reduced mass (in kg mol^{-1}) and the mean relative speed for two gases.

```
>>   reduced_mass__mean_rel_speed_gas_molecules = @(M1,M2,T)   [M1*M2/(M1
+M2); sqrt(8*8.3145*T/pi/(M1*M2/(M1+M2)))]
reduced_mass__mean_rel_speed_gas_molecules =
   function_handle with value:
     @(M1,M2,T)[M1*M2/(M1+M2); sqrt(8*8.3145*T/pi/(M1*M2/(M1+M2)))]
>> format shortE % output numbers into the Command Window in scientific
format: A×10^(B) => AeB which is useful when numbers are many orders of
magnitude apart)
>> mu_c_H2_Cl2 =
reduced_mass__mean_rel_speed_gas_molecules(1.008/1000*2,
35.450/1000*2,298.15)
mu_c_H2_Cl2 =
   1.9603e-03
   1.7945e+03
```

i.e. $\mu = 1.9603 \times 10^{-3}$ kg mol^{-1} and $\bar{c}_{rel} = 1794.5$ m s^{-1} (for the H_2/Cl_2 mixture);

```
>>   mu_c_O2_N2 = reduced_mass__mean_rel_speed_gas_molecules(15.999/
1000*2, 14.007/1000*2,298.15)
mu_c_O2_N2 =
   1.4937e-02
   6.5009e+02
```

i.e. $\mu = 1.4937 \times 10^{-2}$ kg mol^{-1} and $\bar{c}_{rel} = 650.09$ m s^{-1} (for the O_2/N_2 mixture); and so on ...

To put this more into context, imagine trying to calculate the rates of reactions between these gases in the atmosphere throughout a day. That would require regularly computing the mean relative speed for each pair of gas components using the current temperature, which could be conveniently done by calling the `mean_rel_speed_gas_molecules()` function created above.

9.1 Kinetic Modelling

User-defined one-line functions are very useful when we need to evaluate relatively simple simultaneous expressions repeatedly, each time for a set of different values.

(This is what we did above when we were computing the reduced mass and mean relative speed for each pair of gases together in a single step by calling `reduced_mass__mean_rel_speed_gas_molecules()` only once.) In kinetics, calculating the instantaneous rate of each reaction step within a multistep mechanism as time progresses while the concentrations are continuously changing is a prime candidate for using user-defined, one-line functions.

We begin looking into how one-line functions can be used in kinetics by considering the simplest elementary step:

$$A \xrightarrow{k} B$$

where B is produced from A at a $rate = k[A]$. Calculating the *rate* of this single-step reaction multiple times as the concentration of A changes from its initial value, $[A]_0$, is convenient with a one-line function, especially if we want to calculate both the rate of change of [A] and the rate of change of [B] by running a single command.

Let us define a new one-line function that computes the rate of decomposition of A ($d[A]/dt$) and the rate of production of B ($d[B]/dt$) simultaneously. We will only need two input variables for that: k and the concentration of A. We use A in place of [A] as brackets in MATLAB are restricted to defining arrays. First, we write the expressions for the rates of concentration changes of A and B in relation to the $rate = k[A]$:

rate of change of [A]: $\quad \dfrac{d[A]}{dt} = -k[A]$

rate of change of [B]: $\quad \dfrac{d[B]}{dt} = k[A]$

Exercise 9.3

Create a single one-line function containing the two expressions above to calculate the rates of concentration changes of A and B when [A] is 1.00 M, 0.50 M and 0.25 M if $k = 2.5 \times 10^3 \text{ s}^{-1}$.

Answer:

```
>> rates = @(k,A) [-k*A k*A]; % d[A]/dt = -k*A and d[B]/dt = k[A]
>> rates(2.5e3,1)
rates =
   -2500    2500
```

$rate$/mol dm^{-3} s^{-1} : −2500 2500; −1250 1250; −625 625; ($d[A]/dt$ $d[B]/dt$)

Because one molecule of A turns into one molecule of B, the rate of removal of A will be equal to the rate of production of B. Removal refers to a negative concentration change, hence the negative sign in front of the rate of change of [A]. As time passes [A] decreases, and so do the rate of removal of A and the rate of production of B.

To obtain how the concentrations of chemical species change over time during a reaction, integrated rate equations are derived from differential rate equations in introductory physical chemistry textbooks. For example, if a chemical process can be described by a single step A→B, then starting from the differential rate equation $d[A]/dt = -k[A]$, the integrated rate equations for the reactant and product concentrations, $[A] = [A]_0 e^{-kt}$ and $[B] = [A]_0(1 - e^{-kt})$, can be found by solving the differential rate equation *via* integration. The integrated rate equation for a species gives the concentration–time curve for that species, which can be used to predict its concentrations at any time throughout the reaction. For simple reactions (whose mechanisms contain a single elementary step like A→B, 2A→B, A + B→C or very few elementary steps like A ⇌ B or A→B→C), which are in fact rare in chemistry, this treatment is sufficient. However, reactions are usually governed by multi-step mechanisms which do not lend themselves to this analytical approach (*i.e.* integrated rate equations cannot be found for them) and require computational treatment.

The differential rate equations we encounter in chemical kinetics belong to the family of Ordinary Differential Equations (ODE). MATLAB has built-in ODE solvers that are ideal for chemists wanting to establish concentration–time curves without going into the trouble of trying to work out the integrated rate expressions, which are often impossible to find. These ODE solvers, therefore, play a key part in elucidating reaction mechanisms, formulating rate equations and determining rate constants.

MATLAB's ODE solvers require a slightly different format of user-defined one-line functions to the ones we used above. To illustrate how to recast kinetic differential equations to the required MATLAB format, we look at the equilibrium

$$A \underset{k_b}{\overset{k_f}{\rightleftharpoons}} B$$

MATLAB represents variables in ODEs, [A] and [B] in our case, as elements of a vector. If we decide to call that vector c, then c(1) and c(2) are going to serve as [A] and [B], respectively in MATLAB. The kinetic equations need to be changed accordingly to contain c(1) and c(2) before inserting them into the appropriate place within a user-defined one-line function suitable for ODE solvers.

		Scientific/maths notation	MATLAB notation
Net rate of formation of A:	$\dfrac{d[A]}{dt} =$	$k_b[B] - k_f[A]$	kb*c(2) - kf*c(1)
Net rate of formation of B:	$\dfrac{d[B]}{dt} =$	$k_f[A] - k_b[B]$	kf*c(1) - kb*c(2)

One simple way to pass our user-defined one-line function (composed of the two expressions in the third column) to a MATLAB ODE solver is without naming it (*i.e.* as an anonymous function). Time is required to be added as a variable even though it does not appear in the expressions. With these, the kinetic model for the equilibrium $A \rightleftharpoons B$ is cast into a MATLAB one-line function as

```
>> @(t,c) [kb*c(2) - kf*c(1); kf*c(1) - kb*c(2)]
ans =
function_handle with value:
        @(t,c)[kb*c(2)-kf*c(1);kf*c(1)-kb*c(2)]
```

Notice how this single one-line function incorporates all information about the kinetics of the $A \rightleftharpoons B$ equilibrium process and that it has no name (*i.e.*, anonymous). Before we pass this function to an ODE solver, we still need to define:

- *the rate constants*

    ```
    kf=2.4; kb=1.2;
    ```

- *the initial conditions* just as we need to know $[A]_0$ and $[B]_0$ at $t=0$ min for experiments; these are entered as a vector, say c0, so for $[A]_0 = 1$ mol dm^{-3} and $[B]_0 = 0$ mol dm^{-3} we simply write:

    ```
    c0 = [1  0];
    ```

    (Remember the concentration and time units (despite their not being used in the commands) as we will need them for labelling axes later on.)

- *the time points we would like the concentrations to be computed for*; they are defined here as

    ```
    tp = [0:0.01:5];
    ```

    representing instances of time from *0* minute to *5* minutes in the increments of *0.01 minutes*.

We are now ready to solve numerically how the concentrations approach their equilibrium values. We will call ODE solver ode45() for the task. (The full list of ODE solvers and the types of problems they are most suitable for can be found on the MathWorks website.)

```
>> [t,c] = ode45(@(t,c) [kb*c(2) - kf*c(1); kf*c(1) - kb*c(2)], tp, c0);
```

The output will be loaded into two arrays t and c, the former containing the time points we prescribed in tp and the latter storing the concentrations of A and B for each time point in two columns (column 1: $[A]_t$ and column 2: $[B]_t$). Double click on the arrays in *Workspace* to reveal their contents.

Now, we plot the results (Figure 9.2)

```
>> plot(t,c(:,1),'-r', t,c(:,2),'-b')
```

```
>> xlabel('time/min'); ylabel('Concentration/mol dm^{-3}')
>> title('Equilibrium between A and B')
>> legend('[A]','[B]')
```

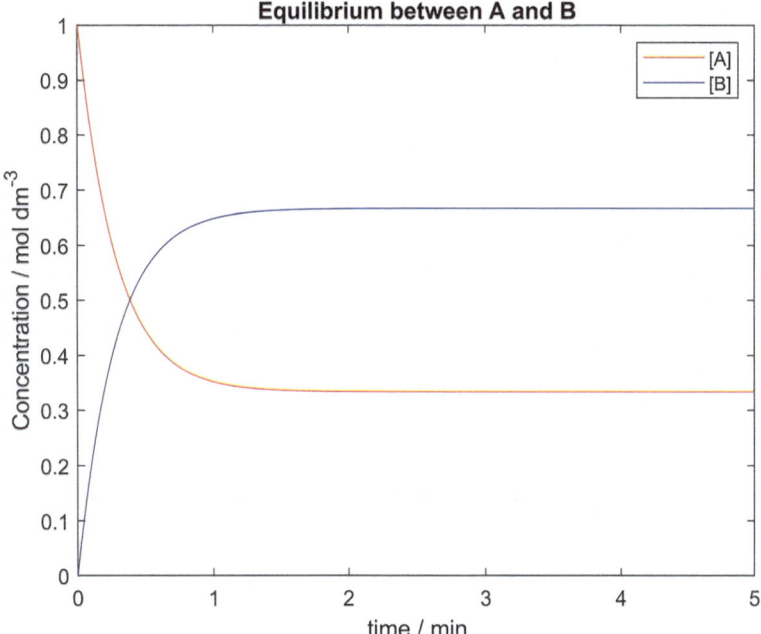

Figure 9.2 Concentration–time plots for a chemical equilibrium between A and B. Initially, only A is present in the reaction vessel.

Exercise 9.4

Consider aniline in the presence of an acid. It partially protonates to form phenylammonium ions as

which can be concisely written as

$$A + H^+ \underset{k_b}{\overset{k_f}{\rightleftharpoons}} P^+$$

For the equilibrium $A + H^+ \rightleftharpoons P^+$, create a one-line function containing the expressions for the rates of change of $[A]$, $[H^+]$ and $[P^+]$. Using ODE solver `ode45()`, predict the *concentration–time* curves of all species within the 0–5 minutes time within the first five minutes of the reaction if the forward and backward rate constants are 3.0 M^{-1}s^{-1} and 1.5 s^{-1}, respectively. Use initial concentrations:

(i)   [A]$_0$ = 1 M,   [H$^+$]$_0$ = 1 M,   [P$^+$]$_0$ = 0 M
(ii) [A]$_0$ = 0 M,   [H$^+$]$_0$ = 0 M,   [P$^+$]$_0$ = 1 M
(iii) [A]$_0$ = 1.8 M,   [H$^+$]$_0$ = 1 M,   [P$^+$]$_0$ = 0 M
(iv) [A]$_0$ = 1.8 M,   [H$^+$]$_0$ = 2 M,   [P$^+$]$_0$ = 0 M
(v)   [A]$_0$ = 1.8 M,   [H$^+$]$_0$ = 4 M,   [P$^+$]$_0$ = 0 M
(vi) [A]$_0$ = 1.8 M,   [H$^+$]$_0$ = 8 M,   [P$^+$]$_0$ = 0 M

How do the equilibrium concentrations of aniline and the phenylammonium ion change as the initial acid concentration is increased, while [A]$_0$ and [P$^+$]$_0$ remain unchanged in (iii–vi)?

(Hint: You need three rate expressions in your one-line function, one for each rate of change d[A]/dt, d[H$^+$]/dt and d[P$^+$]/dt. Use c(1), c(2) and c(3) to represent [A], [H$^+$] and [P$^+$], respectively.)

Answer:
The system of differential equations describing the net rates of formation of the species is

$$\frac{d[A]}{dt} = k_b[P^+] - k_f[A][H^+]$$

$$\frac{d[H^+]}{dt} = k_b[P^+] - k_f[A][H^+]$$

$$\frac{d[P^+]}{dt} = k_f[A][H^+] - k_b[P^+]$$

which translates to the following MATLAB code for solving them to obtain the sought concentration–time curves.

```
tp=0:0.01:5; % minutes
kf=3.0; % M^{-1} s^{-1}
kb=1.5; % s^{-1}
% c(1): [A]; c(2): [H+]; c(3): [P+]
c0=[1  0.9  0]; % (i) [A]0=1 M, [H+]0=0.9 M, [P+]0=0 M
[t1,c1]=ode45(@(t,c) [kb*c(3)-kf*c(1)*c(2); kb*c(3)-kf*c(1)*c(2); kf*c(1)*c(2)-kb*c(3)],tp,c0); % call ODE solver with diff. rate equations
c0=[0  0  1]; % (ii) [A]0=0 M, [H+]0=0 M, [P+]0=1 M
[t2,c2]=ode45(@(t,c) [kb*c(3)-kf*c(1)*c(2); kb*c(3)-kf*c(1)*c(2); kf*c(1)*c(2)-kb*c(3)],tp,c0); % call ODE solver with diff. rate equations
figure % (i)
plot(t1,c1(:,1),'-r', t1,c1(:,2),'-b', t1,c1(:,3),':k','LineWidth',2)
xlabel('time/min'); ylabel('Concentration/mol dm^{-3}')
title(' (i) Equilibrium between aniline (A), H^+ and phenylammonium (P^+)')
legend('[A]','[H^+]','[P^+]','Location','northeast')
figure % (ii)
plot(t2,c2(:,1),'-r', t2,c2(:,2),'-b', t2,c2(:,3),':k','LineWidth',2)
xlabel('time/min'); ylabel('Concentration/mol dm^{-3}')
```

```
title('(ii) Equilibrium between aniline (A), H^+ and phenylammonium (P^+)')
legend('[A]','[H^+]','[P^+]','Location','northeast')
```

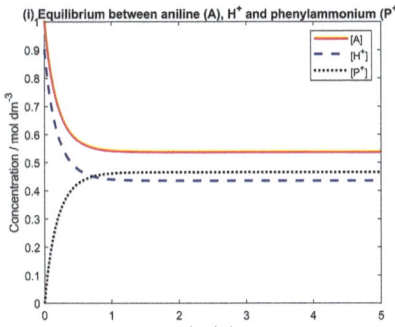
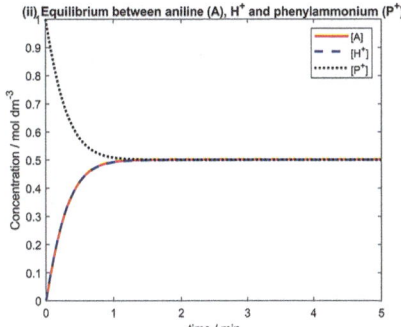

If there is no phenylammonium present initially, it forms from aniline and H^+ over time (i); whereas, if there is only phenylammonium present at the start, aniline and H^+ is produced in equal amounts (ii) until equilibrium is reached.

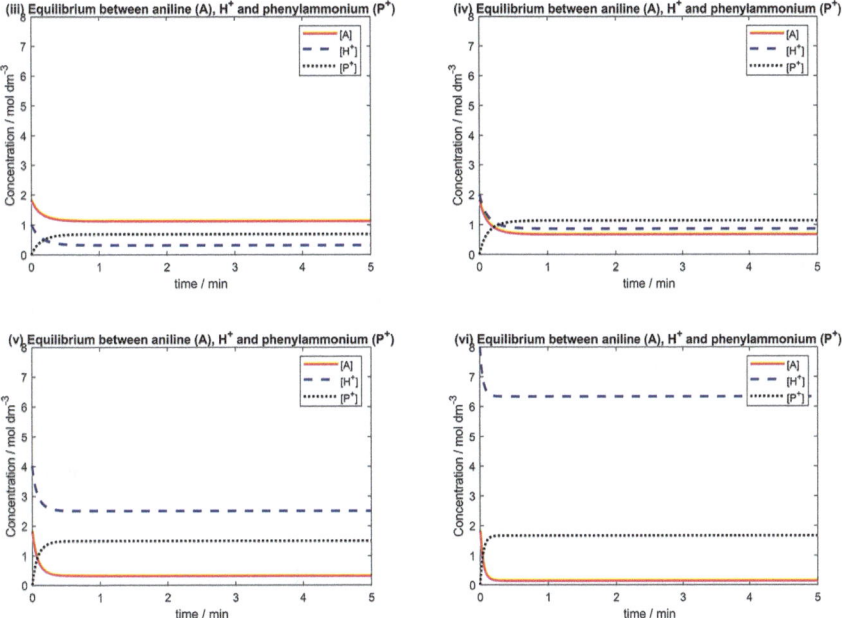

When the initial concentration of aniline is fixed ($[\text{aniline}]_0 = 1.8$ M) while the initial acid concentration is increased (from $[H^+]_0 = 1$ M to $[H^+]_0 = 8$ M), the difference between the equilibrium concentrations of aniline and phenylammonium grows. At very large acid concentrations the equilibrium concentration of aniline is much smaller than that of phenylammonium.

Exercise 9.5

A common mechanism in organic chemistry is where a first reversible step is followed by a second step leading directly to product. For example, during unimolecular nucleophilic substitution, RX dissociates into a carbocation intermediate (R^+) and X^- in a first-order process. The reactive intermediate then either recombines with X^- to form RX or reacts with a nucleophile (:Nuc$^-$) to form the product R–Nuc.

$$RX \underset{k_{-1}}{\overset{k_1}{\rightleftharpoons}} R^+ + X^-$$

$$R^+ + :Nuc^- \xrightarrow{k_2} R-Nuc$$

The S_N1 reaction

follows this mechanism.

Write the differential equations for the net rate of formation of all species in the general scheme and the corresponding MATLAB one-line function with vector c storing their concentrations.

Answer:

The system of differential equations describing the net rates of formation of the species is

$$\frac{d[RX]}{dt} = k_{1b}[R^+][X^-] - k_{1f}[RX]$$

$$\frac{d[R^+]}{dt} = k_{1f}[RX] - k_{1b}[R^+][X^-] - k_2[R^+][:Nuc^-]$$

$$\frac{d[X^-]}{dt} = k_{1f}[RX] - k_{1b}[R^+][X^-]$$

$$\frac{d[:Nuc^-]}{dt} = -k_2[R^+][:Nuc^-]$$

$$\frac{d[R-Nuc]}{dt} = k_2[R^+][:Nuc^-]$$

The corresponding MATLAB function to obtain the concentration–time curves is

```
@(t,c) [k1b*c(2)*c(3) - k1f*c(1); k1f*c(1) - k1b*c(2)*c(3) - k2*c(2)*c(4); k1f*c(1) - k1b*c(2)*c(3); -k2*c(2)*c(4); k2*c(2)*c(4)]
```

where `c(1)`, `c(2)`, `c(3)`, `c(4)`, `c(5)` denote [RX], [R$^+$], [X$^-$], [:Nuc$^-$], [R–Nuc], respectively.

9.2 Case Study: Nitration of Aniline

In the presence of concentrated sulphuric acid, the addition of nitric acid to an aniline–phenylammonium mixture leads to the nitration of the benzene ring through electrophilic substitution. The mechanism in Figure 9.3 captures the main processes taking place in the system. (Elementary steps involve a reactive nitronium ion which we ignore here for simplicity.)

Kinetic measurements have shown that aniline undergoes nitration much faster than phenylammonium (meaning that $k_1 \gg k_2$), which suggests that the main product is *p*-nitroaniline. It was found, however, that the two products form in nearly equal amounts under highly acidic conditions.

This seemingly counterintuitive result can be qualitatively explained based on Exercise 9.4. Under strongly acidic conditions there is much more phenylammonium

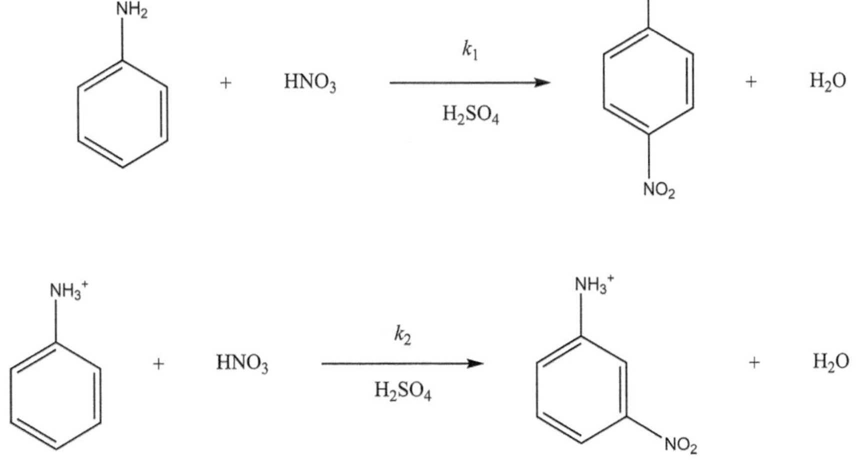

Figure 9.3 Mechanism of aniline nitration.

present than aniline: [aniline] ≪ [phenylammonium]. *The final ratio of p-nitroaniline to m-nitroaniline depends on the relative rates of the two nitration steps and not on the rate constants.* Given that k_1 is large but [aniline] is small and that k_2 is small but [phenyilammonium] is large, the rates of production of *p*-nitroaniline and *m*-nitroaniline become similar; therefore, the two products form in near equal amounts over the same time period.

$$\begin{array}{ccc} \text{rate of aniline nitration} & & \text{rate of phenylammonium nitration} \\ k_1 \times [\text{aniline}] & \approx & k_2 \times [\text{phenylammonium}] \\ \text{Large} \quad \text{small} & & \text{small} \quad \text{large} \end{array}$$

Note that explaining the observed product ratio did not require drawing curly arrows. It only took some knowledge of kinetics and solving the ODEs computationally in Exercise 9.4.

9.3 Case Study: Self-replicating Peptides

Many organic molecules, especially nucleotides and peptides, have a tendency to form secondary structures through hydrogen bonding. The simplest type of such structures is dimers where two molecules align themselves with each other. In this example, when peptides A and B react, a longer peptide, T, forms. T can then serve as a template to bind molecules of A and B which, once bound on the surface of T, react much quicker to form T. Subsequently, this second molecule of T reversibly detaches from the original template and begins acting as a template on its own. This simple self-replicating organic chemical process has been proposed by Wagner *et al.* (*J. Phys. Chem. Lett.* 2015, **6**, 60; *Isr. J. Chem.* 2015, **55**, 880) and is shown in Figure 9.4.

This is a complex mechanism, despite the overall process being A + B → T, as more and more T form, the number of available templates for rapid production of T increases. This self-induced proliferation of the product leading to rate acceleration is

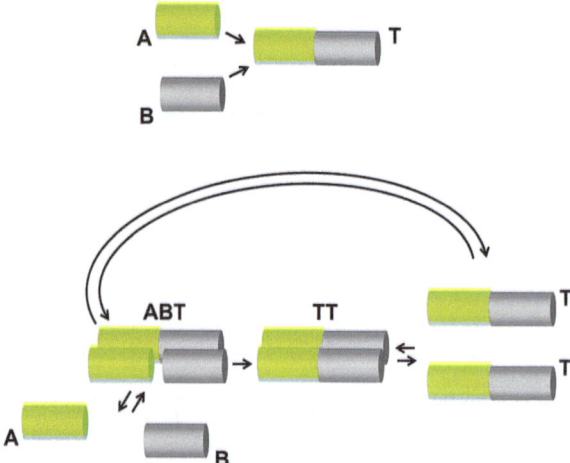

Figure 9.4 Mechanism of a peptide self-replication process, adapted from *Isr. J. Chem.* 2015, **55**, 880.

called *positive feedback* or *autocatalysis*. To model this chemical self-replicator system, we first write down the mechanism consisting of the elementary steps of the reaction.

$$A + B \xrightarrow{k_1} T$$

$$A + B + T \underset{k_{2r}}{\overset{k_2}{\rightleftharpoons}} ABT$$

$$ABT \xrightarrow{k_3} TT$$

$$TT \underset{k_{4r}}{\overset{k_4}{\rightleftharpoons}} T + T$$

Using the mechanism above, we proceed by constructing the equations for the net rate of concentration changes for all species.

$$\frac{d[A]}{dt} = k_{2r}[ABT] - k_1[A][B] - k_2[A][B][T]$$

$$\frac{d[B]}{dt} = k_{2r}[ABT] - k_1[A][B] - k_2[A][B][T]$$

$$\frac{d[T]}{dt} = k_{2r}[ABT] + k_1[A][B] + 2k_4[TT] - k_2[A][B][T] - 2k_{4r}[T]^2$$

$$\frac{d[ABT]}{dt} = k_2[A][B][T] - k_{2r}[ABT] - k_3[ABT]$$

$$\frac{d[TT]}{dt} = k_3[ABT] + k_{4r}[T]^2 - k_4[TT]$$

With `c(1)`, `c(2)`, `c(3)`, `c(4)` and `c(5)` representing [A], [B], [T], [ABT] and [TT], respectively, we write the following code to approximate the concentration–time profiles for all species.

Figure 9.5 shows a *chemical clock* where the reaction appears dormant initially for a while, before a sudden burst of product formation (green curve).

```
k1=1e-5; k2=10; k2r=0.01; k3=5; k4=2; k4r=0.01; % define rate constants

c0=[10 10 0 0 0]; % set initial conditions (initial concentrations)
% ODE solver with nested kinetic model
[t,c]=ode45(@(t,c) [k2r*c(4)-k1*c(1)*c(2)-k2*c(1)*c(2)*c(3);
k2r*c(4)-k1*c(1)*c(2)-k2*c(1)*c(2)*c(3);
k1*c(1)*c(2)+k2r*c(4)+2*k4*c(5)-k2*c(1)*c(2)*c(3)-2*k4r*c(3)^2;
k2*c(1)*c(2)*c(3)-k2r*c(4)-k3*c(4);
k3*c(4)+k4r*c(3)^2-k4*c(5)], 0:0.05:10, c0);
plot(t, c(:,1), 'k', t, c(:,2), '-r', t, c(:,3), 'g', t, c(:,4), 'b', t, c
(:,5), '-c','linewidth',2); % plot all concentrations in one command
xlabel('time/min'); ylabel('concentration/mM'); % label axes
legend('[A]','[B]','[T]','[ABT]','[TT]','location','west');
% add legend
title('Self-replication of peptide T'); % add title
```

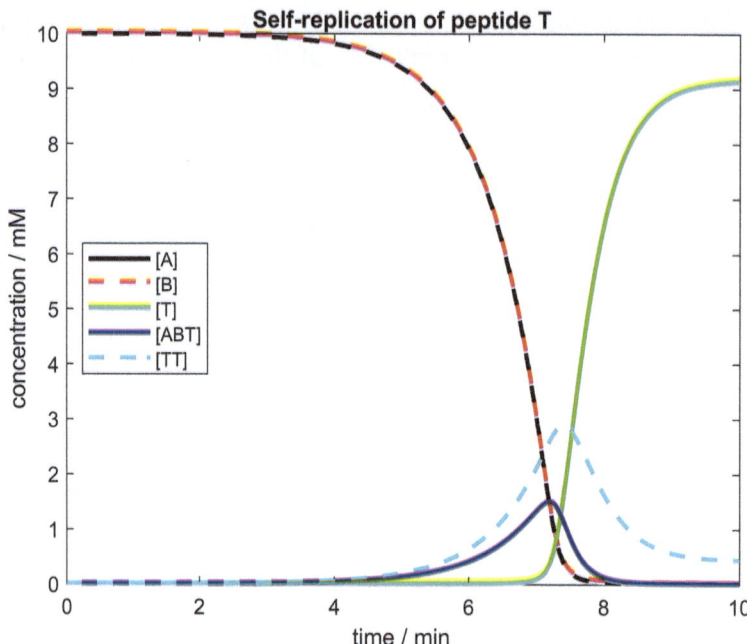

Figure 9.5 Concentration–time profiles for the self-replicating peptide process.

Some may think this kinetic model is too complicated to be handled as a user-defined one-line function. Indeed, as the number of species becomes more than just a few, code readability starts to rapidly decline, which can be remedied by putting the kinetic model (system of ODEs) into a user-defined function and calling it from an ODE solver within the script that does the post-processing of the concentration data. The kinetic model can now be defined with more detail added to increase readability and understanding.

```
function dcdt=kin_mod_srp(t,c)
global k % rate constant array (1:k1 2:k2 3:k2r 4:k3 5:k4 6:k4r)
% net rate of formation equations for the self-replicating peptide model
dAdt = k(3)*c(4)-k(1)*c(1)*c(2)-k(2)*c(1)*c(2)*c(3);
dBdt = k(3)*c(4)-k(1)*c(1)*c(2)-k(2)*c(1)*c(2)*c(3);
dTdt = k(1)*c(1)*c(2)+k(3)*c(4)+2*k(5)*c(5)-k(2)*c(1)*c(2)*c(3)-2*k
(6)*c(3)^2;
dABTdt = k(2)*c(1)*c(2)*c(3)-k(3)*c(4)-k(4)*c(4);
dTTdt = k(4)*c(4)+k(6)*c(3)^2-k(5)*c(5);
dcdt = [dAdt; dBdt; dTdt; dABTdt; dTTdt];
end
```

Save the above function under the name "kin_mod_srp.m" which you will then call from `ode45()` in the script below. Because the rate constants are required by kin_mod_srp(), we need to make sure it is able to access array k. One way of achieving this is by declaring k as a *global array*, which will make k available to all functions even though it is not defined within their scopes.

One-line Functions and Kinetic Modelling

```
global k % initiate global array for rate constants
k1 = 1e−5; k2 = 10; k2r = 0.01; k3 = 5; k4 = 2; k4r = 0.01; % define rate
constants
k = [k1 k2 k2r k3 k4 k4r]; % load rate constants into array k
c0 = [10 10 0 0 0]; % set initial conditions (initial concentrations)
tp = 0 : 0.05 : 10; % time points
[t,c] = ode45(@kin_mod_srp,tp,c0); % ODE solver with kinetic model defined
in function kin_mod_srp(); NOTICE that "(t,c)" has been removed
plot(t, c(:,1), 'k', t, c(:,2), 'r', t, c(:,3), 'g', t, c(:,4), 'b', t, c
(:,5), 'c'); % plot all concentrations in one command
xlabel('time/min'); ylabel('concentration/mM'); % label axes
legend('[A]','[B]','[T]','[ABT]','[TT]','location','west');
% add legend
title('Self-replication of a peptide (T)'); % add title
```

The user-defined function and the script work in tandem. To change parameters (rate constants, initial concentrations, time points) you need to modify the script; if you alter the reaction mechanism, you will have to amend the user-defined function accordingly (then save it before running the script).

It is also possible to stick user-defined functions under the script they are called from to create fully self-contained compartmentalised code. If you were to merge the script and the function into a single code, you would need to copy and paste the function under the script where you would also need to change

`[t,c] = ode45(@kin_mod_srp,tp,c0);`

to

`[t,c] = ode45(@(t,c)kin_mod_srp(t,c),tp,c0);`

Due to the script and the function now being part of the same code, we do not need to declare `k` as a global array, as long as we add it to the list of input arguments of `kin_mod_srp()`. This would mean we can delete "`global k`" from both the script and the function, and change `kin_mod_srp(t,c)` to `kin_mod_srp(t,c,k)` in the function declaration and inside the argument of `ode45()`.[†]

9.4 Appendix: Euler's Method (I)

When we use MATLAB functions to solve kinetic differential equations, like `ode45()` above, the concentration–time curves for the chemical species are approximated without finding the exact expressions linking the concentrations to time. For most chemical reactions, barring a few very simple ones, finding such expressions is an

---

[†] In Octave, move the functions to the top (so they are defined before they are called) preceded only by a command that does not interfere with the rest of the code, for example "`0;`". The latter is required because otherwise Octave would interpret the code as a user-defined function, which this is not, for this is a script that contains function declarations.

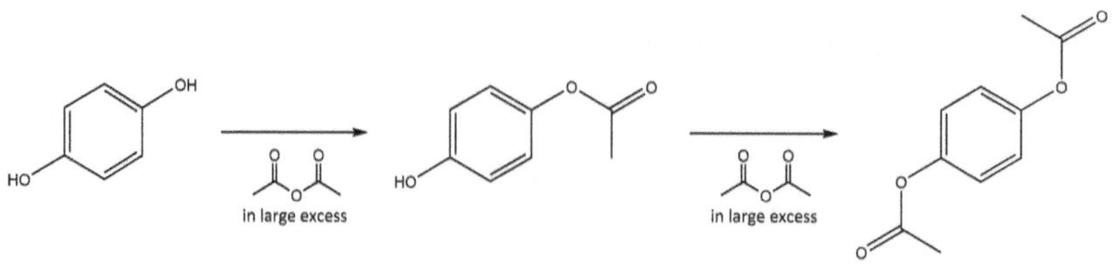

Figure 9.6 Reaction of hydroquinone with acetic anhydride. With a large excess of acetic anhydride, the rates of the steps are independent of the acetic anhydride concentration.

arduous (or most likely impossible) task; thus, we are typically left with no choice but to approximate concentration–time curves. For most applications this is sufficient as we typically only want to visualise the course of the reaction in graphs instead of looking at complicated expressions. Because concentration–time curves vividly capture how concentrations evolve during reactions and because elementary differential rate equations are simple to formulate, approximating concentration–time curves is much more practical than trying to find the exact expressions describing the changes of concentrations over time.

To illustrate how concentration–time profiles can be approximated consider the reaction between hydroquinone and acetic anhydride taking place *via* two consecutive steps under large excess of acetic anhydride (Figure 9.6).

The mechanism with large acetic anhydride excess can be captured by two consecutive first-order steps as

$$A \xrightarrow{k_1} B \xrightarrow{k_2} C$$

The rates at which the concentrations of A, B and C change, called the net rates of formation of the species ($d[A]/dt$, $d[B]/dt$, $d[C]/dt$), can be formulated using the rates of the two steps:

Rate of step 1: $\quad r_1 = k_1[A]$
Rate of step 2: $\quad r_2 = k_2[B]$

As the reaction progresses, the concentrations of all three species change simultaneously and, thus, the rates of both steps are constantly changing as well. At any instance, rates r_1 and r_2 will depend on the current concentrations of species A and B; therefore, the net rates of formation of the compounds will also be continuously changing. For this scheme, the differential rate equations in the left column can be solved mathematically, which yields (for $k_1 \neq k_2$) the integrated rate equations given in the right column. These expressions can be used to calculate the concentrations of the species at any instance (t) and hence draw their exact concentration–time curves.

The tricky part of solving these differential rate equations is finding the integrated rate equation for [B]. Solving the differential rate equation for [A] is straightforward and is discussed in introductory kinetics texts (A→B for which $[A] = [A]_0 e^{-kt}$). The integrated rate expression for [A] (top equation in the right column of Table 9.1) can be then substituted into the differential rate equation for [B] (middle equation in the left column of Table 9.1) which turns it into a differential equation that belongs to a class of differential equations called *linear inhomogeneous differential equations*. There are established ways of solving this class of differential equations, which we are not going to

Table 9.1 Differential and integral rate equations for the consecutive reaction scheme A→B→C.

Net rates of concentration changes	Integrated rate equations giving the concentration–time curves
$\dfrac{d[A]}{dt} = -k_1[A]$	$[A] = [A]_0 e^{-k_1 t}$
$\dfrac{d[B]}{dt} = k_1[A] - k_2[B]$	$[B] = \dfrac{k_1[A]_0}{k_2 - k_1}\left(e^{-k_1 t} - e^{-k_2 t}\right)$
$\dfrac{d[C]}{dt} = k_2[B]$	$[C] = [A]_0\left(1 + \dfrac{k_2 e^{-k_1 t} - k_1 e^{-k_2 t}}{k_1 - k_2}\right)$

look into here. Instead, we just use the result (middle expression in the right column of Table 9.1) to substitute it into the material balance $[C] = [A]_0 - [A] - [B]$, which then yields the integrated rate equation for $[C]$.

The process of finding the integrated rate equations for the chemically simple consecutive reaction scheme is a bit lengthy and cumbersome for most students taking introductory kinetics. Finding the integrated rate equations for more realistic reaction schemes would require even more difficult mathematical treatments, which are in most cases hopeless endeavours as the differential rate equations cannot be solved, *i.e.* there are no established mathematical methods for finding the integrated rate equations corresponding to them. Luckily, formulating equations for net rates of concentration changes is very simple, even for reaction mechanisms composed of many elementary steps. Is it possible to approximate the concentration-time curves, *i.e.* the concentration trajectories describing the course of a reaction, directly from the net rates of concentration changes?

Swiss mathematician Leonhard Euler (1707–1783) realised first that, in general, trajectories starting from known points can be approximated if we have the derivatives of the trajectories. In our chemistry context, this means that if we know the initial concentrations (*i.e.* the starting points of the concentration trajectories of species) and the expressions describing the rates of changes of concentrations (the derivatives of the concentration trajectories), we can approximate concentration–time curves for reactions.

Adapting Euler's method for the consecutive reactions scheme, the rates of concentration changes can be approximated using the expressions in the left column of Table 9.1 if we notice that *time* can be divided into tiny intervals of length Δt and assume that the rates remain constant during these short time periods. Now, instead of changing continuously, the rates will take small jumps from one Δt interval to the next and stay temporarily frozen inside each Δt domain.

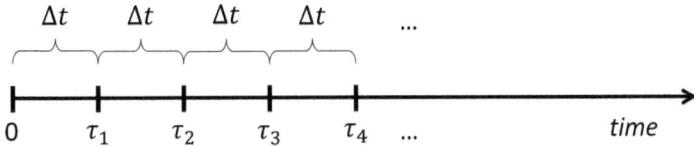

For example, at the start of the reaction ($\tau = 0$), the rate of change of [A] (d[A]/dt) can be readily calculated from the initial concentration of A, $[A]_0$, as

$$\left(\frac{d[A]}{dt}\right)_0 = -k_1[A]_0$$

After the first instance of the reaction, as A begins to be removed, d[A]/dt starts to change (becoming less negative with decreasing [A]). We, however, in our calculations keep it fixed for the duration of the first short Δt period ($0 \rightarrow \tau_1$) on the assumption that Δt is small enough for the value of $-k_1[A]_0$ to remain in the vicinity of the smoothly changing d[A]/dt throughout the entire first Δt period. With that, we can express this fixed rate of change for the first Δt interval ($0 \rightarrow \tau_1$) as

$$\left(\frac{\Delta[A]}{\Delta t}\right)_{0 \rightarrow \tau_1} \approx -k_1[A]_0$$

where $\Delta[A] = [A]_{\tau_1} - [A]_0$ is the change in the concentration of A during the first Δt interval (*i.e.* the concentration of A at $\tau = 0$ subtracted from the concentration of A at τ_1). Multiplying both sides by Δt yields (after Δt in the denominator has been cancelled out) that the estimated change in [A] over the first Δt time span is

$$(\Delta[A])_{0 \rightarrow \tau_1} \approx -k_1[A]_0 \Delta t$$

Once we have computed the estimated initial change in [A], we can approximate [A] at time τ_1, $[A]_{\tau_1}$, by replacing $(\Delta[A])_{0 \rightarrow \tau_1}$ with $[A]_{\tau_1} - [A]_0$

$$[A]_{\tau_1} - [A]_0 \approx -k_1[A]_0 \Delta t$$

which we rearrange for $[A]_{\tau_1}$ yielding

$$[A]_{\tau_1} \approx [A]_0 - k_1[A]_0 \Delta t$$

Having approximated the value for [A] at time τ_1, we can move on to estimating [A] at time τ_2, $[A]_{\tau_2}$. We will do it the same way we obtained $[A]_{\tau_1}$ from $[A]_0$, but now we use $[A]_{\tau_1}$ for estimating the fixed rate for the ($\tau_1 \rightarrow \tau_2$) interval as

$$\left(\frac{\Delta[A]}{\Delta t}\right)_{\tau_1 \rightarrow \tau_2} \approx -k_1[A]_{\tau_1}$$

With $(\Delta[A])_{\tau_1 \rightarrow \tau_2} = [A]_{\tau_2} - [A]_{\tau_1}$, we express the estimated [A] at time τ_2 as

$$[A]_{\tau_2} \approx [A]_{\tau_1} - k_1[A]_{\tau_1} \Delta t$$

The sequence of $[A]_\tau$ values approximating the concentration–time curve (trajectory) for A during the reaction can be given by the recursive formula

$$[A]_{\tau + \Delta t} \approx [A]_\tau - k_1[A]_\tau \Delta t$$

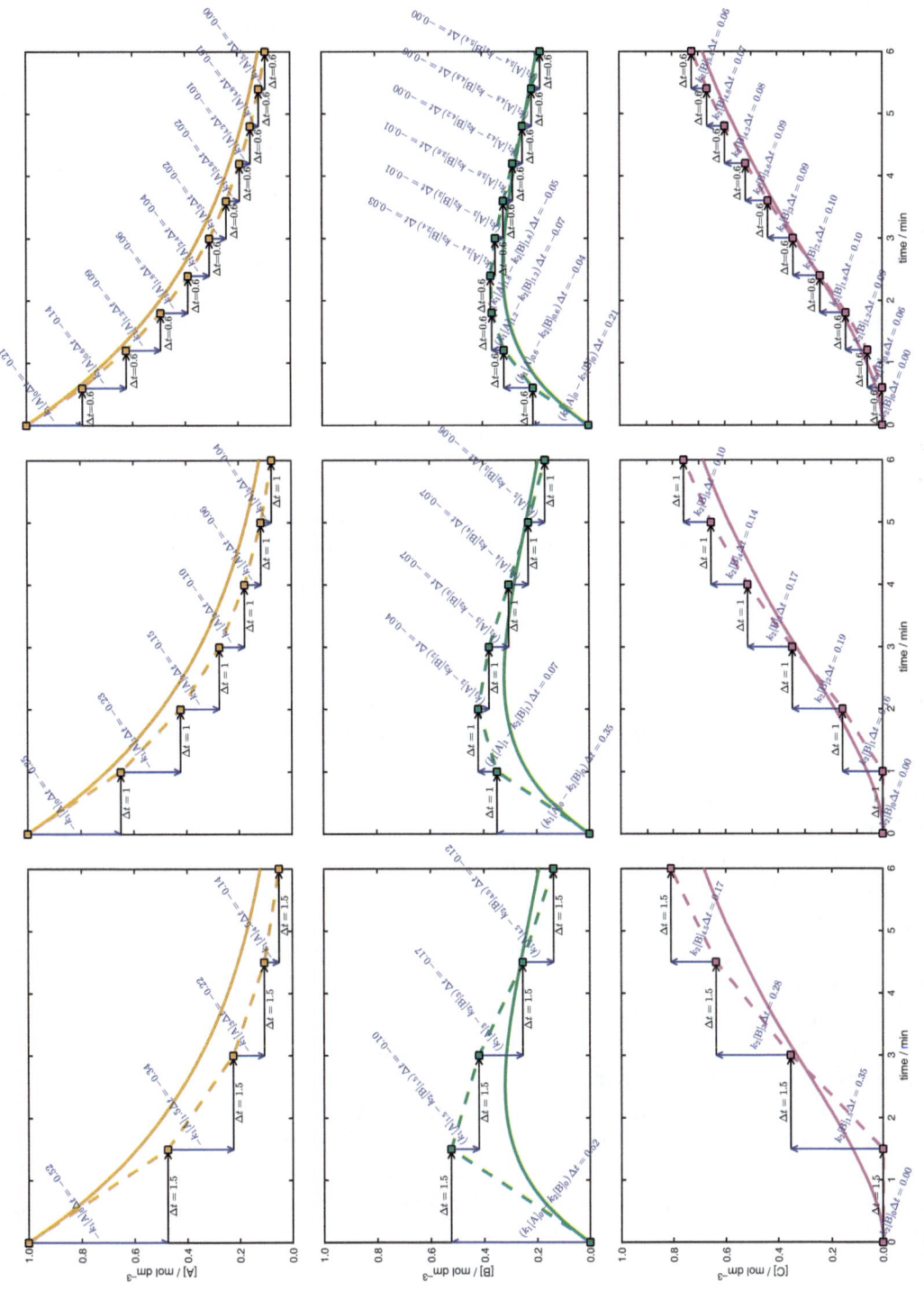

Figure 9.7 Approximating concentration-time profiles using Euler's method.

Table 9.2 Expressions for carrying out Euler's method to approximate the concentration-time profiles for the consecutive reaction scheme A→B→C.

Step	Time	Concentration of A	Concentration of B	Concentration of C
—	0	$[A]_0$	$[B]_0$	$[C]_0$
1	τ_1	$[A]_1 \approx [A]_0 - k_1[A]_0 t$	$[B]_1 \approx [B]_0 + (k_1[A]_0 - k_2[B]_0)\Delta t$	$[C]_1 \approx [C]_0 + k_2[B]_0 \Delta t$
2	τ_2	$[A]_2 \approx [A]_1 - k_1[A]_1 \Delta t$	$[B]_2 \approx [B]_1 + (k_1[A]_1 - k_2[B]_1)\Delta t$	$[C]_2 \approx [C]_1 + k_2[B]_1 \Delta t$
3	τ_3	$[A]_3 \approx [A]_2 - k_1[A]_2 \Delta t$	$[B]_3 \approx [B]_2 + (k_1[A]_2 - k_2[B]_2)\Delta t$	$[C]_3 \approx [C]_2 + k_2[B]_2 \Delta t$
⋮	⋮	⋮	⋮	⋮
i	τ_i	$[A]_i \approx [A]_{i-1} - k_1[A]_{i-1}\Delta t$	$[B]_i \approx [B]_{i-1} + (k_1[A]_{i-1} - k_2[B]_{i-1})\Delta t$	$[C]_i \approx [C]_{i-1} + k_2[B]_{i-1}\Delta t$
⋮	⋮	⋮	⋮	⋮

Similarly, the approximate concentrations of B and C can be computed as

$$[B]_{\tau+\Delta t} \approx [B]_\tau + (k_1[A]_\tau - k_2[B]_\tau)\Delta t$$

$$[C]_{\tau+\Delta t} \approx [C]_\tau + k_2[B]_\tau \Delta t$$

Given that at $\tau = 0$, $[A]_\tau = [A]_0$, $[B]_\tau = [B]_0$ and $[C]_\tau = [C]_0$, the concentrations can be approximated for the entire reaction, as long as the initial concentrations $[A]_0$, $[B]_0$, and $[C]_0$ and the rate constants are known. All we need to do is split time into small Δt intervals so that initially $\tau = 0$, then $\tau_1 = \Delta t$, $\tau_2 = \tau_1 + \Delta t$, $\tau_3 = \tau_2 + \Delta t$ and so on and compute the sequence of calculations given in Table 9.2.

The resulting sequences of estimated concentrations will trace out the predicted concentration–time curves as shown in Figure 9.7 where the smooth solid lines represent the exact concentration–time curves given by the expressions in the right column of Table 9.1. The approximate concentrations of the species are represented by squares connected by dashed lines. Arrows depict the changes in concentrations (blue) and time (black) in each step taken.

When the first 6 minutes of the reaction time is divided into four $\Delta t = 1.5$ min long time spans, the approximate concentrations – obtained by following the calculation sequence provided in Table 9.2 using $\Delta t = 1.5$ – deviate considerably from the smooth solid lines. However, as time is split into shorter and shorter intervals the approximate concentrations move closer and closer to the exact concentration–time curves. The gaps between the solid and dashed curves would continue to reduce, and eventually become undiscernible, if we were to decrease Δt further and further. Notice that the number of steps, and hence the number of calculations, would increase as a result making the process more computationally costly.

Euler's method is rarely used these days, other than to demonstrate how the solutions of linear ordinary differential equations can be approximated. MATLAB's ODE solvers use sophisticated methods which are computationally much more effective than the Euler's method, yet it is very useful in developing an understanding of the basic principles of scientific computational methods.

10 Self-controlling Code

Now that we are becoming more familiar with MATLAB commands, scripts and user-defined functions, we will be looking at automating calculations and data analysis. When analysing data or predicting the outcome of a reaction, we often need to perform the same task multiple times while varying the values of some variables. Also, we commonly encounter situations where we need to execute different sets of commands based on some criteria.

Imagine that we are trying to understand why the electrophilic addition of dienes can result in different products depending on the temperature. To investigate this phenomenon, we consider a reaction between a 1,3-pentadiene derivative and a hydrogen halide (HX) and apply our chemistry knowledge along with our kinetic modelling skills acquired in the previous chapter. In the first step of the reaction, one of the double bonds of the 1,3-pentadiene derivative undergoes regioselective protonation, which yields the more stable carbocation shown below.

This intermediate subsequently reacts with X^- giving two different products. Studies have shown that the 1,2 addition product forms predominantly at low temperatures, and as the temperature is increased the 1,4 addition product gradually becomes dominant. To explain these findings we sketch an energy diagram (Figure 10.1) showing the two energy barriers associated with the two different carbocation······X^- transition states and the products they lead to. Starting from the potential energy well in the middle, the carbocation + X^- system (**A** + **B**) would need to move through either of the barriers to form a product. Because less energy is required to climb over the barrier on the right ($E_{a,1f} < E_{a,2f}$), at low temperatures the 1,2 addition product (**C**) forms predominantly.

A First Look at Coding in Chemistry: Solving Problems Using MATLAB
By Tamas Bansagi
© Tamas Bansagi 2025
Published by the Royal Society of Chemistry, www.rsc.org

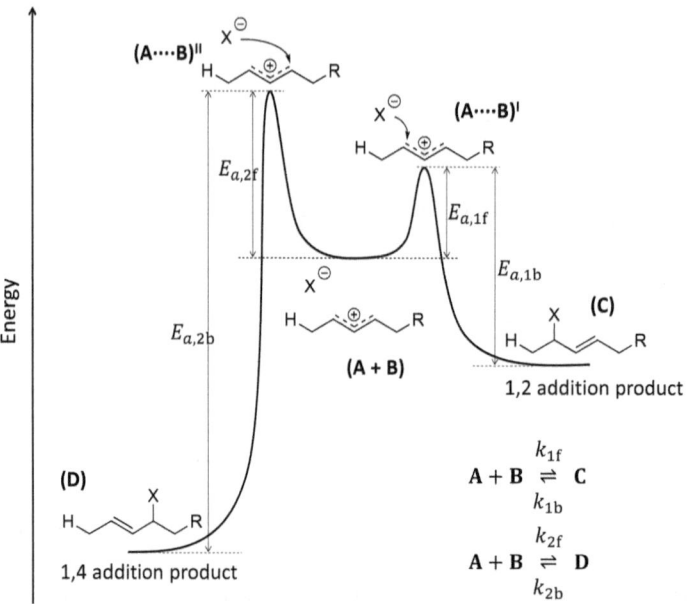

Figure 10.1 Energy diagram for the electrophilic addition of dienes.

As the temperature is raised, more energy becomes available to increase the probabilities (and rates) of the C→A+B and A+B→D processes. These, in turn, will lead to a growing concentration of the 1,4 addition product (**D**) which has the lowest potential energy. (Because $E_{a,2b}$ is the highest activation energy, this product will be less likely to dissociate to carbocation and X^- than the 1,2 addition product.)

These arguments can be tested *via* kinetic modelling, for which rate constants at different temperatures are generated from realistic activation energies that conform to the relative heights of the energy barrier walls. The dependency of rate constants (k) on temperature (T) is given by the Arrhenius equation

$$k = A e^{-\frac{E_a}{RT}}$$

We can predict product ratios using activation energies $E_{a,1f} < E_{a,1b} \sim E_{a,2f} < E_{a,2b}$ while assuming the pre-exponential factors (A) of the reactions to be the same. The activation energies allow us to calculate the rate constants for different temperatures, which in turn enables us to numerically solve the set of kinetic differential equations describing the reaction

$$\frac{d[A]}{dt} = k_{1b}[C] + k_{2b}[D] - k_{1f}[A][B] - k_{2f}[A][B]$$

$$\frac{d[B]}{dt} = k_{1b}[C] + k_{2b}[D] - k_{1f}[A][B] - k_{2f}[A][B]$$

$$\frac{d[C]}{dt} = k_{1f}[A][B] - k_{1b}[C]$$

$$\frac{d[D]}{dt} = k_{2f}[A][B] - k_{2b}[D]$$

The resulting concentration–time curves for the species will then tell us the ratio between the two products as the reactions progress. The kinetic modelling process would involve setting a temperature, calculating the values of the four rate constants, followed by calling an ODE solver to compute the concentration–time curves for the species.

For each temperature, we would need to execute the same sequence of tasks; therefore, making predictions for multiple temperatures may turn into a lengthy and repetitive endeavour, which we would normally like to avoid. (In fact, one of the general principles of software development is the "Don't Repeat Yourself" (DRY) principle which states that duplications should be avoided.) MATLAB (and other programming languages/environments) support the automatic execution of blocks of commands repeatedly to prevent code duplication.

10.1 Loops

Automatically executing blocks of commands multiple times reduces code duplication, which can save us a huge amount of time (on writing/maintaining/running code) while potentially decreasing the number of bugs in our code. Predicting the product ratios for different temperatures would take much less time if we could make MATLAB automatically cycle through a set of temperatures and substitute them into the Arrhenius equation to calculate the rate constants which it would subsequently use to compute the concentration-time data for the species at the different temperatures (red loop). Once the final temperature is reached MATLAB would exit the loop and continue with the last task (blue arrow), displaying the data collected through the cycles (Figure 10.2).

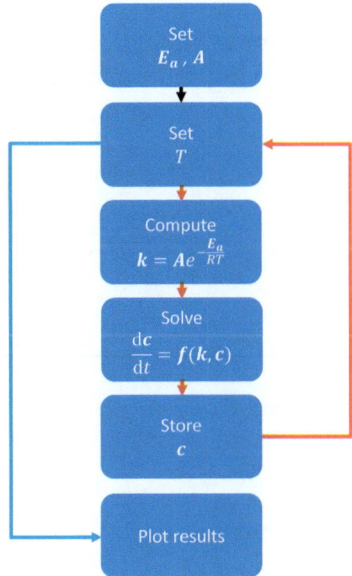

Figure 10.2 Flow diagram of computing concentration–time profiles for a set reaction time at different temperatures and subsequently plotting the final concentrations against temperature.

10.1.1 For Loop

In order to execute this process for multiple temperatures, we are going to be using the so-called *for-loop* where we specify an initial temperature, a temperature increment and a final temperature for MATLAB to automatically cycle through. We define this *for-loop* in MATLAB as seen below. In the *header* starting with for, the specifics of the iterations are outlined: the value of T is going to be automatically changed between 240 and 360 in increments of 20, which means that first MATLAB will set T = 240, then set T = 260 and so on until T is finally set to 360. The block of commands we want MATLAB to execute once in each iteration (also called *body*) is placed between the header and the end statement. For now, the body of the *for-loop* simply outputs the temperature values into the *Command Window* in the format of "Temperature = __ K" with __ being the place where the current value of T is displayed.

```
for T=240:20:360% header of for-loop
    disp(['Temperature = ',num2str(T),' K']) % block of code executed for
each value of temperature T defined in the header
end % end of for-loop
```

```
Temperature = 240 K
Temperature = 260 K
Temperature = 280 K
Temperature = 300 K
Temperature = 320 K
Temperature = 340 K
Temperature = 360 K
```

Now, we include calculating the rate constants for these temperatures using some reasonable activation energies (following the trend $E_{a,1f} < E_{a,1b} \sim E_{a,2f} < E_{a,2b}$ outlined in the energy diagram) and the same pre-exponential factor for simplicity.

```
Ea1f = 80e3; % J/mol
Ea1b = 90e3; % J/mol
Ea2f = 92e3; % J/mol
Ea2b = 130e3; % J/mol
A1f = 1e11; % 1/M/s
A1b = 1e11; % 1/s
A2f = 1e11; % 1/M/s
A2b = 1e11; % 1/s
R = 8.3145; % J/mol/K
for T=240:20:360% header of for-loop
    disp(['Temperature = ',num2str(T),'K'])
    k1f = A1f*exp(-Ea1f/R/T);
    k1b = A1b*exp(-Ea1b/R/T);
    k2f = A2f*exp(-Ea2f/R/T);
    k2b = A2b*exp(-Ea2b/R/T);
```

```
disp(['k1f=',num2str(k1f),';   k1b=',num2str(k1b),';   k2f=',num2str
(k2f),'; k2b=',num2str(k2b)])
end % end of for-loop
```

```
Temperature=240 K
k1f=3.8804e-07; k1b=2.5851e-09; k2f=9.4886e-10; k2b=5.0923e-18
Temperature=260 K
k1f=8.476e-06; k1b=8.3026e-08; k2f=3.2917e-08; k2b=7.6438e-16
Temperature=280 K
k1f=0.00011917; k1b=1.6244e-06; k2f=6.8802e-07; k2b=5.6077e-14
Temperature=300 K
k1f=0.0011779; k1b=2.1379e-05; k2f=9.5887e-06; k2b=2.3202e-12
Temperature=320 K
k1f=0.0087429; k1b=0.00020387; k2f=9.6139e-05; k2b=6.0282e-11
Temperature=340 K
k1f=0.051262; k1b=0.0014911; k2f=0.00073495; k2b=1.0676e-09
Temperature=360 K
k1f=0.24693; k1b=0.0087429; k2f=0.004482; k2b=1.3739e-08
```

In the last step, we add the commands to solve the kinetic ODE of the system and plot the results. Each virtual experiment will last 25 000 s, approximately 7 hours in total.

```
Ea1f=80e3; % J/mol
Ea1b=90e3; % J/mol
Ea2f=92e3; % J/mol
Ea2b=130e3; % J/mol
A1f=1e11;
A1b=1e11;
A2f=1e11;
A2b=1e11;
R=8.3145; % J/mol/K; universal gas constant
tp=0:10:25000; % 25000 s≈7 hours
i=1; % variable to count the number of iterations
for T=240:5:360% temperatures for modelling the reactions
  k1f(i) =A1f*exp(-Ea1f/R/T);
  k1b(i) =A1b*exp(-Ea1b/R/T);
  k2f(i) =A2f*exp(-Ea2f/R/T);
  k2b(i) =A2b*exp(-Ea2b/R/T);
  temp(i) =T; % store current temperature
  % c(1): [A]; c(2): [B]; c(3): [C]; c(4): [D];
  c0=[1 1 0 0]; % set initial concentrations
  [t,c] =ode45(@(t,c) [k1b(i)*c(3) − k1f(i)*c(1)*c(2) +k2b(i)*c(4) − k2f
(i)*c(1)*c(2);    k1b(i)*c(3) − k1f(i)*c(1)*c(2) +k2b(i)*c(4) − k2f(i)*c
(1)*c(2);   k1f(i)*c(1)*c(2) − k1b(i)*c(3);   k2f(i)*c(1)*c(2) − k2b(i)*c
(4)],tp,c0); % call ODE solver with diff. rate equations
   conc(:,:,i) =c(:,:); % store concentrations for the run at current T
   C(i) =c(end,3); % store [1,2-addition product] at the end of run
```

```
    D(i)=c(end,4); % store [1,4-addition product] at the end of run
    i=i+1; % forward counter
end
figure % initiate figure 1
plot(temp,C,'--sr', temp,D,'--db')
xlabel('T/K'); ylabel('concentration/mol dm^{-3}')
legend('[1,2-addition product]_{t=7h}','[1,4-addition product]_
{t=7h}')

figure % initiate figure 2
draw_subplots_for_i=[1  8  17  25] % draw four subplots for runs 1,8,17,25
for f=1:4% for-loop to draw four subplots with concentration-time curves
    subplot(2,2,f)
    plot(t/3600,conc(:,3,draw_subplots_for_i(f)),'-r',t/3600,conc(:,4,
draw_subplots_for_i(f)),'-b')
    xlabel('time/h'); ylabel('concentration/mol dm^{-3}')
    title(['Ad_E of 1,3-pentadiene derivative at
    ',num2str(temp(draw_subplots_for_i(f))),' K']) % generate text for
title featuring temperature which we feed in using num2str() from array
temp
    legend('[1,2-addition product]','[1,4-addition product]','location',
'east')
end
```

This code will produce 2 figures. The first shows the concentrations of the products after 7 hours as the temperature is increased (Figure 10.3), while the second composite figure displays the concentration–time curves for the products (Figure 10.4). At the

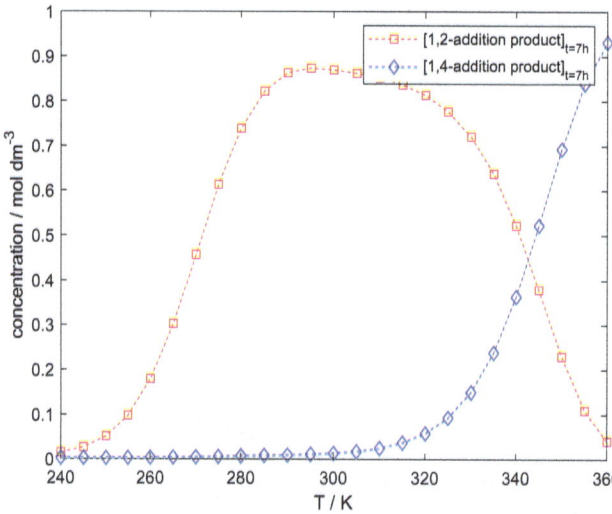

Figure 10.3 Concentrations of 1,2 and 1,4 addition products after 7 hours at different temperatures. The former is called the kinetic, while the latter is called the thermodynamic product of the reaction, as the 1,2 addition product predominantly forms first at low temperatures on this time scale.

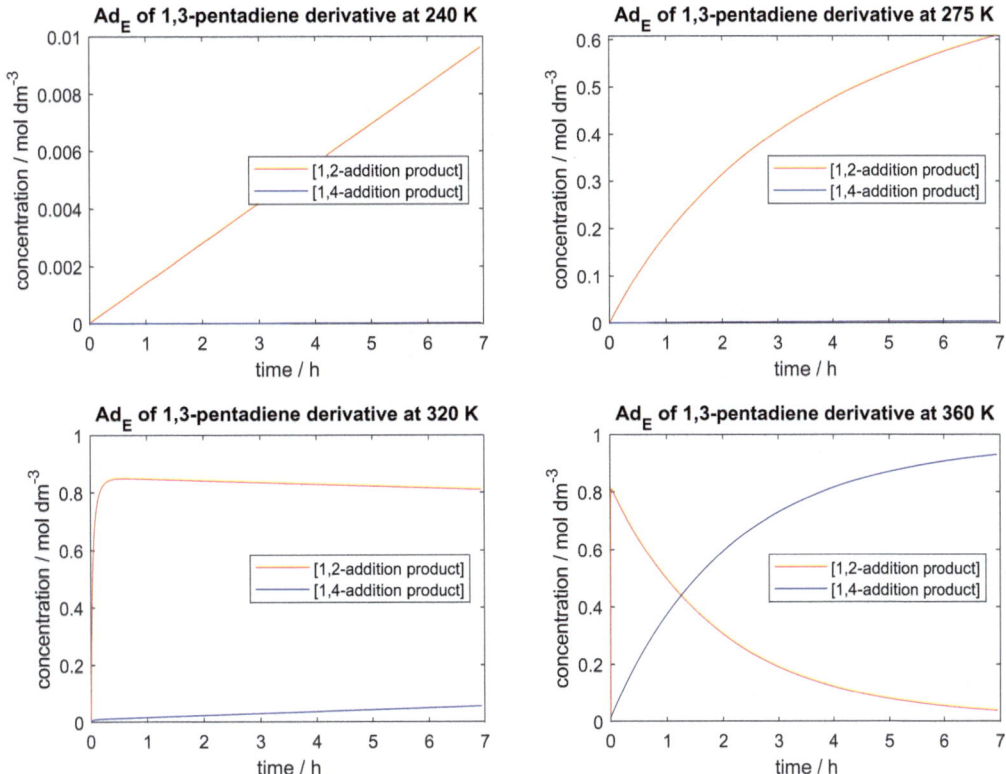

Figure 10.4 Concentration–time profiles for the reaction at some selected temperatures.

lowest temperatures, small amounts of products form during the virtual experiments as reaction rates are low, yet we see that the 1,2 addition product is dominant. At medium temperatures the formation of the 1,2 addition product becomes faster and its concentration reaches high levels within the 7 hour reaction time. As the temperature is further increased, the concentration peak of the 1,2 addition product occurs sooner and sooner so that the 1,4 addition product can eventually become dominant.

It is worthwhile pointing out that the 1,4 addition product will always become dominant eventually but often we cannot wait that long. In these cases, we say that the A+B→C reaction is irreversible (even if we know that all reactions are reversible) and call the 1,2 addition product the *kinetic product* (for it forms at a higher rate at low temperatures on practically feasible timescales) and the 1,4 addition product the *thermodynamic product* (whose production is favoured on practically feasible timescales only at high temperatures).

Exercise 10.1

Calculate the molar masses of alkanes that have 20–30 carbon atoms along with their chemical formulae. Hint: The general chemical formula of alkanes is C_nH_{2n+2}. You will need to construct a *for-loop* where n is changed iteratively between 20 and 30. Then, use each different value of n to compute the mass of the corresponding homologue. The chemical formulae, a sequence of characters of letters and numbers, can be stitched together using square brackets. Do the exercise step by step. Create the *for-loop* first and make sure it works, then construct the command to compute the molar masses and check if the values are correct and so on.

Answer:

```
M_H=1.008; % atomic mass of H
M_C=12.011; % atomic mass of C
for n=20 : 30% change n between 20 and 30
    M_alkane=M_C*n+M_H*(2*n+2); % calculate mass of alkane
    display(['C',num2str(n),'H',num2str(2*n+2),': ',num2str(M_alkane),'
g/mol']) % output string of characters resolving the formula and mass to the
Command Window
end
```

```
C20H42: 282.556 g/mol
C21H44: 296.583 g/mol
C22H46: 310.61 g/mol
C23H48: 324.637 g/mol
C24H50: 338.664 g/mol
C25H52: 352.691 g/mol
C26H54: 366.718 g/mol
C27H56: 380.745 g/mol
C28H58: 394.772 g/mol
C29H60: 408.799 g/mol
C30H62: 422.826 g/mol
```

Exercise 10.2

For a gas, the Maxwell–Boltzmann formula

$$F(c) = 4\pi \left(\frac{m}{2\pi k_B T}\right)^{\frac{3}{2}} c^2 e^{-mc^2/2k_B T}$$

gives the $F(c)$ fraction of the gas molecules/atoms that have a particular speed (c) at temperature T, where k_B is the Boltzmann constant and m is the mass of the gas molecules/atoms. To obtain the speed distribution of the gas molecules/atoms we need to evaluate this expression for a range of speeds.

Compute and display the speed distribution of CO_2 between speeds 0 and 1000 m s^{-1} at 200 K, 300 K...900 K.

Answer:

Notice that we can simplify the calculations by evaluating M/R instead of m/k_B, which is justified because $M = mN_A$ and $R = k_B N_A$, thus $M/R = m/k_B$. We will start by computing a single speed distribution curve for 298 K. To give the impression of a smooth curve, we calculate the fraction of CO_2 molecules at every 10 m s^{-1} starting from 0 m s^{-1} up to 1000 m s^{-1}.

```
% Maxwell-Boltzmann distribution
M=0.044; % molar mass of CO2 in kg
T=298; % temperature in K
c=0:10:1000; % range of speeds in m/s
R=8.3145; % gas constant (J/mol/K)
F=4*pi*(M/2/pi/R/T)^1.5*c.^2.*exp(-M*c.^2/2/R/T); % compute
fractions
plot(c,F)
xlabel('c/m s^{-1}')
ylabel('Fraction')
```

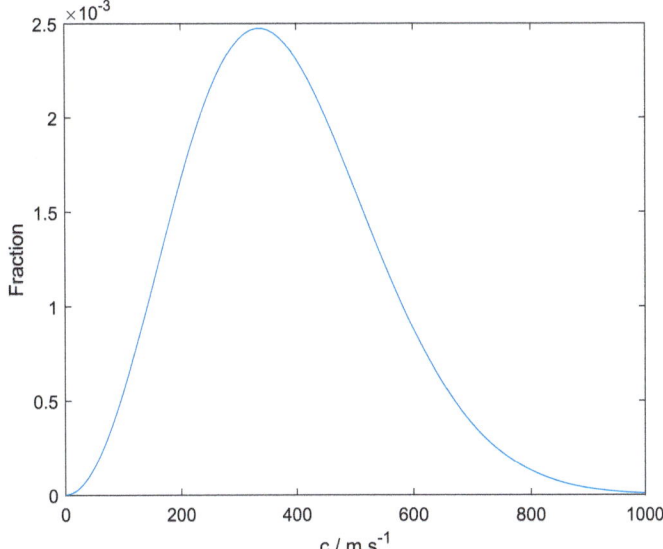

Looking at the Maxwell–Boltzmann distribution curve, we see similarities with the normal distribution curve introduced in Section 5.1. However, because practically there is no upper limit to the speed of gaseous atoms and molecules (although technically the upper limit is the speed of light, 3×10^8 m s^{-1}) and because the lower limit is 0 m s^{-1}, the Maxwell–Boltzmann distribution does not have a perfectly symmetrical bell shape.

Now we add the *for-loop* to compute and plot the speed distribution of CO_2 molecules for temperatures between 200 and 900 K.

```
% Maxwell-Boltzmann distributions between 200 and 900 K
M=0.044; % molar mass of CO2 in kg
R=8.3145; % gas constant (J/mol/K)
c=0:10:1000; % range of speeds in m/s
for T=200:100:900% temperatures in K
```

```
  F=4*pi*(M/2/pi/R/T)^1.5*c.^2.*exp(-M*c.^2/2/R/T); % compute
fractions
  plot(c,F)
  hold on % keep diagram open for more plots to be drawn
end
xlabel('c/m s^{-1}')
ylabel('Fraction')
```

Notice that the first three and the last two lines are outside the *for-loop* as they do not need to be repeated while the temperature is varied. It is good practice to *indent the block of script inside the for-loop for better readability*. The script above will generate the following diagram:

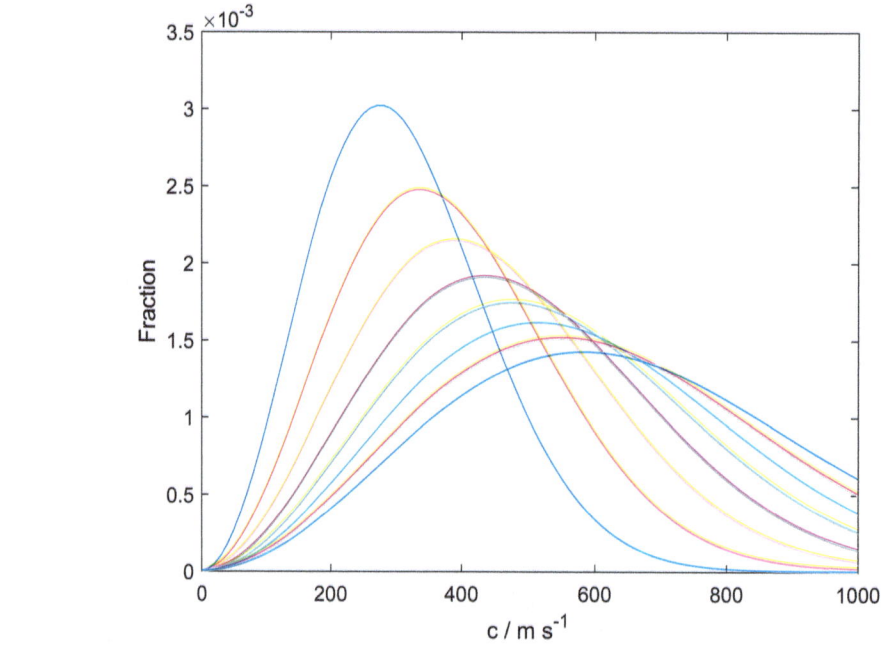

10.1.2 While-loop

This type of loop allows for flexibility in the number of times the loop is rerun. A condition set in the loop header MATLAB evaluates at the beginning of each cycle and the body of the loop is executed if the condition is true.

Suppose we would like to calculate the temperature at which 0.1% of CO_2 molecules reach the speed of 700 m s^{-1} because that is the minimum speed required for collisions with a catalytic surface to result in a particular reaction occurring at a fast enough rate. We could try rearranging the Maxwell–Boltzmann formula for T, but we decide to modify the code in Exercise 10.2 instead. We set the temperature to 200 K and calculate the fraction of molecules with the speed of 700 m s^{-1} before setting up the *while-loop* with condition `F<0.001`. Based on the figure in Exercise 10.2 we are certain that the fraction is less than 0.001; therefore, MATLAB will execute the body of the loop, where the temperature is increased by 1 K and the fraction is recalculated. Then the condition is re-evaluated and if the fraction is still below 0.001, the process is repeated again. This continues until the fraction is no longer smaller than 0.001. At that point MATLAB exits the loop and executes the next command which outputs the last temperature value.

```
% Maxwell-Boltzmann distribution
M=0.044; % molar mass of CO2 in kg
R=8.3145; % gas constant (J/mol/K)
c=700; % speed in m/s
T=200; % temperature in K
F=4*pi*(M/2/pi/R/T)^1.5*c^2*exp(-M*c^2/2/R/T); % compute fraction
while F<0.001% header of while loop (rerun body of loop while F<0.001)
    T=T+1; % increase temperature by 1 K
    F=4*pi*(M/2/pi/R/T)^1.5*c^2*exp(-M*c^2/2/R/T); % compute fraction
end
disp(['The temperature required for CO2 molecules with speed of 700 m/s to make up 0.1% of the population is ',num2str(T),' K.'])
```

When the script is executed, MATLAB returns:

```
The temperature required for CO2 molecules with speed of 700 m/s to make up
0.1% of the population is 497 K.
```

If we require a more precise temperature value, we use a smaller temperature increment in the body of the loop. While-loops are widely employed in science and engineering. An example will be given at the end of Chapter 12, in which we will calculate the H^+ concentration in a solution of a diprotic acid.

When setting the header of a while-loop, we have to carefully consider the condition to avoid the program being stuck in an infinite loop. Should that happen, hold down the "Ctrl" key and press "C" (Ctrl+C) or Ctrl+Break to force MATLAB to stop executing the code – on a Mac use Command+.

10.2 Conditional Statements

Simple decision making is also supported by MATLAB and other programming environments so that the execution of commands can be flexibly altered by the program itself depending on conditions we set. So far we have only written scripts that were executed by MATLAB top to bottom. It would be very advantageous, however, if we could instruct MATLAB to execute certain blocks of commands only if some conditions were met, or not.

Conditional statements can be thought of as questions we instruct MATLAB to ask when it reaches certain points whilst running our script and performing specific sets of tasks depending on the answer. The type of conditional statement we discuss here is the so-called *if-statement*. An *if-statement* can be either TRUE or FALSE, each outcome

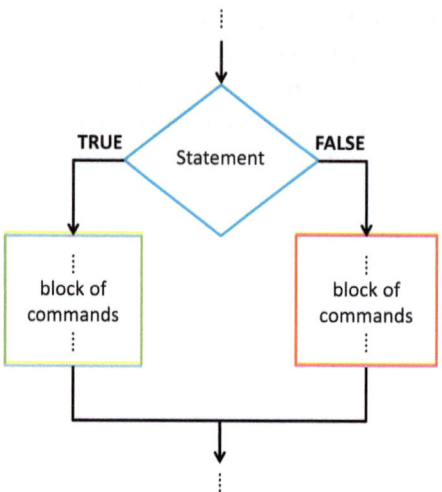

Figure 10.5 Flow diagram of code branching using the *if-statement*.

triggering a different set of actions (block of commands) as illustrated on the flow chart in Figure 10.5.

Let us look at how *if-statements* work through a quick example. We instruct MATLAB to decide if a number (q) is smaller than 10, and to write '$q<10$' or '$q \geq 10$' (q replaced with its the numerical value) in the *Command Window* depending on the outcome. MATLAB will evaluate the expression $q<10$ written as an *if-statement* " if q<10 " and

- **either** carry on until it reaches else, then jump to end – if the answer is TRUE,
- **or** jump to else and continue from there until it reaches end – if the answer if FALSE.

In both cases MATLAB will reach end and continue from there by outputting: 'Evaluation finished.' into the *Command Window*.

```
q=5;
if q<10% statement
   disp([num2str(q),'<10']) % block of commands when statement is TRUE
else
   disp([num2str(q),'≥10']) % block of commands when statement is FALSE
end
disp('Evaluation complete.')
```

5<10
Evaluation complete.

When we write $q=11$; in the first line, the output will be:

11≥10
Evaluation complete.

In the *if-statement*, "<" is the so-called *"less than"* relational operator. MATLAB relational operators are listed below:

Symbol	Meaning	Example (result of evaluation)	
==	Equal to	5 == 5 (TRUE)	6 == 5 (FALSE)
~=	Not equal to	5 ~= 6 (TRUE)	5 ~= 5 (FALSE)
>	Greater than	6 > 5 (TRUE)	4 > 5 (FALSE)
>=	Greater than or equal to	3 >= 3 (TRUE)	3 >= 4 (FALSE)
<	Less than	5 < 8 (TRUE)	8 < 5 (FALSE)
<=	Less than or equal to	5 <= 5 (TRUE)	5 <= 4 (FALSE)

Note that "==" and "=" have different meanings. The former means *"is equal to"* while the latter means *"make equal to"* (or *"assign value to"*). Accordingly:

- a == 10 is interpreted as "the number stored in a is equal to 10" (statement)
- a = 10 is interpreted as "make variable a equal to 10", or "assign 10 to variable a" (command)

```
q=10;
if q==10% statement: q is EQUAL TO 10
   disp([num2str(q),' is equal to 10']) % execute if statement is TRUE
else
   disp([num2str(q),' is not equal to 10']) % execute if statement is FALSE
end
disp('Evaluation complete.')
```

```
10 is equal to 10
Evaluation complete.
```

```
q=10;
if q~=10% statement: q is NOT EQUAL TO 10
disp([num2str(q),' is not equal to 10']) % execute if statement is TRUE
else
disp([num2str(q),' is equal to 10']) % execute if statement is FALSE
end
disp('Evaluation complete.')
```

```
10 is equal to 10
Evaluation complete.
```

Notice that now the statement was *"q is not equal to 10"*, which evaluated to FALSE, therefore MATLAB jumped to `else` and continued from there.

> **Exercise 10.3**
>
> Write a MATLAB script that tells researchers whether the most probable speed of CO_2 molecules is above 700 m s^{-1} at the temperature where 0.1% of CO_2 molecules move with the speed of 900 m s^{-1}.
>
> **Answer:**
>
> ```
> % Maxwell-Boltzmann distribution
> M=0.044; % molar mass of CO2 in kg
> R=8.3145; % gas constant (J/mol/K)
> c=800; % speed in m/s
> T=200; % temperature in K
> F=4*pi*(M/2/pi/R/T)^1.5*c^2*exp(-M*c^2/2/R/T); % compute fraction
> while F<0.001% header of while loop (rerun body of loop while F<0.001)
>      T=T+1; % increase temperature by 1 K
>      F=4*pi*(M/2/pi/R/T)^1.5*c^2*exp(-M*c^2/2/R/T); % compute fraction
> end
> disp(['The temperature required for CO2 molecules with speed of 900 m/s
> to make up 0.1% of the population is ',num2str(T),' K.'])
> c=0:1:1000; % range of speeds in m/s
> F=4*pi*(M/2/pi/R/T)^1.5*c.^2.*exp(-M*c.^2/2/R/T); % compute
> fractions for speed range
> [largest_fraction most_prob_c]=max(F); % find the maximum value of
> F and the corresponding speed
> if most_prob_c<=700% statement
>      disp(['The most probable speed at ',num2str(T),' K is not above
> 700 m/s.']) % execute if statement is TRUE
> else
>      disp(['The most probable speed at ',num2str(T),' K is above
> 700 m/s.']) % execute if statement is FALSE
> end
> ```
>
> When the script is executed, MATLAB returns:
>
> The temperature required for CO2 molecules with speed of 900 m/s to make up 0.1% of the population is 1187 K.
> The most probable speed at 1187 K is not above 700 m/s.
>
> We used the `max()` function to locate the maximum of the speed distribution. More information on the function can be found at https://uk.mathworks.com/help/matlab/ref/max.html.

10.3 Case Study: Monitoring Polymerisation

Applications often require polymers of certain chain lengths. In these cases, chemists and chemical engineers are expected to produce large batches of polymer molecules

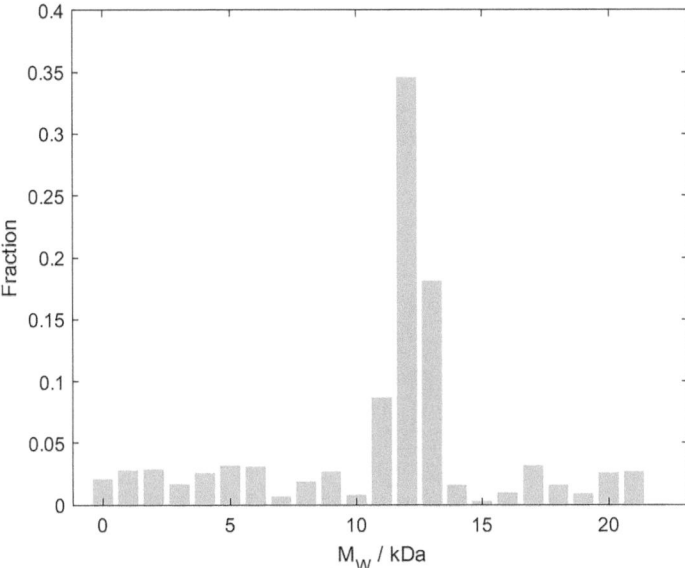

Figure 10.6 Typical molecular weight distribution of a polymer determined by dynamic light scattering. The diagram shows a relatively narrow weight (size) distribution which indicates the high level of monodispersity usually preferred in the polymer industry.

with masses within a narrow range. This could be achieved, for example, by monitoring the polymerisation process and halting it once the required molecular weight has been reached, before the macromolecules could grow too large.

The size of polymers can be estimated by various methods, for example, by light scattering. This sensitive technique requires diluted samples taken from the reactor and yields the size distribution of macromolecules which can be displayed on histograms (see Section 3.6). The molecular weight distribution shown in Figure 10.6 was found 60 minutes after the polymerisation had been initiated.

To determine the mean molecular weight we fit (see Section 4.4) the function

$$y = ae^{-b(x-c)^2}$$

to the data presented in the diagram above, where the mean is represented by fitting parameter c.

Because light scattering measurements require dust-free environments, the samples taken from the reactor during polymerisation can sometimes become contaminated during sample preparation. Whether a sample has become spoiled can be decided only from the molecular weight distribution in most cases. When contamination occurs, the distribution typically has no clear maximum (as shown in Figure 10.8), which results in the fitted function being flat. As a consequence of a poor fit the uncertainties in the fitting parameters are many times higher than those for uncontaminated samples. The stark difference in the uncertainties in the fitting parameters can be exploited in enabling MATLAB to decide if a sample has been spoiled and therefore it should be disregarded.

The following set of data was collected during a test run of a reactor equipped with an automated sample collecting device under development that dilutes and transfers

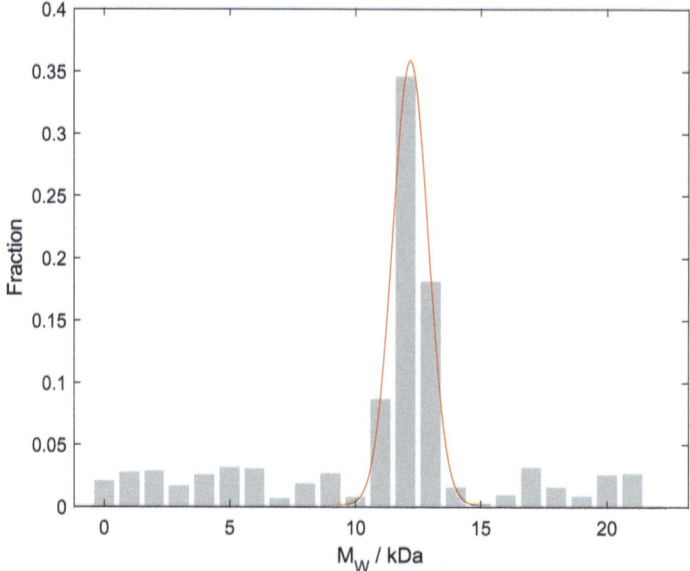

Figure 10.7 Normal distribution curve successfully fitted to the experimentally determined molecular weight distribution shown in Figure 10.6.

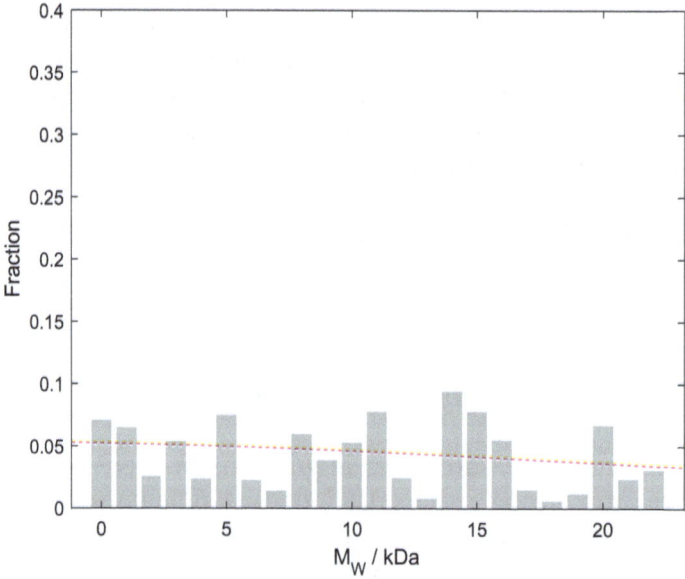

Figure 10.8 Failed determination of the molecular weight distribution leads to a poor fit with high uncertainties in the fitting parameters.

samples directly into the light scattering apparatus. Samples were taken and analysed every 10 minutes. Each row of the data below corresponds to the result of the light scattering measurement performed on a sample. The columns represent molecular masses between 0 and 22 kDa, left to right.

```
data=[
.022 .090 .361 .186 .031 .033 .019 .014 .024 .016 .005 .015 .019
.014 .014 .026 .018 .017 .003 .009 .016 .022 .025;
.038 .036 .004 .048 .219 .381 .073 .010 .003 .011 .031 .003 .014
.005 .006 .002 .027 .003 .024 .007 .025 .019 .011;
.052 .071 .032 .010 .056 .020 .020 .065 .059 .022 .030 .022 .034
.037 .056 .068 .023 .073 .064 .072 .016 .006 .053;
.026 .004 .004 .022 .030 .030 .003 .027 .045 .224 .338 .058 .028
.002 .016 .020 .036 .008 .004 .024 .004 .037 .009;
.021 .028 .029 .017 .026 .032 .031 .007 .019 .027 .008 .087 .346
.181 .016 .003 .010 .032 .016 .009 .026 .027 .001;
.071 .065 .026 .054 .024 .075 .023 .014 .060 .039 .053 .078 .025
.008 .094 .078 .055 .015 .006 .012 .067 .024 .031;
.024 .030 .029 .014 .007 .024 .000 .011 .024 .022 .012 .037 .005
.027 .006 .017 .097 .369 .176 .028 .010 .005 .026;
.018 .007 .004 .007 .005 .012 .035 .020 .009 .016 .025 .020 .023
.032 .014 .028 .006 .002 .024 .237 .347 .085 .023];
```

We are tasked with writing a MATLAB script that can be run as part of the automatic sample collection and analysis process to monitor the polymerisation reaction. For each sample (*i.e.* for each row of array `data`) it is required to

- plot the molecular weight distribution
- fit a distribution function
- calculate the average molecular weight
- disregard the sample if deemed contaminated

We will need a *for-loop* to analyse each row of data separately. In the body of the *for-loop* n will change iteratively between 1 and 8. For each value of n:

I. A row of `data` is loaded into an array named `spl`
II. The expression given above is fitted to `spl`
   (Guesses for the fitting parameters – a, b, c – are provided for the `fit()` function:
   `max_fr` : height of the highest fraction (guess for parameter a)
   `mw_max_fr`: molecular weight of the highest fraction (guess for parameter c)
III. Fit statistics are extracted from `dist_fit`
   `avg_mw`: best estimate for the average molecular weight (value of fitting parameter c)
   `d_avg_mw`: uncertainty in the estimated average molecular weight (from the confidence bounds of b)
IV. An *if-statement* is required to decide if the uncertainty in the estimated average molecular weight is high above its typical level in uncontaminated samples. (0.5 is considered a suitable threshold.)
   If `d_avg_mw<0.5` : create bar chart (histogram), add a graph of fitted expression and output M_W
   If `d_avg_mw≥0.5` : output message that the sample has been disregarded

The output is shown in Figure 10.9.

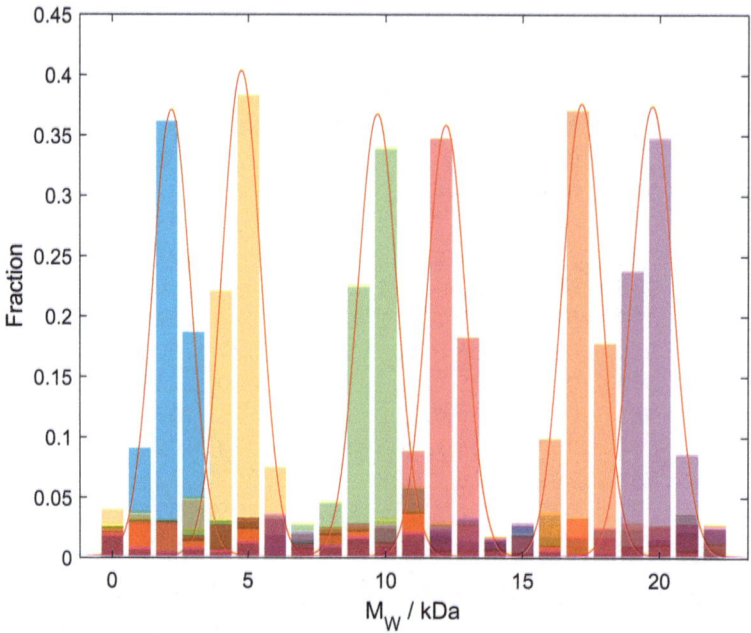

Figure 10.9 Molecular weight distribution with the fitted bell curve for samples successfully collected and analysed throughout the polymerisation process.

```
mw = 0 : 22; % create MW range 0 through 22 kDa
NumOfDisSpl = 0;
for n = 1 : 8
  spl = data(n,:); % I.
  [max_fr mw_max_fr] = max(spl); % II.
  dist_fit = fit(mw',spl','a*exp(-b*(x-c)^2)','startpoint',[max_fr 1 mw_max_fr]); % II.
  fit_stats = confint(dist_fit); % III.
  d_avg_mw = abs(fit_stats(2,3)-fit_stats(1,3))/2; % III.
  avg_mw = dist_fit.c; % III.
  if d_avg_mw<0.5% IV.
    bar(mw,spl,'LineStyle','none','FaceAlpha',0.5)
    hold on
    plot(dist_fit,'-r')
    disp(['Sample ',num2str(n),': MW = ',num2str(avg_mw),' ± ',num2str(d_avg_mw),' kDa'])
  else
    disp(['Sample ',num2str(n),': disregarded'])NumOfDisSpl =
    NumOfDisSpl+1;
  end % end of if-statement
end % end of for-loop
xlabel('M_W/kDa');
ylabel('Fraction')
legend off disp(['Number of discarded samples: ',num2str(NumOfDisSpl)])
```

Self-controlling Code 205

```
Sample 1: MW = 2.1776 ± 0.098602 kDa
Sample 2: MW = 4.7481 ± 0.094038 kDa
Sample 3: disregarded
Sample 4: MW = 9.687 ± 0.11013 kDa
Sample 5: MW = 12.1804 ± 0.11708 kDa
Sample 6: disregarded
Sample 7: MW = 17.1444 ± 0.1047 kDa
Sample 8: MW = 19.717 ± 0.093837 kDa
Number of discarded samples: 2
```

We set the number of discarded samples to zero (NumOfDisSpl = 0) before the *for-loop*, then its value is increased by 1 (NumOfDisSpl = NumOfDisSpl + 1) each time a sample is rejected.

It is important to realise that it took us only 23 lines of MATLAB script to perform this complex task regardless of the number of samples.

Exercise 10.4

You were tasked to synthetize a specific aliphatic alkene. Unfortunately, your reaction did not go to plan and you ended up with a mixture of compounds. From the molecular masses below found on the mass spectrum of the product mixture, you deduced that

- all but two of the first 56 members of the homologue series of aliphatic alkanes were present in the mixture, and that
- instead of the two missing aliphatic alkenes, the mixture contained one aliphatic alkyne and one aliphatic alkyne of the same chain lengths as the missing aliphatic alkenes.

28.0540	42.0810	56.1080	70.1350	84.1620	98.1890	112.2160
126.2430	140.2700	154.2970	168.3240	182.3510	196.3780	210.4050
224.4320	238.4590	252.4860	266.5130	280.5400	294.5670	308.5940
324.6370	336.6480	350.6750	364.7020	378.7290	392.7560	406.7830
420.8100	434.8370	448.8640	462.8910	476.9180	490.9450	504.9720
518.9990	533.0260	547.0530	561.0800	575.1070	589.1340	603.1610
617.1880	631.2150	645.2420	659.2690	671.2800	687.3230	701.3500
715.3770	729.4040	743.4310	757.4580	771.4850	785.5120	799.5390

Which of these molar masses belong to an aliphatic alkane and alkyne?

Answer:

```
data = [28.0540  42.0810  56.1080  70.1350  84.1620  98.1890  112.2160
 126.2430  140.2700  154.2970  168.3240  182.3510  196.3780  210.4050
 224.4320  238.4590  252.4860  266.5130  280.5400  294.5670  308.5940
 324.6370  336.6480  350.6750  364.7020  378.7290  392.7560  406.7830
 420.8100  434.8370  448.8640  462.8910  476.9180  490.9450  504.9720
```

```
    518.9990  533.0260  547.0530  561.0800  575.1070  589.1340  603.1610
    617.1880  631.2150  645.2420  659.2690  671.2800  687.3230  701.3500
    715.3770  729.4040  743.4310  757.4580  771.4850  785.5120  799.5390];
    length_of_data=length(data);
    M_H=1.008; % molar mass of H
    M_C=12.011; % molar mass of C
    for m=1:length_of_data
        n=m+1; % 1st element of data (m=1) corresponds to n=2 in the
    homologue series and so on
        M_alkene=n*M_C+(2*n)*M_H; % calculate M_r of alkane from n
        % Always use integers when deciding if they are equal or not!
        % Multiplying the masses by 1000 ensures that all non-zero decimal
    places become part of the integers to be compared.
        int_M_alkene=round(M_alkene*1000); % generate integer
        int_data=round(data(m)*1000); % generate integer
        if int_M_alkene~=int_data % statement: the integers are NOT equal
            disp(['M_r ',num2str(data(m)),' does not belong to an alkene.'])
            int_M_alkane=round((n*M_C+(2*n+2)*M_H)*1000); % generate integer
            if int_M_alkane==int_data % statement: the integers are equal
                disp(['M_r ',num2str(data(m)),' belongs to an alkane.'])
            else % when the integers are not equal
                disp(['M_r ',num2str(data(m)),' belongs to an alkyne.'])
            end
        end
    end
```

The code generates the output:

```
M_r 324.637 does not belong to an alkene.
M_r 324.637 belongs to an alkane.
M_r 671.28 does not belong to an alkene.
M_r 671.28 belongs to an alkyne.
```

Exercise 10.5

You made another attempt. This time the outcome was even more disappointing. The product mixture contained multiple aliphatic alkanes and alkynes, with each missing aliphatic alkene in the homologue series replaced by either its alkane or alkyne analogue. Given the molar masses identified on the mass spectrum

```
30.0700   44.0970   54.0920   70.1350   82.1460   96.1730   110.2000
124.2270  140.2700  152.2810  168.3240  182.3510  196.3780  210.4050
226.4480  236.4430  252.4860  266.5130  278.5240  294.5670  306.5780
322.6210  334.6320  352.6910  364.7020  378.7290  390.7400  404.7670
```

420.8100 434.8370 448.8640 464.9070 476.9180 492.9610 504.9720
518.9990 533.0260 549.0690 561.0800 577.1230 589.1340 601.1450
619.2040 629.1990 647.2580 661.2850 671.2800 687.3230 699.3340
713.3610 729.4040 743.4310 755.4420 769.4690 785.5120 799.5390
815.5820 825.5770 841.6200 855.6470 867.6580 883.7010 895.7120
911.7550 923.7660 939.8090

how many different alkanes, alkenes and alkynes does the product mixture contain? (Use $M_r(C) = 12.011$ and $M_r(H) = 1.008$)

Answer:

```
data = [30.0700  44.0970  54.0920  70.1350  82.1460  96.1730  110.2000
124.2270 140.2700 152.2810 168.3240 182.3510 196.3780 210.4050
226.4480 236.4430 252.4860 266.5130 278.5240 294.5670 306.5780
322.6210 334.6320 352.6910 364.7020 378.7290 390.7400 404.7670
420.8100 434.8370 448.8640 464.9070 476.9180 492.9610 504.9720
518.9990 533.0260 549.0690 561.0800 577.1230 589.1340 601.1450
619.2040 629.1990 647.2580 661.2850 671.2800 687.3230 699.3340
713.3610 729.4040 743.4310 755.4420 769.4690 785.5120 799.5390
815.5820 825.5770 841.6200 855.6470 867.6580 883.7010 895.7120
911.7550 923.7660 939.8090];
length_of_data = length(data);
M_H = 1.008;
M_C = 12.011;
num_of_alkanes = 0;
num_of_alkynes = 0;
for m = 1:length_of_data
   n = m+1;
   M_alkene = n*M_C+(2*n)*M_H;
   int_M_alkene = round(M_alkene*1000);
   int_data = round(data(m)*1000);
   if int_M_alkene ~= int_data
     int_M_alkane = round((n*M_C+(2*n+2)*M_H)*1000);
     if int_M_alkane == int_data
       num_of_alkanes = num_of_alkanes+1;
     else
       num_of_alkynes = num_of_alkynes+1;
     end
   end
end
disp(['Number of alkanes found: ',num2str(num_of_alkanes)])
disp(['Number of alkynes found: ',num2str(num_of_alkynes)])
```

The code generates the output:

```
Number of alkanes found: 12
Number of alkynes found: 23
```

Exercise 10.6

In Section 2.3, we said we will revisit displaying environmental data on interactive maps using pins with colours determined by the levels of the monitored compound found in samples. Write a MATLAB function that allows users to divide the overall concentration range of the monitored compound across the environmental samples into a set number of tiers and automatically assign a different pin colour to each. Assume that the geocoordinates of samples and the corresponding levels of the compound monitored are stored in a single array (GPS_Conc).

Answer:

We will need to add the number of concentration tiers (NumOfTiers) as an additional input argument to the function declaration used in Exercise 2.2. This will determine how many different colours will be set for the pins to indicate the local levels of the monitored substance. To sort samples into concentration tiers, we use the histcounts() function which divides the range the environmental values stored in GPS_Conc(:,3) span across into a number of equally-sized bins (concentration tiers) defined in NumOfTiers and returns three arrays: the number of samples in each tier (TC), the edges of the tiers (TE) and the bin index for each sample (TI), *i.e.*, which tier contains the sample.

We also need to define as many different colours as there are concentration tiers that we require. The quickest way of generating a set of different colours is through calling a colour map which should not be confused with a map used for presenting environmental samples. A colour map is a colour palette resolved as an ($N \times 3$) array, where each row represents the intensity of the red (R), green (G) and blue (B) components a specific colour, with N being the number of different colours making up the palette. Command "PinColPal = parula(NumOfTiers)" creates array PinColPal (colour palette for the pins) while loading a different set of three numbers (RGB values) into each of the rows whose total number is determined by NumOfTiers. The RGB values will be picked from a colour map (here, from the in-built colour map called "Parula") whose name will be passed to our function as input argument ColMap. Because neither the name of the colour map, nor NumOfTiers is defined within the scope of our function, we need to build and execute the above command from inside the function. We will create the character array exactly matching the command above and have it executed by the eval() function, just as it would be typed into the *Command Window* followed by pressing "ENTER/RETURN/↵" on the keyboard.

Now with the colour palette (PinColPal) defined, we just need to assign the correct colour from the palette to each pin before the interactive map is created. The array storing the colour for each pin individually (PinCols) will be created first as an empty array, then the rows of RGB values will be added one-by-one to the bottom of the latest iteration of PinCols in a *for-loop* from the first sample to the last in GPS_Conc. A new row will always be added by vertically concatenating it to the PinCols by the vertcat() function.

The function then converts the numerical values of the concentrations (stored in GPS_Conc(:,3)) to their character equivalents which are loaded into ConcText, before the map can be generated in the final step.

Self-controlling Code

```
function
env_data_tiers_map(GPS_Conc,Analyte,NumOfTiers,PinSize, ColMap,
MapType)
[TC,TE,TI]=histcounts(GPS_Conc(:,3),NumOfTiers); % sort
concentrations in samples into bins
% combine character arrays using syntax [...,...,...] to create the string
of characters spelling out the command run for generating our colour
palette
cmd_str=['PinColPal=',ColMap,'(',num2str(NumOfTiers),');'];
eval(cmd_str); % run command string defined above to create colour
palette
PinCols=[]; % create empty array for receiving RGB colours as rows
for m=1:length(GPS_Conc(:,3)) % for loop to assign RGB colours to pins
  PinCols=vertcat(PinCols,PinColPal(TI(m),:));
end
ConcText=string(GPS_Conc(:,3)); % turn numbers in column GPS_Conc
(:,3) into text characters for displaying on map
webmap(MapType); % Map types: www.mathworks.com/help/map/ref/webmap.
html
wmmarker(GPS_Conc(:,1),GPS_Conc(:,2),'FeatureName',Analyte,
'Description',ConcText,'IconScale',PinSize,'Color',PinCols)
end
```

```
>> GPS_Conc=[−41.865  146.746  12.200; −41.982  146.946  45.300;
−42.103  147.171  51.900;  −42.157  146.643  38.700;  −42.066
146.168  25.200;  −42.223  146.243  29.300; -42.175  145.644  32.500];
```

We now call `env_data_tiers_map()` first with four, then with only two concentration tiers, which will result in the maps below. Try using different colour maps (mathworks.com/help/matlab/ref/colormap.html) and number of tiers.

```
>> env_data_tiers_map(GPS_Conc,'Nitrate (mg/l)',4,1,'parula',
'openstreetmap')
>> env_data_tiers_map(GPS_Conc,'Nitrate (mg/l)',2,1,'parula',
'openstreetmap')
```

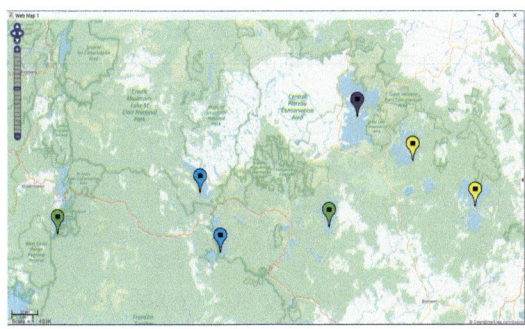 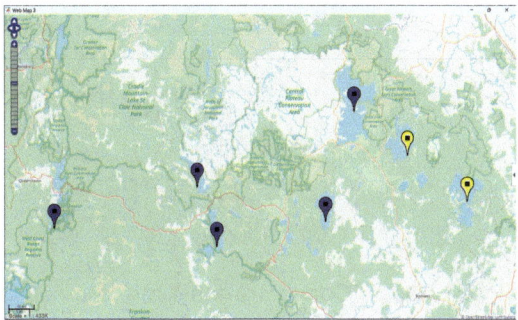

10.4 Appendix: Euler's Method (II)

In the previous chapter we discussed how concentration–time plots can be approximated for a reaction using Euler's method. Now, we are going to put what we have learned about *for-loops* into action and estimate the concentrations of the species (seen in Figure 10.10) involved in the consecutive reactions scheme discussed in Section 9.4.

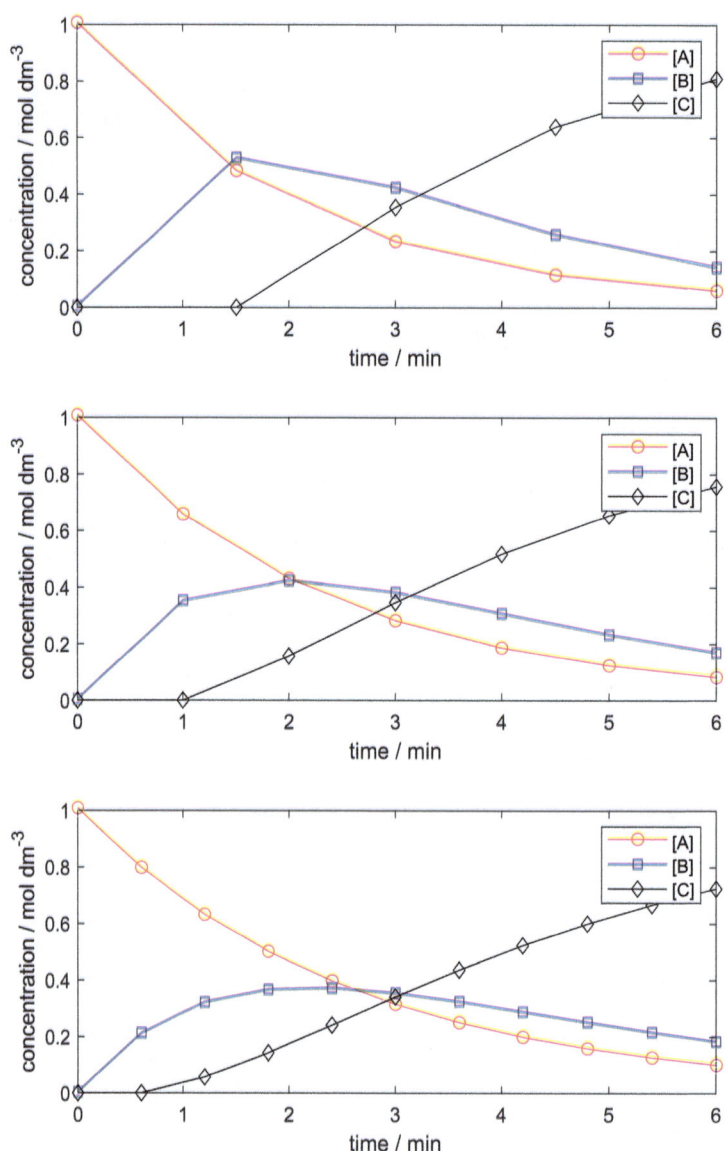

Figure 10.10 Concentration–time profiles computed using *for-loop* to implement Euler's method.

```
k1=0.35; % (1/min) rate constant of step 1
k2=0.45; % (1/min) rate constant of step 2
dt=1; % (min) time step
t_fin=6; % final time
c0=[1  0  0]; % (M) assign initial concentrations: [A]0, [B]0, [C]0
i=1; % counter
time(i)=0; % initial time (1st item in array "time" will be 0)
conc(i,1:3)=c0; % load initial concentrations into 1st row of array "conc"
for t=dt:dt:t_fin % concentrations for t=0 known→loop starts on t=dt
  i=i+1; % increase counter by 1
  time(i)=t; % store current time
  % compute approximate concentrations at time t
  conc(i,1)=conc(i-1,1) − k1*conc(i-1,1)*dt; % [A]
  conc(i,2)=conc(i-1,2) + (k1*conc(i-1,1) − k2*conc(i-1,2))*dt; % [B]
  conc(i,3)=conc(i-1,3) + k2*conc(i-1,2)*dt; % [C]
end
plot(time,conc(:,1),'-or',time,conc(:,2),'-sb',time,conc(:,3),'-dk')
xlabel('time/min')
ylabel('concentration/mol dm^{-3}')
legend('[A]', '[B]', '[C]')
```

Concentration–time curves become smoother as the time step is reduced. Time step: top (1.5 min), middle (1 min), bottom (0.6 min) as in Section 9.4.

11 Maths for Chemistry with MATLAB I

Solving chemistry problems often requires performing mathematical operations. If we are lucky, we can do the maths involved on a piece of paper. However, as we gradually move away from the simplest chemical systems and problems, sooner or later there comes a point where the maths needed becomes prohibitively cumbersome, even when the problem can be solved exactly. Knowing the correct sequence of mathematical operations does not mean we should always attempt to carry them all the way through. The solution may simply take too long to work out or it may prove too difficult to find with insufficient mathematical training. Often the exact solution cannot even be found, thus it can only be approximated. In the last two chapters we will be looking at how we can do maths with MATLAB to solve problems in chemistry.

11.1 Introduction to Symbolic Maths

When manipulating mathematical expressions seems too challenging, or if we would like to check whether we have done them correctly, we can use MATLAB to do the maths for us. It has vast capabilities (once the Symbolic Math toolbox is added) to deal with those mathematical manipulations we often shy away from. MATLAB can perform algebraic computations, differentiation, integration, and it can solve equations too.

Imagine that we are in a lab and we have two beakers in front of us. Each contains a solution of the same compound. The volumes and the concentrations are different. What will be the concentration of the compound after combining the two solutions in a third beaker? The amounts of the compound will add up and we can reasonably assume that the volumes would also add up, unless the concentrations are vastly different and/or different solvents were used to prepare the solutions (Figure 11.1).

First, we recall the definition of concentration which states that the concentration of a compound in a solution (c) is the amount of the compound (n) divided by the volume of the solution (V), $c = n/V$. The amounts of the compound in the two beakers (n_1 and n_2)

A First Look at Coding in Chemistry: Solving Problems Using MATLAB
By Tamas Bansagi
© Tamas Bansagi 2025
Published by the Royal Society of Chemistry, www.rsc.org

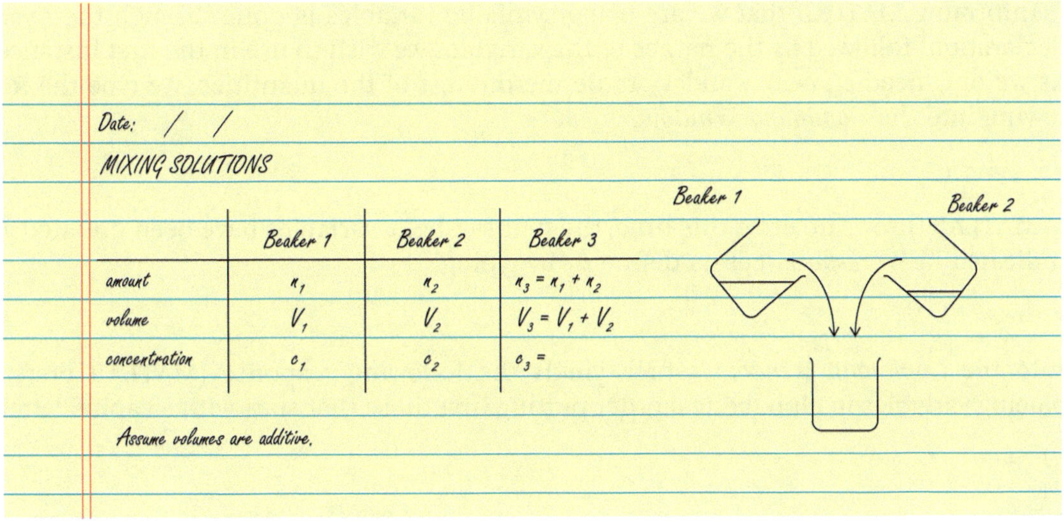

Figure 11.1 Excerpt of a lab notebook.

can be calculated by rearranging this formula for n ($n = cV$) and applying it to the two solutions as

$$n_1 = c_1 V_1 \quad n_2 = c_2 V_2$$

When we pour the solutions together into beaker 3 the total amount will be

$$n_3 = n_1 + n_2 = c_1 V_1 + c_2 V_2$$

Now we return to the definition of concentration, $c = n/V$, and apply it to the combined solution to obtain the expression for calculating its concentration which becomes

$$c_3 = \frac{n_3}{V_3} = \frac{c_1 V_1 + c_2 V_2}{V_1 + V_2}$$

At this point, we substitute the concentration and volume data into this expression which yields the final answer.

Let us try to do the same procedure in MATLAB. (Of course, we would not rush to start up MATLAB to solve this problem, but it is always worthwhile practicing on some simple problems to see how things work before deploying MATLAB to solve problems that could be rather cumbersome to work out by hand.) We need to first tell MATLAB that the variables (c_1, c_2, c_3, V_1, V_2, V_3, n_1, n_2, and n_3) we are using must be interpreted symbolically. This means that MATLAB should not expect numerical values to be assigned to them, just treat them as symbols representing quantities whose values are not specified. If we were to simply type in c1 into the *Command Window*, MATLAB would not be able to interpret what we mean and return an error message as seen below.

```
>> c1
Unrecognized function or variable 'c1'.
```

Informing MATLAB that we are using symbolic variables is done through the `syms` declaration[†] followed by the names of the variables we wish to use in the first instance. As we only need c_1, c_2, V_1, and V_2 to define the rest of the quantities, we type the following into the *Command Window*.

```
>> syms c1  c2  V1  V2
```

MATLAB throws no error this time; the four symbolic variables have been initiated as indicated in *Workspace*. Let us define n_1 by typing

```
>> n1 = c1*V1
```

into the *Command Window*, which yields the following response (MATLAB prefers placing variables in alphabetical order putting first those that start with a capital letter)

```
n1 =
V1*c1
```

If we look in *Workspace* we will see that MATLAB created a new symbolic variable n1 without us having to initiate it. MATLAB assumes that variables derived from symbolic variables could only be symbolic variables themselves too. Now we define n_2 similarly, and then n_3 and V_3

```
>> n2 = c2*V2
n2 =
V2*c2
>> n3 = n1 + n2
n3 = V1*c1 + V2*c2
>> V3 = V1 + V2
V3 =
V1 + V2
```

With the expressions for n_3 and V_3 defined, all we need to do is to type in the definition expression of concentration applied to the combined solution, and MATLAB will give us the full expression for c_3 including all the variables needed for the calculation:

```
>> c3 = n3/V3
c3 =
(V1*c1 + V2*c2)/(V1 + V2)
```

If we find it difficult to read in-line expressions, we can ask MATLAB to make them more easily readable by using the `pretty()` function[‡] as follows

```
>> pretty(c3)
V1 c1 + V2 c2
--------------
    V1 + V2
```

[†] In Octave, the Symbolic package has to be loaded in by issuing "pkg load symbolic" before `syms` is called.
[‡] Octave generates plain-text expressions resembling typeset mathematics automatically eliminating the need for using `pretty()`.

At this point, you may think that working this out with MATLAB took just as long, or even longer, than by hand. You are probably right. But what if we were not told the volumes but instead given the masses and densities of the two solutions. The task is still manageable without MATLAB but we perhaps start to feel a bit more inclined to use it. Let us clear *Workspace* and declare the primary symbolic variables – concentrations (c_1, c_2), masses (m_1, m_2) and densities (ρ_1, ρ_2) as

```
>> clear all
>> syms c1 c2 m1 m2 p1 p2
```

Now we enter the expressions for volumes calculated from the masses and densities $(V = m/\rho)$:

```
>> V1 = m1/p1
V1 =
m1/p1
>> V2 = m2/p2
V2 =
m2/p2
```

As before, we write the expressions for the amounts of the compound in the two beakers, n_1 and n_2, followed by combining them. (There is no need to retype these, just keep pressing ↑ on your keyboard until you find the expression you need amongst those you recently entered, then press "ENTER/RETURN/↵" to select it):

```
>> n1 = c1*V1
n1 =
(c1*m1)/p1
>> n2 = c2*V2
n2 =
(c2*m2)/p2
>> n3 = n1 + n2
n3 =
(c1*m1)/p1 + (c2*m2)/p2
```

Let us combine the volumes and enter the expression for c_3 without re-typing anything:

```
>> V3 = V1 + V2
V3 =
m1/p1 + m2/p2
>> c3 = n3/V3
c3 = ((c1*m1)/p1 + (c2*m2)/p2)/(m1/p1 + m2/p2)
>> pretty(c3)
 c1 m1    c2 m2
 ----- + -----
  p1       p2
-----------------
    m1    m2
    -- +  --
    p1    p2
```

This looks a bit more complicated than anticipated, so we wonder if MATLAB could simplify this expression for us. The function `simplify()` does just that. Before running `simplify(c3)`, which would give us a long in-line expression, nest `simplify(c3)` inside `pretty()` to obtain a more readable expression straight away:

```
>> pretty(simplify(c3))
c1 m1 p2 + c2 m2 p1
-------------------
   m1 p2 + m2 p1
```

which indeed looks easier to substitute into once we have the concentrations, masses and densities. Better yet, MATLAB can even do the substitution into the expression for us. First we need to ask MATLAB to transform the expression we have obtained for c_3 into a one-line function, which we shall name c3_func, by using the function `matlabFunction()`:

```
>> c3_func = matlabFunction(c3)
c3_func =
  function_handle with value:
    @(c1,c2,m1,m2,p1,p2)((c1.*m1)./p1+(c2.*m2)./p2)./(m1./p1+m2./p2)
```

Now we can simply run c3_func() with the concentrations, masses and densities of the solutions in Beaker 1 and Beaker 2 and obtain the concentration of the combined solution. We note the order of variables in the argument of c3_func() because we must supply them in that order to obtain the correct result. With $c_1 = 0.053$ M, $c_2 = 0.075$ M, $m_1 = 235$ g, $m_2 = 369$ g, $\rho_1 = 1.015$ g cm^{-3}, and $\rho_2 = 1.022$ g cm^{-3}, we enter

```
>> c3 = c3_func(0.053,0.075,235,369,1.015,1.022)
```

which yields:

```
c3 =
    0.0664
```

mol dm^{-3}.

Exercise 11.1

Use symbolic maths to obtain the formula for the pressure (p) of a three-component mixture of perfect gases based on the masses of the gases (m_1, m_2, m_3,), and the temperature (T) and volume (V) of the mixture. Hint: We will need the ideal gas equation (rearranged for pressure: $p = nRT/V$), the fact that the total amount of gases is the sum of the individual amounts: $n = n_1 + n_2 + n_3$) and that the number of moles of a component can be calculated by dividing its mass by its molar mass ($n_i = m_i/M_i$).

Answer:
```
>> syms m1 m2 m3 M1 M2 M3 T V R
>> n1 = m1/M1;
>> n2 = m2/M2;
>> n3 = m3/M3;
>> n = n1 + n2 + n3;
>> p = n*R*T/V;
```

```
>> pretty(p)
        / m1     m2     m3 \
  R T  |  -- +  -- +  --   |
        \ M1     M2     M3 /
  ------------------------------
                V
```

To avoid making mistakes while substituting into this expression using your calculator, simply create a MATLAB one-line function for the expression.

Exercise 11.2

Create a MATLAB function using symbolic maths that calculates the bond length of a diatomic molecule from the positions of two consecutive peaks on the molecule's rovibrational spectrum which typically looks like this:

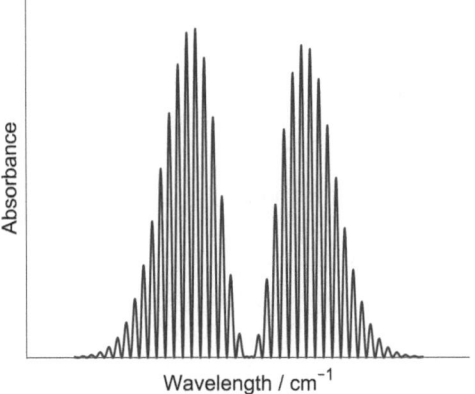

Take peak positions 2123.72 cm^{-1} and 2127.58 cm^{-1} recorded for $^{12}C^{16}O$, and calculate the bond length between the carbon and oxygen atoms in the molecule.

Answer:

We assume that the bond length of a rotating diatomic molecule, AB, is constant. For such rigid rotors, the quantised angular momenta correspond to quantised rotational energies linked by the rotational constant of the molecule, $B = h(8\pi^2 cI)^{-1}$, where h is Plank's constant, c is the speed of light and I is the moment of inertia of the molecule. The latter determines how much torque is needed for a certain acceleration in the rotation of the molecule. The longer the bond and the heavier the atoms at its ends, the greater the molecule's moment of inertia which is calculated as $I = \mu r^2$, where r denotes the bond length and $\mu = m_A m_B/(m_A + m_B)$ is the reduced mass of the molecule, with m_A and m_B being the masses of the two atoms making up the molecule.

From the equally-spaced peaks on the rovibrational spectrum of a molecule, its rotational constant can be calculated as half the distance between adjacent peak positions ($\tilde{v}_i$ and $\tilde{v}_{i+1}$ on either side of the centre, where the gap size is $4B$). With the rotational constant determined, the moment of inertia is obtained from the first expression above, followed by calculating the bond length using the reduced mass which we compute directly from the atomic masses. All the way through we have to pay attention to the units: the peak positions are typically given in cm^{-1}, the reduced mass is in kilograms and the speed of light is in cm s^{-1} to yield the bond length in metres.

This may seem like a lengthy process, especially if you do the calculations by hand. We, however, know how to create a MATLAB function that will do all the number crunching for us. With v1: wavenumber of peak i in cm^{-1}; v2: wavenumber of peak $(i+1)$ in cm^{-1}; m1: atomic mass of atom 1; m2: atomic mass of atom 2; B: rotational constant in cm^{-1}; mu: reduced mass of (a single) molecule in kg; I: Inertia of molecule in kg m^2; h: Planck's constant $h = 6.626 \times 10^{-34}$ J s; c: speed of light $c = 2.998 \times 10^{10}$ cm s^{-1} (in cm s^{-1} because wavenumbers are given in cm^{-1} and this way cm conveniently cancels out) and using the unified atomic mass unit, $u = 1.661 \times 10^{-27}$ kg, to convert atomic numbers (12 for C and 16 for O) to atomic masses (m1, m2), the process is as follows:

```
>> syms v1 v2 m1 m2 h c
>> B = (v2 - v1)/2;
>> mu = m1*m2/(m1 + m2);
>> I = h/(8*pi^2*c*B);
>> r = sqrt(I/mu);
>> BondLength = matlabFunction(r);
BondLength =
  function_handle with value:
    @(c,h,m1,m2,v1,v2)sqrt((h.*(m1+m2).*(-1.266514795529222e-2))./(c.*m1.*m2.*(v1./2.0-v2./2.0)))
```

Now, with the function generated, we can call BondLength() paying attention to passing the input arguments in the required order (c,h,m1,m2,v1,v2) while supplying the atomic masses as (atomic number $\times u$) for each atom.

```
>> CO_bond_length = BondLength(2.998e10, 6.626e-34, 12*1.661e-27, 16*1.661e-27, 2123.72, 2127.58)
CO_bond_length =
   1.1284e-10
```

Because we used SI units throughout, except for $\tilde{v}_i, \tilde{v}_{i+1}$ (cm^{-1}) and c (cm s^{-1}) with cm cancelling out, the sought bond length is 1.128×10^{-10} m = 112.8 pm.

11.2 Symbolic Maths: *Unit Algebra*

Expressions in science and engineering typically contain quantities with units. When formulating an expression, it is important that we check that it is correct. A quick and rudimentary way to do this is by looking at all the units in the expression and work out if they simplify to the expected units. For example, in the expression obtained earlier for the concentration of a compound in solution prepared by mixing two solutions

$$\frac{\dfrac{c_1 m_1}{p_1} + \dfrac{c_2 m_2}{p_2}}{\dfrac{m_1}{p_1} + \dfrac{m_2}{p_2}}$$

the units must simplify to moles/volume (n/V), otherwise the expression is incorrect. (We have to be cautious though because having the right units does not necessarily mean that an expression is correct, although if the units are wrong we can be certain that the expression is wrong. Many wrong answers in exams and lab reports result from not watching the units.)

It is quick and easy to do unit algebra with MATLAB.[§] Let us re-visit the example in which two solutions were mixed. First, we clear *Workspace* and declare the symbolic variables as before

```
>> clear all
>> syms c1 c2 m1 m2 p1 p2
```

however, now we continue by defining a symbolic array, u, that will contain all the units needed

```
>> u = symunit;
```

To assign units to each of our symbolic quantities (concentrations, masses and densities) use the following format:

```
>> c1 = u.mol/u.dm^3
>> c2 = u.mol/u.dm^3
>> m1 = u.kg
>> m2 = u.kg
>> p1 = u.g/u.cm^3
>> p2 = u.g/u.cm^3
```

As a result, MATLAB will not only remember, for example, that c1 is a symbolic variable but it will also carry its units through algebraic manipulations.

Note that MATLAB stores symbolic variables with units in the format of "1 × units" internally if the value of the symbolic variable is not defined. When a value is assigned to a symbolic variable, "1" is replaced by the value; so, for example, the gas constant would be stored as "8.3145 × J mol^{-1} K^{-1}". The "1" is carried and displayed only when the "units" cannot be simplified to a single unit. Consequently, for c1 and c2 MATLAB

[§] As of mid-2024, symunit is not yet implemented in Octave.

puts 1 in front of their units, (`[mol]/[dm]^3`); whereas for `m1`, `m2` and `V1`, `V2` it does not.

When we proceed by defining the volume of the solution in Beaker 1 as

```
>> V1=m1/p1
```

MATLAB will simplify the units and display the unit of `V1` automatically as

```
V1 =
[cm]^3
```

which is very useful (and convenient) when we are trying to stay on top of our units. Let us re-enter now the rest of the expressions as before – keep pressing the ↑ key on the keyboard until the relevant expression comes up, then press "ENTER/RETURN/↵" to select the ones we need.

```
>> V2=m2/p2;
>> n1=c1*V1;
>> n2=c2*V2;
>> n3=n1 + n2;
>> V3=V1 + V2;
>> c3=n3/V3;
c3=1*([mol]/[dm]^3)
```

which is what we expect for a correct expression for concentration. The "1" is included because these units do not simplify to a single unit.

Exercise 11.3

Use symbolic maths to verify that the pressure formula derived in Exercise 11.1 produces the correct units. (Assign units as "R=u.J/u.mol/u.K", "M1=u.kg/u.mol", *etc.*)

Answer:

```
>> syms m1 m2 m3 M1 M2 M3 T V R
>> u=symunit;
>> R=u.J/u.mol/u.K;
>> T=u.K;
>> V=u.m^3;
>> m1=u.g;
>> m2=u.g;
>> m3=u.g;
>> M1=u.g/u.mol;
>> M2=u.g/u.mol;
>> M3=u.g/u.mol;
>> n1=m1/M1;
```

```
>> n2 = m2/M2;
>> n3 = m3/M3;
>> n = n1 + n2 + n3;
>> p = n*R*T/V
p =
3*([J]/[m]^3)
```

where the "3" in front of the units comes from computing the "n = n1 + n2 + n3" expression with n1, n2, n3 being all stored internally as "1 × mol"; therefore, the "3" should be neglected. This result may seem a bit unusual if we expected pressure to be given in Pa or PSI (Pa = N m^{-2} and PSI = pound per inch^2).

MATLAB's Symbolic Math Toolbox contains a list of units, including SI units, prefixes, and various unit systems along with functions to perform conversions between them.

1. Here we will use the rewrite() function to convert both J m^{-3} and Pa to SI base units to check if they give the same SI base units, proving that they are indeed the same. First we convert J m^{-3} as

    ```
    >> rewrite(u.J/u.m^3,'SI')
    ans =
    1*([kg]/(1*[m]*[s]^2))
    ```

    i.e. kg (m s^2)$^{-1}$. Next we convert Pa into SI base units

    ```
    >> rewrite(u.Pa,'SI')
    ans =
    1*([kg]/(1*[m]*[s]^2))
    ```

    which is identical to the answer above confirming that J m^{-3} is the same as Pa.

2. Another approach can be asking MATLAB to convert one set of units into another set of units which contains a particular unit we prescribe. These can be done by using unitConvert(). Let us instruct MATLAB to take "J m^{-3}" and convert it to a set of units containing "N" to check if it results in "N m^{-2}".

    ```
    >> unitConvert(u.J/u.m^3,u.N)
    ans =
    1*([N]/[m]^2)
    ```

    which confirms again that J m^{-3} is the same as Pa.

3. We can also use that Pa is a derived SI unit included in the Symbolic Math Toolbox, hence we can ask MATLAB to associate J m^{-3} with the appropriate derived SI units by

    ```
    >> rewrite(u.J/u.m^3,'SI','derived')
    ans =
    [Pa]
    ```

Exercise 11.4

The Nernst equation is one of the key equations of electrochemistry. Briefly, it is used to calculate the potential of an electrode (E) from the activities, a_i, (or concentrations in dilute solution) of the species participating in the electron transfer, given the potential of an electrode under standard conditions ($E°$) and the number of electrons exchanged in the redox process (z).

$$E = E° - \frac{RT}{zF} \ln \prod a_i$$

The expression RT/F is called thermal voltage (V_T) and it gives how much the potential of an electrode deviates from its standard value with temperature. Use MATLAB to show that the unit of V_T is Volt.

Quantity	Unit(s)
R:	$J\,mol^{-1}\,K^{-1}$
T:	K
F:	$C\,mol^{-1}$

Answer:

Assign units to quantities, then enter expression for thermal voltage:

```
>> syms R T F
>> u = symunit;
>> R = u.J/u.mol/u.K;
>> T = u.K;
>> F = u.C/u.mol;
>> VT = R*T/F
VT =
1*([J]/[C])
>> rewrite(u.J/u.C,'SI','derived')
ans =
[V]
```

The same can be achieved by passing VT directly to `rewrite()`

```
>> rewrite(VT,'SI','derived')
ans =
[V]
```

If you need quick information within MATLAB about a particular unit that you are not entirely sure about, use the `unitInfo()` function. Let us ask MATLAB to display information on the pressure unit Pa.

```
>> unitInfo(u.Pa)
pascal - a physical unit of pressure. ['SI']
```

Get all units for measuring 'Pressure' by calling unitInfo('Pressure').
SI units accept all SI prefixes. For details, see SI Unit Prefixes List.

MATLAB even adds that we can have all pressure units it knows displayed by calling `unitInfo('Pressure')`.

Finally, let us look at how the units of more complicated expressions can be conveniently handled in MATLAB. The formula for the Bohr radius, a_0, used in Quantum Mechanics as the atomic unit of length (5.229×10^{-11} m), is

$$a_0 = \frac{\varepsilon_0 h^2}{\pi m_e e^2}$$

where ε_0 is the permittivity of free space (8.854×10^{-12} F m^{-1}), h is the Planck constant (6.626×10^{-34} J s), m_e is the rest mass of the electron (9.109×10^{-31} kg) and e is the charge of the electron (-1.602×10^{-19} C). After substituting these into the expression we obtain

$$a_0 = \frac{(8.854 \times 10^{-12} \text{ F m}^{-1}) \times (6.626^2 \times 10^{-2 \times 34} \text{ J}^2 \text{s}^2)}{3.142 \times (9.109 \times 10^{-31} \text{ kg}) \times (1.602^2 \times 10^{-2 \times 19} \text{ C}^2)}$$

which may not look very appealing. The most important thing is that we understand this expression, and not that we work out the value of a_0 every time we need it, for which we can conveniently use MATLAB. First, we initiate the symbolic variables and type in the constants together with their units.

```
>> syms me qe e0 h
>> u = symunit;
>> me = 9.109e-31*u.kg;
>> qe = -1.602e-19*u.C;
>> h = 6.626e-34*u.J*u.s;
>> e0 = 8.854e-12*u.F/u.m;
```

Now, we type in the expression for a_0

```
>> a0 = h^2*e0/(pi*me*qe^2);
```

with a semicolon at the end of the command to supress the output which is a very long symbolic number followed by the units. For clarity, it is good practice to separate the *symbolic numeric value* (snv) from the units and display them individually. The function `separateUnits()` severs the symbolic numeric value from the units and we can load them into two separate variables: numeric value of $a_0 \rightarrow$ a0_snv and the units of $a_0 \rightarrow$ a0_u

```
>> [a0_snv,a0_u] = separateUnits(a0)
a0_snv =
3289202269825916565639048849513548015098499905/
(1978080338065917729167508829851492779116101311337070592*pi)
a0_u =
1*((1*[F]*[J]^2*[s]^2)/(1*[C]^2*[kg]*[m]))
```

The symbolic numeric value (a0_snv) is far too long (and written as a quotient of two whole numbers) which makes it inconvenient to work with; therefore, we turn it into *standard numeric value* by using the `double()` function.

```
>> a0_nv=double(a0_snv)
a0_nv=
5.2929e-11
```

To put it into a convenient format, we output the result as an in-line string of characters using `display()`. While we convert a0_nv into a string for displaying by `num2str()`, we round a_0 to four significant figures (same as the precisions of the input quantities) by passing `'%.4g'` to `num2str()`.

```
>> display(['a_0=', num2str(a0_nv,'%.4g'),' ',symunit2str(a0_u)])
a_0=5.293e-11 (F*J^2*s^2)/(C^2*kg*m)
```

We used the `symunit2str()` function to convert the symbolic units into strings for displaying. If we would like to have (F*J^2*s^2)/(C^2*kg*m) simplified before outputting a_0, we should use `rewrite()` prior to separating the units from the numerical value:

```
>> a0_SI=rewrite(a0,'SI','Derived');
>> [a0_SI_snv,a0_SI_u]=separateUnits(a0_SI);
display(['a_0=', num2str(a0_nv,'%.4g'),' ',symunit2str(a0_SI_u)])
a_0=5.293e-11 m
```

which is exactly what we were looking for.
MATLAB stores the seven defining constants for the SI base units. Once

```
>> u=symunit;
```

is executed, the constants can be accessed as pre-defined units (listed in Table 11.1).

MATLAB also supports user-defined constants. For more information visit mathworks.com/help/symbolic/units-list.html

Table 11.1 Pre-defined units of fundamental physical constants in MATLAB.

Constant	Access symbolic value as
Speed of light in vacuum (c)	u.c_0
Planck constant (h)	u.h_c
Elementary charge (e)	u.e
Hyperfine transition frequency of Cesium-133 ($\delta\nu_{Cs}$)	u.dv_Cs
Boltzmann constant (k_B)	u.k_B
Avogadro constant (N_A)	u.N_A
Luminous efficacy of monochromatic radiation of frequency 540×10^{12} Hertz (K_{cd})	u.K_cd

Exercise 11.5

Rydberg is a unit of energy commonly used in atomic physics. It is given by the expression

$$R = \frac{m_e e^4}{8\varepsilon_0^2 h^2}$$

(i) Show that the Rydberg energy is 2.179×10^{-18} J *via* accessing the defining constants as given in the table above (Table 11.1).

Electronic transitions of atoms produce spectra composed of sharp lines. The spacing between spectral lines can be given using the Rydberg constant, R_∞, defined as

$$R_\infty = \frac{R}{hc}$$

where h and c denote the Planck constant and the speed of light, respectively.

(ii) Show that the Rydberg constant is equal to 109 700 cm^{-1} (cm^{-1} is a common unit in spectroscopy).

Answer:

```
>> syms qe e0 me h c
>> u = symunit;
>> qe = -u.e;
>> h = u.h_c;
>> c = u.c_0;
>> e0 = 8.854e-12*u.F/u.m;
>> me = 9.109e-31*u.kg;
>> R = me*qe^4/8/e0^2/h^2;
>> R = rewrite(R,'SI','derived');
>> [R_snv R_u] = separateUnits(R);
>> disp(['R= ',num2str(double(R_snv),'%.4g'),' ',
symunit2str(R_u)])
R = 2.179e-18 J
R_inf = R/h/c;
R_inf = rewrite(R_inf,'SI','derived')
R_inf = 1.097e+07 1/m
R_inf_cm = unitConvert(R_inf,u.cm); % convert 1/m to 1/cm
[R_inf_cm_snv R_inf_cm_u] = separateUnits(R_inf_cm);
disp(['R_inf= ',num2str(double(R_inf_cm_snv),'%.4g'),' ',
symunit2str(R_inf_cm_u)])
R_inf = 1.097e+05 1/cm
```

11.3 Appendix: Intermolecular Forces between Two Dipoles

The alignment of molecular dipoles is opposed by the effect of thermal energy, which leads to the rotation of the dipoles. At temperatures where the thermal energy ($k_B T$) is greater than the electrostatic dipole–dipole interaction energy (U_{DD}), the *rotationally averaged energy of interaction between two dipoles* μ_1 and μ_2, separated by distance r, can be obtained from a Boltzmann-weighted average of the energies of orientations:

$$U_{DD}(r) = -\frac{2}{3k_B T}\left(\frac{\mu_1 \mu_2}{4\pi\varepsilon_0}\right)^2 \left(\frac{1}{r^6}\right)$$

This expression gives the interaction energy between two dipoles in Joules which is on the order of 10^{-21}. Working with these energies is often not convenient as we are more used to handling energies in kJ mol^{-1}. The expression can be rewritten to

$$U_{DD}(r)/\text{kJ mol}^{-1} = -\frac{(\mu_1/\text{D})^2 (\mu_2/\text{D})^2}{1025 \,(r/\text{nm})^6}$$

which gives interaction energies for room temperature ($T = 298$ K) in kJ mol^{-1} if the dipole moments are in Debye (1 D = 3.336×10^{-30} C m) and the distance between dipoles is in nanometres (10^{-9} m).

We will use MATLAB to derive the second expression from the first with e0: $\varepsilon_0 = 8.854 \times 10^{-12}$ F m^{-1}, kB: $k_B = 1.381 \times 10^{-23}$ J K^{-1}, mu_D: $\mu = 1$D $= 3.336 \times 10^{-30}$ C m, r_nm: $r = 1$ nm $= 10^{-9}$ m, NA: $N_A = 6.022 \times 10^{23}$ mol^{-1}, T: 298 K. We substitute in 1D for the size of each electric dipole moment, 1 nm for the distance between the two dipoles and 298 K into the first equation, and multiply the resulting interaction energy by the Avogadro constant while carrying the units of all quantities.

```
>> syms e0 kB mu_D r_nm NA T
>> u=symunit;
>> e0=8.854e-12*u.F/u.m;
>> kB=1.381e-23*u.J/u.K;
>> mu_D=3.336e-30*u.C*u.m;
>> r_nm=1e-9*u.m;
>> NA=6.022e23/u.mol;
>> T=298*u.K;
>> U_dd=-2/3/kB/T*(mu_D*mu_D/4/pi/e0)^2*1/r_nm^6*NA;
>> U_dd_SI=rewrite(U_dd,'SI','Derived');
>> [U_dd_SI_snv U_dd_SI_u]=separateUnits(U_dd_SI);
>> disp(['U_dd=',num2str(double(U_dd_SI_snv),'%.4g'),' ',symunit2str(U_dd_SI_u)])
U_dd=-0.976 J/mol
```

This corresponds to $U_{DD} = -0.000976$ kJ mol^{-1}, where $-0.000976 = -1/1025$; thus, using MATLAB we were able to conveniently justify the correctness of the second expression which is much easier to use than the first one.

12 Maths for Chemistry with MATLAB II

Having covered the basics of symbolic manipulations and unit algebra, we are now ready to tackle more complex problems, for example how to calculate the typical speed of gas molecules. The rates of reactions in the atmosphere are related to the speeds of molecules; the faster they move the higher their kinetic energy and the more frequently they collide with each other and particulates in the air. Therefore, not only the number of solid particles (soot, dust, *etc.*), the levels of unburned hydrocarbons and nitrogen oxides (coming primarily from car exhaust), and the intensity of UV radiation, but also temperature (which affects the speed of gas molecules) play a role in the formation of toxic ozone in the air in big cities and industrial centres. As we saw in Exercise 10.2, the speed distribution of gas molecules changes with temperature with the typical speed increasing as the temperature grows. The typical speed of gas molecules can refer to the *most probable speed*, the *average speed* or the so-called *root-mean-square speed* which can all be calculated from the Maxwell–Boltzmann formula

$$f(c) = 4\pi \left(\frac{m}{2\pi k_B T} \right)^{3/2} c^2 e^{-mc^2/2k_B T}$$

where the $f(c)$ is the fraction of the gas molecules of mass m, having a particular speed (c) at temperature T, with k_B denoting the Boltzmann constant. The distribution of speeds of gas molecules can be computed by setting m and T and then substituting a range of speeds into the expression above, as seen in Exercise 10.2.

From the speed distribution shown in Figure 12.1 we can read off what percent of the gas molecules possess a particular speed; for example, molecules with a speed of 200 m s^{-1} make up about 0.1% (1×10^{-3}) of the molecules, and the largest fraction of the gas molecules (approx. 0.2%) have speeds of about 400 m s^{-1}.

- The most probable speed is the speed we most likely obtain when observing a randomly chosen gas molecule. When randomly picking a molecule for observation we are most likely to come across a molecule that belongs to the largest fraction of molecules in terms of their speed. Therefore, the most probable speed is the speed at which the distribution curve peaks.

A First Look at Coding in Chemistry: Solving Problems Using MATLAB
By Tamas Bansagi
© Tamas Bansagi 2025
Published by the Royal Society of Chemistry, www.rsc.org

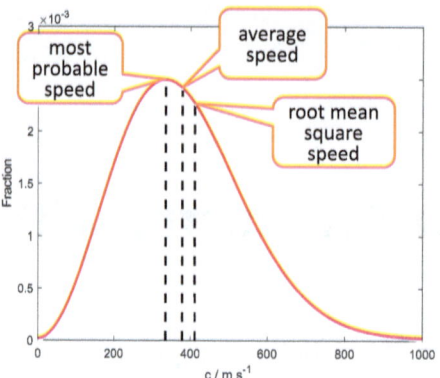

Figure 12.1 Typical speed distribution of gaseous atoms and molecules. The most probable, average and root-mean-square speeds can be calculated form the Maxwell–Boltzmann formula.

- The average (or mean) speed is the speed we obtain when we add up the speeds of all the molecules and divide the sum by the total number of molecules. While calculating the sum of speeds, when a certain speed is measured for multiple molecules, we can multiply that speed by the number of molecules with that speed instead of adding the speed separately each time we obtain it; for example, instead of

$$\bar{c} = (c_1 + c_3 + c_2 + c_2 + c_3 + c_1 + c_1 + c_2 + c_3 + c_1)/10$$

we can write

$$\bar{c} = (4c_1 + 3c_2 + 3c_3)/10 = c_1 \frac{4}{10} + c_2 \frac{3}{10} + c_3 \frac{3}{10}$$

i.e. we multiply the fractions of molecules having a specific speed by the corresponding speeds and add up the results to obtain the average molecular speed as

$$\bar{c} = \sum c_i f(c_i)$$

- The root-mean-square speed is the square root of the mean of the squared speeds

$$c_{rms} = \sqrt{(c_1^2 + c_3^2 + c_2^2 + c_2^2 + c_3^2 + c_1^2 + c_1^2 + c_2^2 + c_3^2 + c_1^2)/10}$$

which rearranges to

$$c_{rms} = \sqrt{(4c_1^2 + 3c_2^2 + 3c_3^2)/10} = \sqrt{c_1^2 \frac{4}{10} + c_2^2 \frac{3}{10} + c_3^2 \frac{3}{10}}$$

i.e. we multiply the fractions of molecules with a specific speed by the corresponding speeds squared and add up the results, followed by taking the square root of the sum as

$$c_{rms} = \left(\sum c_i^2 f(c_i) \right)^{1/2}$$

Now, we are going to discuss how these can be determined from the Maxwell–Boltzmann distribution of gases while learning to differentiate, integrate and solve equations using MATLAB.

12.1 Symbolic Maths: *Differentiation*

Differentiating a function yields an expression describing how the slope of the function depends on the independent variable. We know from calculus that at the extrema (minima and maxima) of a function the slope is zero. If we want to determine the most probable speed of gas molecules, we need to determine the location at which the Maxwell–Boltzmann distribution peaks. In the language of calculus, this task involves differentiating the Maxwell–Boltzmann distribution function, and then finding the speed at which its derivative is zero.

The MATLAB function for differentiation is `diff()`. It takes a symbolic function and the variable we wish to differentiate with respect to. For example, consider the function $y = f(x) = 5x^2 + 2x + 10$, and use MATLAB to obtain its derivative dy/dx or df/dx

```
>> syms x
>> f = 5*x^2 + 2*x + 10
f =
5*x^2 + 2*x + 10
>> dfdx = diff(f,x)
dfdx =
10*x + 2
```

MATLAB can compute partial derivatives which follows the same syntax. The derivative of $z = 5x^2y^3 + 2x + 10y$ with respect to x and y can be obtained as

```
>> syms x y
>> f = 5*x^2*y^3 + 2*x + 10*y
f =
5*x^2*y^3 + 2*x + 10*y
>> pretty(f)
     2  3
  5 x  y  + 2 x + 10 y
>> dfdx = diff(f,x)
dfdx =
10*x*y^3 + 2
>> pretty(dfdx)
           3
  10 x y  + 2
>> dfdy = diff(f,y)
dfdy =
15*x^2*y^2 + 10
>> pretty(dfdy)
       2  2
  15 x  y  + 10
```

Now we return to differentiating the Maxwell–Boltzmann distribution function with respect to c as

```
>> syms c k T m
f = 4*pi*(m/(2*pi*k*T))^1.5*c^2*exp(-m*c^2/(2*k*T));
```

```
pretty(f)
              /       2   \
     2        |    m c    | /    m      \3/2
    c  pi exp| - -----   | |  -------   |      4
              \   2 T k /  \ 2 T k pi /
```

MATLAB has rearranged the terms but this expression is equivalent to what we have typed in.[†]

```
>> dfdc=diff(f,c)
pretty(dfdc)
                                                      /    2  \
                                           3          |  m c  | /    m     \3/2
                                        m c pi exp|  - ----- | |  ------- |      4
           /    2  \                              \    2 T k /  \ 2 T k pi /
           |  m c  | /    m     \3/2
  c pi exp | - ----- | |  ------- |    8 - --------------------------------
           \   2 T k /  \ 2 T k pi/                         T k
```

which looks a lot worse than what it is. Notice that the composite constants $(m/2kT)$ and $\pi(m/2kT\pi)^{3/2}$ appear twice each throughout the expression. Replacing them by two constants A and B would collapse the expression considerably, which is a common technique to help focus on the key characteristics of expressions and reduce the risk of making mistakes.

To find the most probable speed, we need to find the value of c for which this expression is equal to zero. In other words, we need to solve the $df/dc = 0$ equation for c. We will come back to this problem later in this chapter.

To replace the above composite expressions, we will use the subs() function. It requires three input arguments, (i) the expression in which the replacements need to be made, (ii) the sub-expression we want swapped, and (iii) the replacement expression. Because $\pi(m/2kT\pi)^{3/2}$ contains $(m/2kT)$, we first make the replacement $\pi(m/2kT\pi)^{3/2} \to B$.

```
>> syms A B
>> simp_dfdc=subs(dfdc,pi*(m/(2*T*k*pi))^(3/2),B);
>> pretty(simp_dfdc)
                                    /   2  \
                               3    |  c m |
                 /    2  \    B c m exp| - ----- |    4
                 |  c m  |             \  2 T k /
        B c exp | - ------ | 8 - -----------------------
                 \  2 T k /                T k
```

Now, we proceed to swap $(m/2kT)$ to A.

```
>> simp_dfdc=subs(simp_dfdc,m/(2*T*k),A);
>> pretty(simp_dfdc)
```

[†] Octave may prefer rendering the expression differently, for example simplifying $4\pi(1/2\pi)^{3/2}$ to $\sqrt{2/\pi}$.

```
                              /      2  \
                          3   |   c  m  |
         /    2  \     B c  m exp| - ----- | 4
         |  c  m  |            \   2 T k  /
B c exp| - ----- | 8 - ------------------------
         \  2 T k /                T k
```

which shows that this replacement failed. This may have happened because $(m/2kT)$ is part of a larger term $(-c^2m/2kT)$, which is a part of an even larger term $\exp(-c^2m/2kT)$, and MATLAB could not recognise it hidden under two layers. On second attempt, we remove one layer of obstacle by requesting the replacement $(c^2m/2kT)\to(Ac^2)$ as

```
>> simp_dfdc=subs(simp_dfdc,c^2*m/(2*T*k),A*c^2);
>> pretty(simp_dfdc)
                         3              2
             2       4 B c  m exp(-A c )
  8 B c exp(-A c ) - ------------------
                            T k
```

which results in the sought output.[‡]

Exercise 12.1

Simplify expression `dfdc` by making replacements $(m/2kT)\to A$; $\pi(m/2kT\pi)^{3/2}\to B$; $(m/kT)\to D$ in a *single step* by using arrays for the expressions.

Answer:

We can pass arrays as the second and third input arguments of `subs()` to replace multiple expressions in one step. Using the format `[old_exp1 old_exp2 ...]`, `[new_exp1 new_exp2 ...]` for these arrays, we make the single-step swap as

```
>> syms A B D
>> simp_dfdc=subs(dfdc,[pi*(m/(2*T*k*pi))^(3/2) c^2*m/(2*T*k)
   m/(T*k)], [B  A*c^2  D]);
>> pretty(simp_dfdc)
               2                  3          2
   8 B c exp(-A c ) - 4 B D c  exp(-A c )
```

Exercise 12.2

Using MATLAB, show how the pressure of an ideal and a real gas change with either temperature or volume. The pressure of an ideal gas is given by

$$p = nRT/V$$

[‡] As Octave may render the expressions slightly differently, alternative expressions may be chosen for substitutions.

The pressure of a real gas can be expressed by rearranging the van der Waals equation to

$$p = nRT/(V - nb) - an^2/V^2$$

The ideal gas equation only works well if the overall volume of the gas molecules is much smaller than the volume of the vessel containing the gas (nb is negligible) and when the attraction between gas molecules is not reducing the pressure considerably (an^2/V^2 is negligible). These assumptions are reasonable at high temperatures and/or low pressures. When these assumptions do not hold (for example, close to the boiling temperature of the substance at a particular pressure) the van der Waals equation should be used.

Answer:

Partial derivatives express how the pressure changes with either temperature or volume while the other is kept constant. The partial derivatives of the expressions above with respect to T and V are obtained as

```
>> syms p n R T V
>> p=n*R*T/V;
>> dpdT=diff(p,T);
>> pretty(dpdT)
R n
---
 V
>> dpdV=diff(p,V);
>> pretty(dpdV)
  R T n
- -----
    2
   V
>> syms p n R T V a b
>> p=n*R*T/(V-n*b) - n^2*a/V^2;
>> dpdT=diff(p,T);
>> pretty(dpdT)
  R n
-------
V - b n
>> dpdV=diff(p,V);
>> pretty(dpdV)
        2
2 a n       R T n
------ - --------
   3         2
   V     (V - b n)
```

12.2 Symbolic Maths: *Integration*

Finding the area under a peak on a chromatogram to determine the amount of an analyte, establishing ratios between ^{1}H NMR peaks to resolve the relative amounts of hydrogen atoms in different environments in an organic molecule, to calculate the probability of certain outcomes in quantum mechanics, or to work out expressions to link quantities from their differential relationships (integrated kinetic rate laws, formulae in thermodynamics for macroscopic changes) are a few examples where integration is used in chemistry.

Integration in MATLAB is straightforward and can be done by using the `int()` function. Consider the expression often appearing in introductory thermodynamics: $1/x$. We find its indefinite integral (antiderivative or primitive function) as

```
>> syms x
>> f=1/x;
>> F=int(f)
F =
log(x)
```

Note that MATLAB uses `log(x)` for $\ln(x)$ and `log10(x)` for $\log_{10}(x)$. To calculate a definite integral between limits we specify the limits in square brackets as

```
>> F=int(f,[0.1 0.2])
F =
log(2)
```

Notice that the definite integral of the same expression between limits is an expression that has a value. For finding the numeric value of an indefinite integral call the `double()` function.

```
>> double(F)
ans =
    0.6931
```

To illustrate integration through a chemistry example we consider the isothermal expansion of an ideal gas. The work done (dW) by an ideal gas expanding isothermally against external pressure (p_{ext}) can be given as

$$dW = -p_{ext}\, dV$$

where dV is a differential (infinitesimal, vanishingly small or tiny) change in the volume of the ideal gas. If the expansion were performed under reversible conditions, the external pressure would be slowly adjusted throughout the process and kept always infinitesimally below the internal pressure of the gas, p. As a result of $p \cong p_{ext}$, we can replace p_{ext} in the above equation with p, and by recalling the ideal gas equation ($p = nRT/V$) we obtain

$$dW = -\frac{nRT}{V}\, dV$$

The work (W) done while expanding from the initial volume, V_i, to the final volume, V_f, is obtained by adding up all dW contributions, which is mathematically captured by integration as

$$W = \int_{V_i}^{V_f} -\frac{nRT}{V}\, dV$$

The expression to be integrated (the integrand) is the same type, a constant divided by the independent variable, as $1/x$ which we have seen already. Here, however, we need to tell MATLAB that we wish to integrate with respect to V because it does not know that n, R and T are constant. Also, it is good practice when dealing with physical quantities to pass our assumptions on symbolic variables upon declaration. Here, n, R, T, V, V_i and V_f are all positive real numbers, which is useful information for MATLAB as this will limit the scope of the search for the answer.

```
>> syms n R T V Vi Vf positive real
>> f=-n*R*T/V;% the integrand does not contain the differential dV
>> F=int(f,V,[Vi Vf])
F =
-R*T*n*(log(Vf) - log(Vi)) % definite integral of the integrand
```

which is equivalent to the familiar expression found in thermodynamics textbooks:

$$W = -nRT(\ln V_f - \ln V_i) = -nRT \ln\left(\frac{V_f}{V_i}\right)$$

Please note that the same result is achieved without separately defining f first before integrating.

```
>> syms n R T V Vi Vf positive real
>> W=int(-n*R*T/V,V,[Vi Vf])
W =
-R*T*n*(log(Vf) - log(Vi))
```

Exercise 12.3

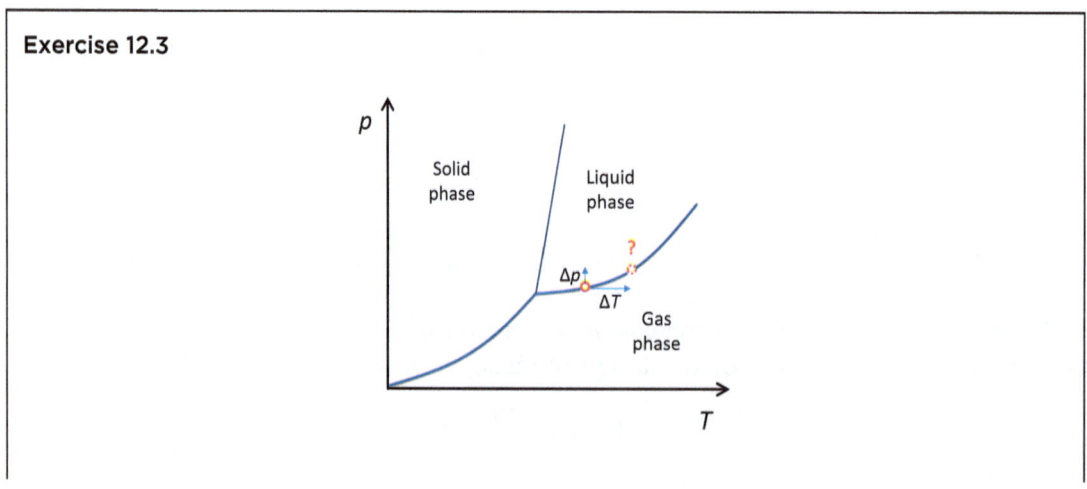

On the phase diagram of a substance, the slope of the boundary curve separating the liquid and gas phases can be approximated as

$$\frac{dp}{dT} = \frac{p\Delta_{vap}H}{RT}$$

Using MATLAB, solve this differential equation and hence express how much the pressure changes in response to a temperature change.

Answer:

To put this question into context, imagine putting enough water into a pressure cooker so that it cannot boil off completely. How much is the pressure inside the pressure cooker if the temperature of the pressure cooker is 120 °C? The solid red circle on the diagram represents a known pressure and temperature point at which the vapour/gas and liquid phases of a substance are in equilibrium, *i.e.* their amounts do not change over time. The dotted red circle represents the position of the new equilibrium pressure and temperature point which we are estimating from the position of the solid circle. At atmospheric pressure, water boils at 100 °C, which gives the location of the solid red circle for water; therefore, how much will the pressure change (*i.e.* $\Delta p = ?$) in response to increasing the temperature to 120 °C (*i.e.* for $\Delta T = 20$ °C)?

We rearrange the expression to

$$\frac{1}{p}dp = \frac{\Delta_{vap}H}{RT^2}dT$$

and integrate both sides to establish the formula connecting the macroscopic changes in pressure and temperature

$$\int_{p_i}^{p_f} \frac{1}{p}dp = \int_{T_i}^{T_f} \frac{\Delta_{vap}H}{RT^2}dT$$

We assume that the molar enthalpy of vaporisation of the substance, $\Delta_{vap}H$, remains constant if ΔT is small. First, we integrate the left-hand side (LHS):

```
>> syms p pI pF positive real
>> int_LHS = int(1/p,p,[pI pF])
int_LHS =
(log(pF) - log(pI))
```

The RHS is integrated as

```
>> syms R DvH T TI TF positive real
>> int_RHS = int(DvH/R/T^2,T,[TI TF]);
>> pretty(RHS)
       / 1     1 \
  DvH | --  - -- |
       \ Tf    Ti /
  - ---------------
           R
```

Bringing the two sides together yields

$$\ln p_f - \ln p_i = -\frac{\Delta_{vap}H}{R}\left(\frac{1}{T_f} - \frac{1}{T_i}\right)$$

from which p_f can be calculated if p_i, T_i, T_f and $\Delta_{vap}H$ are known. (Using the water in the pressure cooker example, $p_i = 101\,325$ Pa, $T_i = 273.15$ K, $T_f = 293.15$ K and $\Delta_{vap}H = 40.67$ kJ mol^{-1} enables us to calculate p_f which together with T_f would define the position of the dotted circle on the phase diagram of water i.e. estimating how the phase boundary curve continues from the solid red circle towards higher temperatures.)

Now we return to the Maxwell–Boltzmann distribution function and determine the average speed of gas molecules. We know from statistics that the average of a distribution ($\bar{x}$) is calculated as $\bar{x} = \sum x_i f(x_i)$. For a continuous distribution (where x_i are not discrete values) the sum is calculated by integrating as

$$\bar{x} = \int_{-\infty}^{\infty} x f(x) dx$$

where $f(x)$ is a distribution function. In our case, $f(x)$ is the Maxwell–Boltzmann speed distribution function defined above, noting that we only need to integrate between 0 and positive infinity as speeds cannot be negative. Therefore, we compute the integral

$$\bar{c} = \int_0^{\infty} c 4\pi \left(\frac{m}{2\pi k_B T}\right)^{3/2} c^2 e^{-mc^2/2k_B T} \, dc$$

which MATLAB can handle easily. Remember that it is often more important that we understand concepts and the steps of developing solutions to problems than spending time solving them manually. (Unless you are enthusiastic about maths and physical chemistry; in which case working out how important formulae are derived is an excellent way to apply your maths to chemistry.)

```
>> syms k m positive real
>> syms c T
>> assume([c>=0 T>=0]); % another way of passing assumptions to MATLAB§
>> f = 4*pi*(m/(2*pi*k*T))^1.5*c^2*exp(-m*c^2/(2*k*T));
>> c_avg = int(c*f,c,[0 inf])
c_avg =
(2*2^(1/2)*T^(1/2)*k^(1/2))/(m^(1/2)*pi^(1/2))
```

Combine square roots by calling the `combine()` function,¶ before making the expression more readable

```
>> c_avg = combine(c_avg)
c_avg =
```

---

§ In Octave, algebraic assumptions cannot be made. Use "`syms k m c T positive real`" in place of the first three commands.
¶ As of mid-2024, `combine()` is not yet available in Octave.

```
2*((2*T*k)/(m*pi))^(1/2)
>> pretty(c_avg)
     / 2 T k \
sqrt| ----- | 2
     \ m pi /
```

This result is identical to the average speed of gas molecules expression found in kinetics textbooks $(2^{1/2}2 = 2^{1/2}4^{1/2} = 8^{1/2})$

$$\bar{c} = \left(\frac{8kT}{\pi m}\right)^{1/2}$$

Obtaining the expression for the root-mean-square average requires multiplying the Maxwell–Boltzmann speed distribution function by c^2, integrating the resulting expression between 0 and positive infinity, then taking the square root of the integral as shown below

$$\bar{c}_{rms} = \left(\int_0^\infty c^2 \, 4\pi \left(\frac{m}{2\pi k_B T}\right)^{3/2} c^2 e^{-mc^2/2k_B T} \, dc\right)^{1/2}$$

which may sound complicated. Fortunately, it takes only a few seconds with MATLAB:

```
>> c_rms = sqrt(int(c^2*f,c,[0 inf]));
>> combine(c_rms);
>> pretty(ans)
     / 3 T k \
sqrt| ----- |
     \ m /
```

which is indeed the expression

$$\bar{c}_{rms} = \left(\frac{3kT}{m}\right)^{1/2}$$

we are familiar with from our kinetics studies.

12.3 Symbolic Maths: Solving Equations

In MATLAB, the function for solving algebraic equations (and inequalities) is `solve()`. Setting up the equations is intuitive; the only thing we need to remember is to use two equal signs "==" to connect the two sides of the equations. Solving the equation $8x + 4 = 32$ with MATLAB is as follows

```
>> syms x
solve(8*x + 4 == 32)
ans =
     7/2
```

MATLAB can solve fully algebraic equations. The above equation recast as $ax+b=c$ and solved as

```
>> syms x a b c
>> solve(a*x + b==c,x)
ans =
    -(b - c)/a
```

where we instructed MATLAB to solve the equation for x by writing ", x" after the equation inside `solve()`. For evaluating this solution with the numbers above, we turn the expression stored in ans into a MATLAB one-line function and substitute in $a=8$, $b=4$ and $c=32$.

```
>> func=matlabFunction(ans)
func =
    function_handle with value:
    @(a,b,c)-(b-c)./a
>> func(8,4,32)
ans =
    3.5000
```

`solve()` is ideal for rearranging equations, as above where we essentially used `solve()` to rearrange $ax+b=c$ for x yielding $x=-(b-c)/a$.

To demonstrate the usefulness of `solve()` as a rearrangement tool in chemistry, we consider enzyme reactions. The equilibrium between the enzyme, E, substrate, S, and the enzyme–substrate complex, ES, can be viewed through the expressions

$$K_M = \frac{[E][S]}{[ES]} = \frac{([E_0]-[ES])[S]}{[ES]}$$

where $[E_0]$ denotes the initial enzyme concentration (before complex formation could have taken place) and K_M is the Michaelis constant. During the derivation of the Michaelis–Menten equation, it becomes necessary to rearrange the above expression for [ES]. With MATLAB, it is an easy task as shown below.

```
>> syms KM S ES E0 positive
>> ES=solve(KM==(E0-ES)*S/ES,ES)
ES =
(E0*S)/(KM + S)
```

Exercise 12.4

Using MATLAB, rearrange the van der Waals equation

$$(p+an^2/V^2)(V-nb)=nRT$$

for pressure to obtain the expression used in Exercise 12.1.

Answer:
```
>> syms p V T R n a b
>> p=solve((p + a*n^2/V^2)*(V − n*b) == n*R*T,p)
p =
(R*T*n − (a*n^2*(V − b*n))/V^2)/(V − b*n)
>> pretty(p)
                2
           a n  (V - b n)
  R T n - --------------
                2
                V
  ---------------------
         V - b n
>> pretty(simplify(p,4))
                 2
  R T n      a n
  ------- - ---
  V - b n    2
             V
```

In the last step, we asked MATLAB to try simplifying the expression in 4 steps,[¶¶] which results in a more compact expression than the one obtained using `simplify(p)`.

With MATLAB we can also solve systems of equations. For example, having determined the rate constant of a reaction at two different temperatures, $k(T_1)$ and $k(T_2)$, we can calculate the activation energy of the reaction (E_a) and the pre-exponential factor (A) using the Arrhenius equation. With $k_1 = k(T_1)$ and $k_2 = k(T_2)$ the system of equations we need to solve for E_a and A is

$$\ln(k_1) = \ln(A) - \frac{E_a}{RT_1}$$

$$\ln(k_2) = \ln(A) - \frac{E_a}{RT_2}$$

This problem can be fed into MATLAB as

```
>> syms k1 k2 A R T1 T2 Ea real positive
>> S=solve([log(k1) ==log(A)-Ea/R/T1, log(k2) ==log(A)-Ea/R/T2],
[Ea A]);
```

with the solution loaded into array S. Notice that the equations and the variables to solve for are in square brackets. (Within square brackets, equations and variables are

---

¶¶ As of mid-2024, only single-step simplification is available in Octave. Please note that simplifying expressions is a non-exact area; therefore simplified expressions may differ across MATLAB and Octave editions, which might result in failing to obtain the expressions given here.

typically comma-separated, but it is not a strict requirement.) We obtain the expressions for the activation energy and the pre-exponential factor as

```
>> pretty(simplify(S.Ea))
R T1 T2 (log(k1) - log(k2))
---------------------------
         T1 - T2
>> pretty(simplify(S.A))
   / T1 log(k1) - T2 log(k2) \
exp| ----------------------- |
   \         T1 - T2         /
```

It is worthwhile noting that once we have calculated the value of E_a, rearranging either the Arrhenius equation for A and substituting into that is a bit quicker than substituting into the expression above. However, turning these expressions into MATLAB functions for substituting in our values determined in the lab is even quicker.

Exercise 12.5

When determining the composition of a mixture using light absorption, we often encounter the problem of having more than one chemical species absorb light at a wavelength. For example, when we would like to follow the progress of a reaction $\alpha + \beta \rightarrow \gamma$ by determining concentrations c_α, c_β, and c_γ through measuring the absorbance of the reaction mixture – with all three components absorbing light in the range of our spectrophotometer – we need to measure the absorbance at three different wavelengths (A_1, A_2, and A_3) and solve the system of equations

$$A_1 = \varepsilon_{\alpha 1} c_\alpha l + \varepsilon_{\beta 1} c_\beta l + \varepsilon_{\gamma 1} c_\gamma l$$
$$A_2 = \varepsilon_{\alpha 2} c_\alpha l + \varepsilon_{\beta 2} c_\beta l + \varepsilon_{\gamma 2} c_\gamma l$$
$$A_3 = \varepsilon_{\alpha 3} c_\alpha l + \varepsilon_{\beta 3} c_\beta l + \varepsilon_{\gamma 3} c_\gamma l$$

where $\varepsilon_{\alpha 1}$, $\varepsilon_{\beta 1}$, and $\varepsilon_{\gamma 1}$ are the molar absorption coefficients of species α, β and γ, respectively at wavelength 1 and so on, with $l = 3$ cm being the path length of light through the reaction vessel.

Date: / /

DETERMINING MULTIPLE CONCENTRATIONS USING UV-VIS SPECTROMETRY

Absorbance	molar absorption coefficient, ε / m^2 mol^{-1}		
	α	β	γ
$A_1 = 1.204$	10 520	2402	589
$A_2 = 1.533$	4267	8958	892
$A_3 = 0.536$	376	1302	3594

(Watch your units; molar absorption coefficients are given in m² mol⁻¹, whereas l is given in cm.)

What other method could you use to solve this problem? Hint: the problem is described by a system of linear equations.

Answer:

```
>> syms ea1 eb1 eg1 ea2 eb2 eg2 ea3 eb3 eg3 ca cb cg A1 A2 A3 l
real positive
>> S=solve([A1==ea1*ca*l + eb1*cb*l + eg1*cg*l, A2==ea2*ca*l +
eb2*cb*l + eg2*cg*l, A3==ea3*ca*l + eb3*cb*l + eg3*cg*l],
[ca cb cg]);
>> A2c_func=matlabFunction([S.ca S.cb S.cg],'Vars',{l A1 A2 A3 ea1
eb1 eg1 ea2 eb2 eg2 ea3 eb3 eg3}); % order of input variables specified
in {...}
>> A2c_func(0.03,1.204,1.533,0.536,10520,2402,589,4267,8958,
892,376,1302,3594)
ans =
    0.0027    0.0041    0.0032
```

which are in mol m⁻³.

Now we return again to the Maxwell–Boltzmann distribution function and determine the most probable speed of gas molecules building upon what we have learnt about differentiation and solving equations with MATLAB.

```
>> syms k m positive real
>> syms c T real
>> assume([c>=0 T>=0]);
>> f=4*pi*(m/(2*pi*k*T))^1.5*c^2*exp(-m*c^2/(2*k*T));
```

Differentiate the Maxwell–Boltzmann distribution function to obtain the expression for its derivative (df/dc):

```
>> dfdc=diff(f,c); % differentiate the Maxwell-Boltzmann distr. func.
```

Find the location(s) where the Maxwell–Boltzmann distribution function has minima or maxima

```
>> extrema=solve(dfdc==0,c)
extrema =
                        0
(2^(1/2)*T^(1/2)*k^(1/2))/m^(1/2)
```

The distribution has a minimum at `extrema(1)` (at $c=0$ as seen in the Figure 12.1) and a maximum at `extrema(2)` which is the most probable speed.[‖] Combine roots in the expression giving the most probable speed, then display:

```
>> c_most_prob=combine(extrema(2))
c_most_prob =
((2*T*k)/m)^(1/2)
>> pretty(c_most_prob)
     / 2 T k \
sqrt| ----- |
     \   m   /
```

which is the expression found in kinetics textbooks as

$$c^* = \left(\frac{2kT}{m}\right)^{1/2}$$

12.4 Case Study: Propagation of Uncertainties

In Chapter 6, we left off our discussion on error propagation with the expression

$$\delta y \approx \sqrt{\left(\frac{\partial}{\partial u}f\right)^2 (\delta u)^2 + \left(\frac{\partial}{\partial v}f\right)^2 (\delta v)^2 + \left(\frac{\partial}{\partial w}f\right)^2 (\delta w)^2}$$

which we can now return to. This formula tells us how to estimate the uncertainty in quantity y, δy, from the uncertainties in quantities u, v, w, δu, δv, and δw, respectively. This is a general formula applicable to all relationships, including the specific relationships $y = u \pm v$, $y = u \times v$ and $y = u/v$ we have already derived the error propagation formulae for:

Relationship between quantities	Propagation of uncertainties	Example
$y = u \pm v$	$\delta y = \sqrt{(\delta u)^2 + (\delta v)^2}$	Uncertainty in volume when pipetting twice (or more) into a receiving container
$y = u \times v$	$\delta y = \sqrt{\left(\frac{\delta u}{u}\right)^2 + \left(\frac{\delta v}{v}\right)^2}\, y$	Uncertainty in the amount of moles of a solute calculated by multiplying its concentration by the volume of the solution
$y = \dfrac{u}{v}$		Uncertainty in the concentration of a solute calculated by dividing its number of moles by the volume of the solution

These propagation formulae have limited usefulness, for not all relationships we encounter are combinations of them. Therefore, we often need to resort to looking up error propagation formulae specific to the particular relationships we are dealing with (involving powers, logs, trigonometric functions and their various combinations).

[‖] Following on from footnote on page 230, Octave will return one extremum.

Instead of building ever larger tables of error propagation formulae to suit everyone, we could just use our knowledge of MATLAB to conveniently generate the correct error propagation formula, whatever the relationship between the quantities may be. We only have to apply the general formula using MATLAB for each specific relationship we encounter. The process will simply involve

1. computing the partial derivatives of the relationship $(\partial f/\partial u)$, $(\partial f/\partial v)$, ...
2. multiplying each squared partial derivative by its corresponding squared uncertainty
3. adding the resulting terms together and taking the square root of the sum

all of which we have seen to be straightforward in MATLAB.

Consider the simple yet fundamental relationship of chemistry: $[H^+] = 10^{-pH}$. Using the above notation, this relationship becomes $y = 10^{-u}$. Applying the general error propagation expression to this relationship is done as

```
>> syms u du real  % u: pH value measured; du: accuracy of pH meter
>> assume(du>0)  % uncertainty is a positive quantity, but pH could be
negative
>> f=10^(-u);  % define relationship: H = f(pH) = 10^(-pH)
>> dfdu=diff(f,u);  % compute partial derivative df/du equivalent to
dH/d(pH)
>> dfdu2=dfdu^2;  % square df/du equivalent to (dH/d(pH))^2
>> dy=sqrt(dfdu2*du^2)  % substitute into general error prop. formula
dy =
log(10)*(1/10^(2*u)*du^2)^(1/2)
>> pretty(simplify(dy))
du log(10)
----------
     u
   10
```

Therefore, the sought expression to propagate the uncertainty in pH values measured into the corresponding calculated $[H^+]$ values is

$$\delta[H^+] \approx \ln(10)\frac{\delta pH}{10^{pH}}$$

which can be re-written into a more convenient form $\delta[H^+] \approx \ln(10)10^{-pH}\delta pH = 2.303 [H^+]\delta pH$. (This means that if the accuracy of a pH meter is the same across its pH range (*i.e.* if δpH is constant) $\delta[H^+]$ will depend on the pH; it will increase as the pH is lowered.)

Of course, we can write the above MATLAB script using H and pH:

```
>> syms pH dpH real
>> assume(dpH>0)
>> H=10^(-pH);
>> dH=sqrt(diff(H,pH)^2*dpH^2);  % Nested functions for brevity
>> pretty(simplify(dH))
```

$$\frac{dpH \; \log(10)}{pH^{10}}$$

To make this script more practical for lab use, we could convert it into one-line function format so that we can directly substitute in our measured pH values. For the accuracy of the pH meter used, δpH, we should consult the manual.

```
>> dH_func=matlabFunction(dH,'Vars',{pH dpH})
dH_func =
  function_handle with value:
    @(pH,dpH)log(1.0e+1).*sqrt(1.0e+1.^(pH.*-2.0).*dpH.^2)
```

Now, let us say our pH reading was 5.68 ± 0.01 which can be passed to our dH_func() as

```
>> dH_func(5.68,0.01)
```

which generates the result

```
ans =
    4.8108e-08
```

meaning that $\delta[H^+] = 5 \times 10^{-8}$ mol dm^{-3} (rounded up to 1 significant figure). We may prefer to have the $[H^+]$ displayed along with $\delta[H^+]$. For that, we double click on dH_func() in *Workspace* and copy the one-line function into a composite one-line function we will define as H_dH_func(). We put in square brackets the expression 10^{-pH} (to calculate $[H^+]$) first, followed by the one-line function for $\delta[H^+]$

```
H_dH_func=@(pH,dpH) [10^(-pH) log(1.0e+1).*sqrt(1.0e+1.^(pH.*-2.0).
*dpH.^2)]
```

Now, when we pass the accuracy of our pH meter and the measured pH to H_dH_func() as

```
H=H_dH_func(5.68,0.01)
```

we obtain

```
H =
   1.0e-05 *
    0.2089    0.0048
```

By default, MATLAB tries to display the results on the same scale so it multiplies both $[H^+]$ and $\delta[H^+]$ by 10^{-5}. If we prefer each to be displayed in its most compact format, we change the output format to

```
>> format shortG
```

then re-run, which outputs the slightly more readable

```
H =
   2.0893e-06   4.8108e-08
```

Therefore pH $= 5.68 \pm 0.01$ corresponds to (after rounding) $[H^+] = 2.09 \pm 0.05 \times 10^{-6}$ mol dm^{-3}.

Exercise 12.6

Imagine you are interested in a particular reaction and you would like to determine its activation energy. You set up and run the reaction several times at two different temperatures (T_1, T_2). No matter how hard you try there are always some temperature variations about the desired two different temperatures $(\delta T_1, \delta T_2)$. Having determined the rate constant and its uncertainty (standard error) at two different temperatures, $k(T_1) = k_1 \pm \delta k_1$ and $k(T_2) = k_2 \pm \delta k_2$, you recall the expression for calculating the activation energy

$$E_a = \frac{RT_1 T_2 (\ln k_1 - \ln k_2)}{T_1 - T_2}$$

we obtained earlier in this chapter. (Note that the uncertainty in the gas constant is much lower than the uncertainties in your rate constants and temperatures, thus you do not need to include the uncertainty in R in the analysis.) Using MATLAB, apply the general error propagating formula to this expression with

$$T_1 = 293.1 \pm 0.2 \text{ K}$$

$$T_2 = 303.2 \pm 0.3 \text{ K}$$

$$k_1 = 58 \pm 7 \text{ mol}^{-1} \text{ dm}^3 \text{ s}^{-1}$$

$$k_2 = 110 \pm 12 \text{ mol}^{-1} \text{ dm}^3 \text{ s}^{-1}$$

The error propagation expression generated by MATLAB will probably look complicated. We should not worry about that though because once it is turned into a one-line function, MATLAB will handle substituting the values in and computing the uncertainty in the activation energy. Place the expression above and the corresponding MATLAB generated error propagation expression into a single composite one-line function.

(You could propagate the uncertainties by applying the rules for $y = u \pm v$, $y = u \times v$ and $y = u/v$ along with expression $\delta y = \delta u/u$ for $y = \ln(u)$; however, this involves more steps which could lead to more mistakes than directly applying the general error propagation expression.

Answer:
```
>> syms k1 k2 A R T1 T2 dT1 dT2 dk1 dk2 Ea real positive
>> S=solve([log(k1) == log(A)-Ea/R/T1, log(k2) == log(A)-Ea/R/T2],
[Ea A]); % solve system of Arrhenius equations for Ea and A
```

```
>> dEadT1 = diff(S.Ea,T1);  % compute partial derivative dEa/dT1
>> dEadT2 = diff(S.Ea,T2);  % compute partial derivative dEa/dT2
>> dEadk1 = diff(S.Ea,k1);  % compute partial derivative dEa/dk1
>> dEadk2 = diff(S.Ea,k2);  % compute partial derivative dEa/dk2
>> dEadT1_2 = dEadT1^2;  % (dEa/dT1)^2
>> dEadT2_2 = dEadT2^2;  % (dEa/dT2)^2
>> dEadk1_2 = dEadk1^2;  % (dEa/dk1)^2
>> dEadk2_2 = dEadk2^2;  % (dEa/dk2)^2
>> dEa = sqrt(dEadT1_2*dT1^2 + dEadT2_2*dT2^2 + dEadk1_2*dk1^2 +
dEadk2_2*dk2^2);  % substitute into general error propagation formula
>> Ea_dEa_func = matlabFunction([S.Ea dEa],'Vars',{T1 dT1 T2 dT2 k1
dk1 k2 dk2 R});  % convert symbolic equations for Ea and dEa into a one-
line function
>> Ea_dEa = Ea_dEa_func(293.1,0.2,303.2,0.3,58,7,110,12,8.3145)
Ea_dEa =
   1.0e+04 *
      4.6824      1.2016
```

meaning that $E_a = 50 \pm 20$ kJ mol^{-1} (having rounded the uncertainty up to one significant figure to avoid underestimating the error and the activation energy to the same decimal place).

12.5 Appendix A: Exact Solutions, Approximations, and Simplifications

Most science and engineering problems cannot be solved exactly. This may come as a surprise to you because so far in your chemistry (and other science and maths) topics you have most likely only seen problems that can be solved by hand. In reality, these (introductory) problems simple enough to be solved exactly (or analytically) on a piece of paper are quite rare, nevertheless they are very important in developing your problem-solving skills early on. They are often the result of (vast) simplifications, without which they could not be solved exactly, or the exact solution would require so much effort that it would distract students from the key scientific points of the discussion.

What can we do if we want to solve a problem that cannot be solved – or is too difficult to be solved – exactly? This is a very important question indeed, as most problems fall into this category. What we do in these cases is either *simplify the problem* to something we can solve exactly, or *approximate the solution*. Now, you may ask – *is a simplified problem solved exactly or an approximate solution good enough?* The short answer is: it is, as long as it is close enough to the exact solution (if it exists) and/or it is sufficient for the intended purpose. The vague terms "close enough" and "sufficient" always depend on the acceptable tolerance decided on a case-by-case basis.

Algorithms approximating solutions work iteratively. Starting from some initial guess solutions, algorithms keep refining the approximate solution step by step, from one iteration to the next. The iterative refinement process is stopped when the approximated solution improves so little during the iteration that a further iteration step would not be worth computing, or we do not require that level of accuracy. To briefly illustrate how simplifications and approximations work in relation to finding exact solutions we use MATLAB to find the solution to two familiar chemistry problems. We will first simplify the problems so that we can solve them with intermediate maths knowledge. Then we use symbolic maths to find the exact solutions and finally we will approximate the solutions by using the `vpasolve()` function.

Imagine that we work for a household products company and our task is to create environmentally friendly window cleaners, dishwashing liquids and mildew stain removers. A key ingredient in all these products is acetic acid (vinegar) so when we are working on these formulations, we are essentially making dilute acetic acid solutions and investigating their effectiveness. Calculating the pH of an acetic acid solution is therefore an integral part of the formulation development process. For the pH we need to calculate $[H^+]$ first, which involves solving the following simultaneous equations for a weak monoprotic acid like acetic acid:

$$K_w = [H^+][OH^-] \qquad H_2O \rightleftharpoons H^+ + OH^-$$

$$K_a = \frac{[H^+][A^-]}{[HA]} \qquad HA \rightleftharpoons H^+ + A^-$$

$$[A^-] + [HA] = [HA]_0$$

$$[H^+] = [OH^-] + [A^-]$$

where the first two equations represent the chemical equilibria on the right. The third equation is the material balance for the weak acid stating that the concentration of the initially added weak acid, before any dissociation could have taken place, $[HA]_0$ (known as its analytical concentration), will be equal to the sum of the equilibrium concentrations of the undissociated weak acid, $[HA]$, and the conjugate base, $[A^-]$. The last equation represents the charge balance: the total number of positive charges must be equal to the total number of negative charges in the solution at all times. As eager formulation chemists, we would like to be able to calculate the exact value of $[H^+]$ for any initial acetic acid added, $[HA]_0$, using the ionic product of water, K_w, and the acid dissociation constant of acetic acid, K_a, which we can find in the CRC Handbook of Chemistry and Physics.

We make the following rearrangements and substitutions

$$[OH^-] = \frac{K_w}{[H^+]}$$

$$[HA] = [HA]_0 - [A^-]$$

$$K_a = \frac{[H^+][A^-]}{[HA]_0 - [A^-]} \Rightarrow K_a[HA]_0 = [H^+][A^-] + K_a[A^-] \Rightarrow [A^-] = \frac{K_a[HA]_0}{[H^+] + K_a}$$

$$[\mathrm{H^+}] = \frac{K_\mathrm{w}}{[\mathrm{H^+}]} + \frac{K_\mathrm{a}[\mathrm{HA}]_0}{[\mathrm{H^+}]+K_\mathrm{a}} \Rightarrow [\mathrm{H^+}] = \frac{K_\mathrm{w}[\mathrm{H^+}]+K_\mathrm{w}K_\mathrm{a}+K_\mathrm{a}[\mathrm{HA}]_0[\mathrm{H^+}]}{[\mathrm{H^+}]^2+K_\mathrm{a}[\mathrm{H^+}]}$$

$$[\mathrm{H^+}]^3 + K_\mathrm{a}[\mathrm{H^+}]^2 = (K_\mathrm{w}+K_\mathrm{a}[\mathrm{HA}]_0)[\mathrm{H^+}] + K_\mathrm{w}K_\mathrm{a}$$

which rearranges to

$$[\mathrm{H^+}]^3 + K_\mathrm{a}[\mathrm{H^+}]^2 - (K_\mathrm{w}+K_\mathrm{a}[\mathrm{HA}]_0)[\mathrm{H^+}] - K_\mathrm{w}K_\mathrm{a} = 0$$

This is a cubic equation given by the general form

$$ax^3 + bx^2 + cx + d = 0$$

where $[\mathrm{H^+}]$ plays the role of x, $a=1$, $b=K_\mathrm{a}$, $c=-(K_\mathrm{w}+K_\mathrm{a}[\mathrm{HA}]_0)$ and $d=-K_\mathrm{w}K_\mathrm{a}$. Just like how the familiar quadratic equations ($ax^2+bx+c=0$) can be solved exactly using the quadratic formula, cubic equations can also be solved exactly by using the formulae

$$x_1 = t - \frac{s}{3a} - \frac{q}{3as}$$

$$x_2 = t + \frac{s(1+\sqrt{3}i)}{6a} + \frac{q(1-\sqrt{3}i)}{6as}$$

$$x_3 = t + \frac{s(1-\sqrt{3}i)}{6a} + \frac{q(1+\sqrt{3}i)}{6as}$$

where

$$p = 2b^3 - 9abc + 27a^2d \qquad q = b^2 - 3ac \qquad r = \sqrt{p^2-4q^3} \qquad s = \sqrt[3]{\frac{r+p}{2}} \qquad t = -\frac{b}{3a}$$

with i called the imaginary number** defined as $\sqrt{-1}$.

Before solving the above equation cubic in $[\mathrm{H^+}]$, we approach the problem *via* finding graphically the values of $[\mathrm{H^+}]$ for which the cubic equation holds by looking at the graph of the expression

$$[\mathrm{H^+}]^3 + K_\mathrm{a}[\mathrm{H^+}]^2 - (K_\mathrm{w}+K_\mathrm{a}[\mathrm{HA}]_0)[\mathrm{H^+}] - K_\mathrm{w}K_\mathrm{a}$$

We substitute in the acidity constant of acetic acid, $K_\mathrm{a} = 1.8 \times 10^{-5}$, an analytical concentration, for example $[\mathrm{CH_3COOH}]_0 = 0.0001$ mol dm^{-3}, and the ionic product of water, $K_\mathrm{w} = 10^{-14}$, and then we put the resulting expression into a loop. The loop allows us to effortlessly change $[\mathrm{H^+}]$ in a broad range while computing the corresponding values of the expression. Subsequently, we plot these values against their respective $[\mathrm{H^+}]$ values, which will trace out the graph of the cubic expression as $[\mathrm{H^+}]$ varies (Figure 12.2).

** To limit the extent of this introductory book, we are not going to discuss imaginary numbers. Their elusive nature puzzled mathematicians for centuries, but they were put on a firm footing eventually. Despite their mysterious name, imaginary numbers have wide-ranging real-world applications, far beyond finding the roots of polynomials. For example, quantum mechanics, which underpins all chemistry, would not be possible without them.

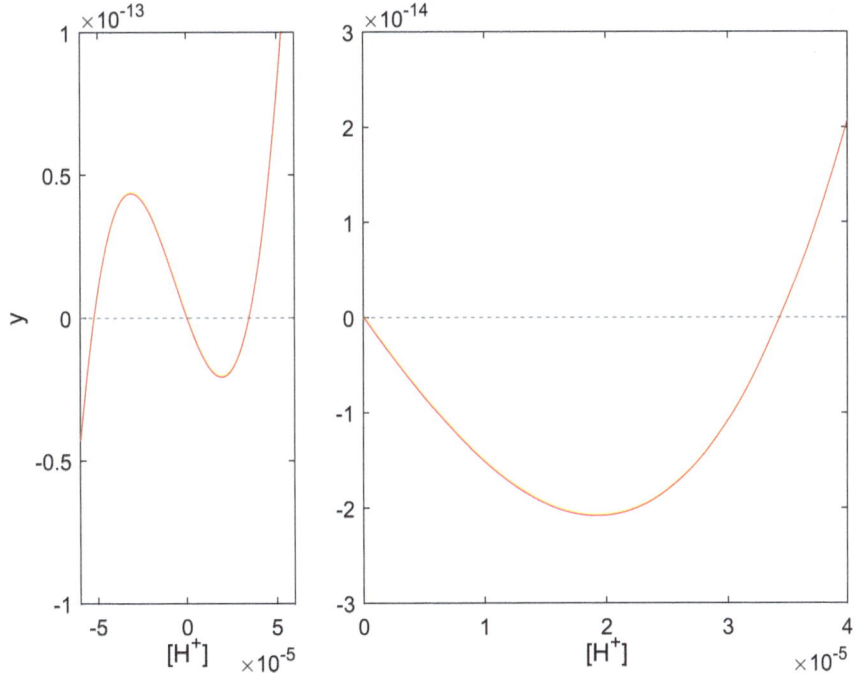

Figure 12.2 Graph of the cubic expression derived for the acid concentration in an acetic acid solution with analytical concentration $[CH_3COOH]_0 = 0.0001$ mol dm^{-3}.

We set up our loop to start on a negative $[H^+]$ because we would like to gain insight into how the expression behaves in general (left), before focusing on the chemically relevant domain (right). For some $[H^+]$ values the cubic expression is negative while for others it is positive. We also see that for $[CH_3COOH]_0 = 0.0001$ mol dm^{-3} the graph goes through zero three times as $[H^+]$ is changed, which means that mathematically our cubic equation has three distinct roots for this particular acetic acid analytical concentration. (In other words, there are three different $[H^+]$ values at which our cubic equation holds, where the expression is exactly zero.) Now looking along the range of chemically feasible $[H^+]$ values (right), we see that the graph goes through zero only once – somewhere around 3.5×10^{-5} mol dm^{-3} – and that is going to be the root of our cubic equation, the H^+ concentration we are looking for when $[CH_3COOH]_0 = 0.0001$ mol dm^{-3}.

MATLAB does not need to produce graphs like this to find the roots of cubic equations, for it uses sophisticated algorithms to numerically approximate the roots of polynomials. There is even no need to set up the cubic equation for MATLAB to solve, we can just give MATLAB the four equations we used along with the values for the parameters ($[HA]_0$, K_a, K_w) and its advanced algorithms will find the equilibrium concentrations of $[H^+]$, $[OH^-]$, $[HA]$ and $[A^-]$.

Exact Solution

Now, we look at how MATLAB can be used to derive the cubic equation we obtained above. (As discussed earlier, the Symbolic Math Toolbox is there to do the heavy lifting, saving us from having to do the often daunting algebraic manipulations so that we can focus on the chemistry. Of course, if you are particularly interested in maths, physical, theoretical and/or computational chemistry you can use the Symbolic Math Toolbox only to check whether you have done the maths right.) The step-by-step derivation of cubic formula is as follows.

```
>> clear all % clear Workspace
>> syms H OH HA A HA0 Ka Kw positive real % declare symbolic variables
>> OH = solve(Kw==H*OH,OH)   % rearrange Kw = H * OH for OH
OH =
Kw/H
>> HA = solve(A + HA==HA0,HA) % rearrange A + HA = HA0 for HA
HA =
HA0 - A
>> A = solve(Ka==H*A/HA,A) % rearrange Ka = H*A/HA for A
A =
(HA0*Ka)/(H + Ka)
>> eq1 = simplify(H - A - OH==0) % charge balance eq. with OH, HA,
A defined above
eq1 =
Kw*(H + Ka) + H*HA0*Ka==H^2*(H + Ka)
```

We now rearrange eq1 by taking its right-hand side and left-hand side using the rhs() and lhs() functions, and subtracting the latter from the former.

```
>> eq2 = rhs(eq1) - lhs(eq1) ==0% rearrange for zero
eq2 =
H^2*(H + Ka) - Kw*(H + Ka) - H*HA0*Ka==0
```

In the final step, we collect the powers of [H^+] from all terms by utilising the collect() function.

```
>> eq3 = collect(eq2,H) % rewrite eq2 into powers of H
eq3 =
H^3 + Ka*H^2 + (- Kw - HA0*Ka)*H - Ka*Kw==0
```

which is the same as the cubic equation we derived above. These few lines of script illustrate how easy it is to do maths with **MATLAB** for deriving formulae used in chemistry.

To use the above cubic formulae to solve our cubic equation exactly, we need to execute

```
>> Ka=1.8e-05; Kw=1e-14; HA0=0.0001;
>> a=1;
>> b=Ka;
>> c=-(Kw + Ka*HA0);
>> d=-Kw*Ka;
>> p=2*b^3 - 9*a*b*c + 27*a^2*d;
>> q=b^2 - 3*a*c;
```

```
>> r=(p^2 - 4*q^3)^(1/2);
>> s=((r + p)/2)^(1/3);
>> t=-b/(3*a);
>> format long
>> x=[t - s/(3*a) - q/(3*a*s); t + s*(1 + 3^(1/2)*i)/(6*a) + q*(1
 - 3^(1/2)*i)/(6*a*s); t + s*(1 - 3^(1/2)*i)/(6*a) + q*(1 + 3^(1/2)*i)/
(6*a*s)]
x =
  -5.237057254585508e-05   - 3.049318610115481e-20i
  -9.999934444288139e-11   + 7.792703114739563e-20i
   3.437067254519952e-05   - 4.404571325722362e-20i
```

The three solutions are expressed in two parts with the second parts being negligibly small. (In this case, they exist only because during the calculations the parts containing the imaginary number failed to round exactly to zero, thus being eliminated, which would have been the technically correct answer.) We can remove the second parts in x by

```
>> real(x)
ans =
  -5.237057254585508e-05
  -9.999934444288139e-11
   3.437067254519952e-05
```

As expected from the graphs of the cubic expression, we have three roots, two negative and one positive, which means that $[H^+] = 3.437067254519952 \times 10^{-5}$ mol dm^{-3} (pH = 4.463811969759449) in the acetic acid solution when $[CH_3COOH]_0 = 0.0001$ mol dm^{-3}.

It is important to emphasise that this *exact solution*

- is only available to the given number of digits
- was computed using $[HA]_0$, K_a and K_w whose values are not exact

which shows the weakness of exact solutions produced *via* finite-precision calculations from measured and/or estimated, *i.e.* non-exact, quantities.

The same result is achieved when we simply give MATLAB the four equations and the values for $[HA]_0$, K_a and K_w.

```
>> syms H OH HA A positive real
>> Ka=1.8e-05; Kw=1e-14; HA0=0.0001;
% declare all equations in a single array:
>> eq=[Kw==H*OH, Ka==H*A/HA, A + HA==HA0, H==A + OH];
% solve equations and load solutions for H, OH, HA, A into an array
>> S=solve(eq);
>> format long
>> H=double
(S.H)  % symbolic numeric value of H => standard numeric value[††]
H =
   3.437067254519952e-05
```

[††] In Octave, use "real(double(S{2}.H))" here.

If we want MATLAB to use the cubic formula to solve our cubic equation analytically, we need to write[‡‡]

```
>> syms H positive real
>> Ka = 1.8e-05; Kw = 1e-14; HA0 = 0.0001;
>> H = root(H^3 + Ka*H^2 + (- Kw - HA0*Ka)*H - Ka*Kw == 0);
>> format long
>> double(H)
ans =
   1.0e-04 *
  -0.000000999993444
   0.343706725451995
  -0.523705725458551
```

Approximation

MATLAB has powerful algorithms to approximate solutions which are most useful when a solution like the above is difficult to obtain or not possible to find analytically. Here we look at the same weak acid problem and approximate the solution (which is often called numerically solving a problem) using `vpasolve()`:

```
>> syms H OH HA A positive real
>> Ka = 1.8e-05; Kw = 1e-14; HA0 = 0.0001;
>> eq = [Kw == H*OH, Ka == H*A/HA, A + HA == HA0, H == A + OH];
>> format long
% approximate solution if finding exact/analytical solution fails[§§]
>> S_approx = vpasolve(eq);
% solution found symbolically; still useful to write out all steps
for future reference
>> H_approx = double(S_approx.H)
>> H_approx =
     3.437067254519952e-05
```

Simplification

The [H^+] of a weak acid is a familiar problem from secondary education and general chemistry textbooks, where simplifications are made in order to keep the maths manageable at the cost of reduced precision. Neglecting [OH^-] enables us to crack the problem without higher maths.

[‡‡] As of mid-2024, `root()` is not implemented in Octave. `root()` is not to be confused with `roots()` which also calculates the roots of polynomials; however, it does not take symbolic expressions as its input. It requires the coefficients (from highest to lowest order) of the terms passed as an array. We will use `roots()` in the appendix to predict titration curves.

[§§] In Octave, use instead `S_approx = vpasolve(eq, [H OH A HA], [1 1 1 1]*1e-10); H_approx = double(S_approx(1))`.

$$K_a = \frac{[\text{H}^+][\text{A}^-]}{[\text{HA}]} \qquad \text{HA} \rightleftharpoons \text{H}^+ + \text{A}^-$$

$$[\text{A}^-] + [\text{HA}] = [\text{HA}]_0$$

$$[\text{H}^+] = [\text{A}^-]$$

with

$$[\text{HA}] = [\text{HA}]_0 - [\text{A}^-]$$

$$K_a = \frac{[\text{H}^+]^2}{[\text{HA}]_0 - [\text{H}^+]} \Rightarrow K_a[\text{HA}]_0 = [\text{H}^+]^2 + K_a[\text{H}^+]$$

which rearranges to the quadratic equation

$$[\text{H}^+]^2 + K_a[\text{H}^+] - K_a[\text{HA}]_0 = 0$$

As before, we put the left-hand side of the expression into a loop to trace out its graph by varying $[\text{H}^+]$. For $K_a = 1.8 \times 10^{-5}$ and analytical concentration $[\text{CH}_3\text{COOH}]_0 = 0.0001 \text{ mol dm}^{-3}$, we obtain the graph seen in Figure 12.3.

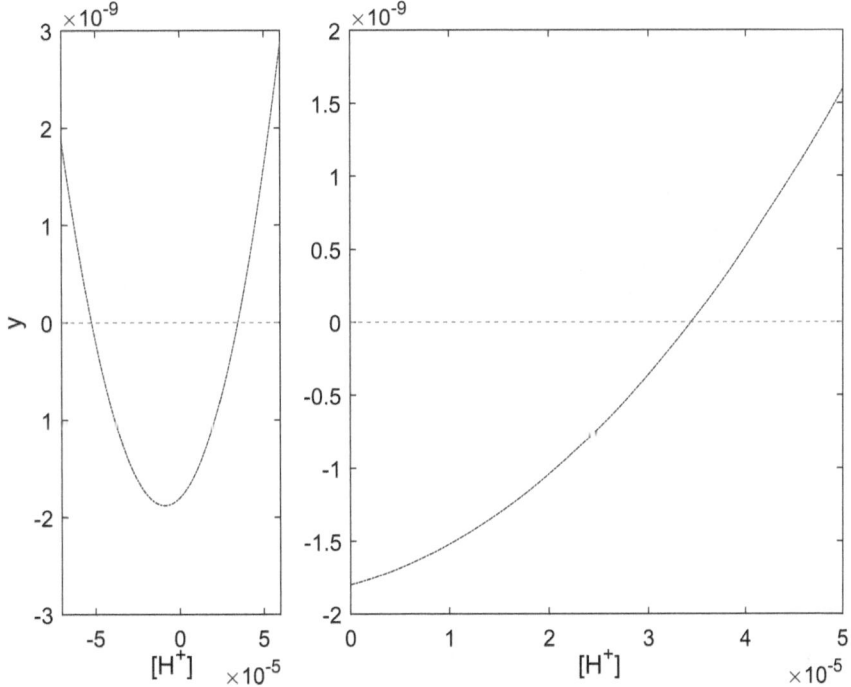

Figure 12.3 Graph of the quadratic expression derived (while making one simplification) for the acid concentration in an acetic acid solution with analytical concentration $[\text{CH}_3\text{COOH}]_0 = 0.0001 \text{ mol dm}^{-3}$.

If we want MATLAB to use the quadratic formula and solve exactly our quadratic equation resulting from the simplification made by neglecting [OH$^-$], we need to write

```
>> syms H HA HA0 Ka positive real
>> Ka=1.8e-05; HA0=0.0001;
>> H=root(H^2 + Ka*H - HA0*Ka==0);
>> format long
>> double(H)
ans =
   1.0e-04 *
   0.343704968844029
  -0.523704968844029
```

meaning that [H$^+$] = 3.43704968844029 × 10^{-5} mol dm^{-3} (pH = 4.463814189346787), which is very close to the exact solution of the non-simplified problem.

Further simplification can be made by assuming that [HA] $\cong$ [HA]$_0$ if the degree of dissociation is small. This, along with [H$^+$] = [A$^-$], leads to

$$K_a = \frac{[H^+]^2}{[HA]_0}$$

which yields the formula for estimating [H$^+$]:

$$[H^+] = \sqrt{K_a [HA]_0}$$

With $K_a = 1.8 \times 10^{-5}$ and analytical concentration [CH$_3$COOH]$_0$ = 0.0001 mol dm^{-3}, we obtain [H$^+$] = 4.24264068711929 × 10^{-5} mol dm^{-3} (pH = 4.372363747448347) which is still reasonably close to the exact [H$^+$] concentration.

The choice between trying to find the exact solution, approximating it, or making simplifications needs to be made depending on the nature of the problem and how accurately we need to have the solution. The pH of a household cleaning product can safely vary in a much broader range than the pH of an intravenous fluid; therefore, it is likely that different approaches are required during the formulation process to predict the pH of the final product in these cases.

Exercise 12.7

Calculate [H$^+$] if we were to use a diprotic acid in the formulation. The equations describing this chemical system are given below. The 2 appearing in the charge balance equation accounts for each A^{2-} anion carrying two negative charges.

$$K_w = [H^+][OH^-] \qquad H_2O \rightleftharpoons H^+ + OH^-$$

$$K_{a1} = \frac{[H^+][HA^-]}{[H_2A]} \qquad H_2A \rightleftharpoons H^+ + HA^-$$

$$K_{a2} = \frac{[H^+][A^{2-}]}{[HA^-]} \qquad HA^- \rightleftharpoons H^+ + A^{2-}$$

$$[A^{2-}] + [HA^-] + [H_2A] = [H_2A]_0$$

$$[H^+] = [OH^-] + [HA^-] + 2[A^{2-}]$$

a. Using MATLAB, show that the concentration of H^+ is given by the quartic equation:

$$[H^+]^4 + K_{a1}[H^+]^3 + (K_{a1}K_{a2} - K_{a1}[H_2A]_0 - K_w)[H^+]^2 - (2K_{a1}K_{a2}[H_2A]_0 + K_wK_{a1})[H^+]$$

$$- K_wK_{a1}K_{a2} = 0$$

(You are encouraged to derive this by hand as well.)

b. Vary $[H^+]$ and plot the graph of the expression for a potential formulation proposed to contain carbonic acid at analytical concentration $[H_2CO_3]_0 = 0.0001$ mol dm^{-3} ($K_{a1} = 4.2 \times 10^{-7}$, $K_{a2} = 5.6 \times 10^{-11}$).

c. Calculate the roots of the quartic equation and the H^+ concentration of the formulation if the analytical concentration of carbonic acid is $[H_2CO_3]_0 = 0.0001$ mol dm^{-3}.

d. (i) Approximate the solution of, and (ii) solve exactly the equations describing the carbonic acid system for the above analytical concentration and hence obtain the H^+ concentration.

Answer:

a. This derivation is similar to how we obtained the cubic formula for $[H^+]$ of a solution of a weak monoprotic acid.

```
>> clear all  % clear Workspace
>> syms H OH H2A HA A H2A0 Ka1 Ka2 Kw positive real  % declare symbolic
variables
>> OH = solve(Kw == H*OH, OH)    % rearrange Kw = H * OH for OH
OH =
Kw/H
>> HA = solve(Ka2 == H*A/HA, HA)  % rearrange Ka2 = H*A/HA for HA
HA =
(A*H)/Ka2
>> H2A = solve(Ka1 == H*HA/H2A, H2A)  % rearrange Ka1 = H*HA/H2A for H2A
H2A =
(A*H^2)/(Ka1*Ka2)
>> A = solve(A + HA + H2A == H2A0, A)  % rearrange A + HA + H2A == H2A0
for A
A =
```

```
(H2A0*Ka1*Ka2)/(H^2 + Ka1*H + Ka1*Ka2)
>> eq1=OH + HA +2*A - H==0% input charge balance using H, A, H2A0
eq1 =
Kw/H - H + (A*H)/Ka2 + (2*H2A0*Ka1*Ka2)/(H^2 + Ka1*H + Ka1*Ka2) ==0
```

Because we have a new expression for A containing H2A0 and H, we re-substitute the expressions for the variables (most importantly the new expression for A) into eq1 using subs().

```
>> eq2=subs(eq1) % substitute into eq1 to contain A now defined with H2A0 and H
eq2 =
Kw/H - H + (H*H2A0*Ka1)/(H^2 + Ka1*H + Ka1*Ka2) + (2*H2A0*Ka1*Ka2)/(H^2 + Ka1*H + Ka1*Ka2) ==0
```

As this equation does not contain A, we continue by multiplying out all expressions using the expand() function, which usually yields better outcomes in the subsequent simplification step.

```
>> eq3=simplify(expand(eq2)) % carry out all multiplications then simplify
eq3 =
Kw*(H^2 + Ka1*H + Ka1*Ka2) + H^2*H2A0*Ka1 + 2*H*H2A0*Ka1*Ka2 ==H^2*(H^2 + Ka1*H + Ka1*Ka2)
>> eq4=rhs(eq3) - lhs(eq3) ==0% rearrange for zero
eq4 =
H^2*(H^2 + Ka1*H + Ka1*Ka2) - Kw*(H^2 + Ka1*H + Ka1*Ka2) - H^2*H2A0*Ka1 - 2*H*H2A0*Ka1*Ka2 ==0
>> eq5=collect(eq4,H) % rewrite eq4 into powers of H
eq5 =
H^4 + Ka1*H^3 + (Ka1*Ka2 - H2A0*Ka1 - Kw)*H^2 + (- Ka1*Kw - 2*H2A0*-Ka1*Ka2)*H - Ka1*Ka2*Kw ==0
```

which is the same as the quartic equation given in the problem.

b.

```
>> Ka1=4.2e-7; Ka2=5.6e-11; Kw=1e-14; H2A0=0.0001;
% constants
>> H=-1e-5 : 1e-7 : 1e-5; % create values for H
>> y=H.^4 + Ka1.*H.^3 + (Ka1*Ka2 - H2A0*Ka1 - Kw).*H.^2 + (- Ka1*Kw - 2*H2A0*Ka1*Ka2).*H - Ka1*Ka2*Kw; % compute quartic expression for H values
>> subplot(1,3,1) % define figure layout and move to left panel (subplot 1)
>> plot([H(1) H(end)],[0 0],'-k',H,y) % plot dashed baseline and quartic exp.
```

```
>> axis([-1e-5 1e-5 -2e-21 2e-21]) % set axis limits
>> xlabel('[H^+]'); ylabel y % set axis labels
>> subplot(1,3,[2 3]) % in figure, move to right panel spanning across subplot areas 2 and 3
>> plot([H(1) H(end)],[0 0],'-k',H,y) % plot dashed baseline and quartic exp.
>> axis([0 7e-6 -4.5e-22 4.5e-22]) % set axis limits
>> xlabel('[H^+]') % set axis label
```

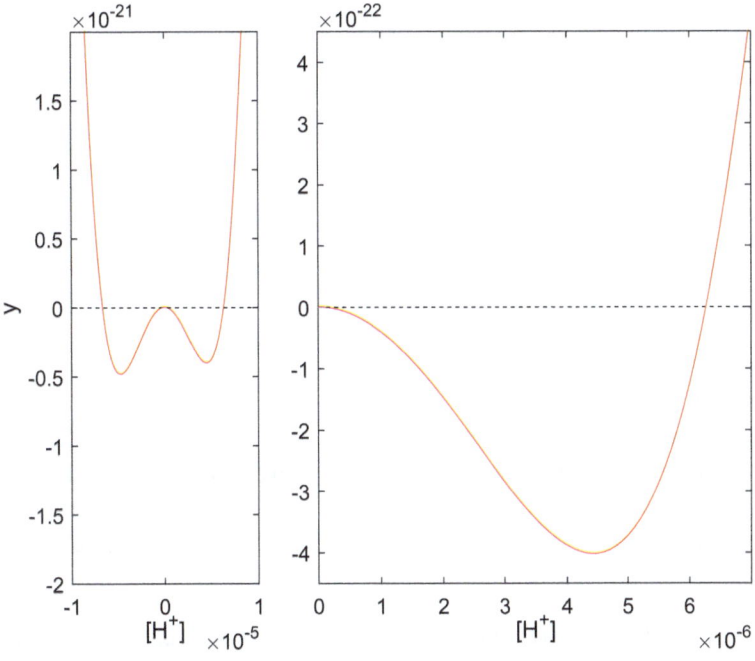

c.
```
>> clear all % clear Workspace
>> syms H positive real % declare symbolic variable
>> Ka1=4.2e-7; Ka2=5.6e-11; Kw=1e-14; H2A0=0.0001; % constants
>> H=root(H^4 + Ka1*H^3 + (Ka1*Ka2 - H2A0
*Ka1 - Kw)*H^2 + (- Ka1*Kw - 2*H2A0*Ka1*Ka2)*H - Ka1*Ka2*Kw==0);
>> format long
>> H=double(H)
H =
  1.0e-05 *
 -0.000003092821776
 -0.000018102104375
  0.627502085157800
 -0.669480890231649
```

$[H^+] = 6.275020851578003 \times 10^{-6}$ mol dm^{-3} (pH = 5.202384826705853)

d. (i)

```
>> clear all % clear Workspace
>> syms H OH H2A HA A positive real % declare symbolic variables
>> Ka1 = 4.2e-7; Ka2 = 5.6e-11; Kw = 1e-14; H2A0 = 0.0001; % constants
>> eq = [Kw == H*OH, Ka1 == H*HA/H2A, Ka2 == H*A/HA, A + HA + H2A == H2A0, H == OH + HA + 2*A];
>> S_approx = vpasolve(eq);
>> H_approx = double(S_approx.H)
H_approx =
    6.275020851578002e-06
```

d. (ii)
Same as the above but replace

```
>> S_approx = vpasolve(eq);
```

with

```
>> S = solve(eq);
```

When trying to directly solve the equations describing the carbonic acid system, MATLAB returns a warning that basically tells us that the exact solution is too complicated to be condensed into a set of straightforward expressions. MATLAB essentially fails to formulate and find the exact solution of this problem without some guidance. This is, however, not an issue because we were able to obtain the approximate solution.

We have seen that the exact [H$^+$] for a monoprotic (weak) acid is calculated by solving a cubic equation. For a diprotic acid, a quartic equation has to be formulated and solved, which is possible but would be far more cumbersome than solving a cubic equation, because the quartic formula is so large that it would probably take up an entire page. A triprotic acid would require a quintic polynomial to be solved, and so on. The trouble is that quintic polynomials cannot be solved analytically; no general formulae exist to solve them or higher order polynomials. Consequently, the value of [H$^+$] in such important systems as phosphoric acid solutions (H$_3$PO$_4$; pK_1 = 1.96, pK_2 = 7.12, pK_3 = 12.3; present in a multitude of biological systems including us) and citric acid solutions (HOC(CO$_2$H)(CH$_2$CO$_2$H)$_2$; pK_1 = 3.13, pK_2 = 4.76, pK_3 = 6.39, pK_4 = 14.4; naturally occurring in citrus fruits and a widely-used ingredient in food and cleaning products) can only be obtained through numerical approximation. This puts finding the exact concentration of H$^+$ in a diprotic acid solution at the edge of what is mathematically possible *via* analytical methods. Therefore, MATLAB's failure to find the exact [H$^+$] for a carbonic acid solution should not surprise us. Luckily, MATLAB has sophisticated algorithms to approximate solutions numerically, which are routinely used to calculate the equilibrium concentrations of chemical species in complex real systems such as blood, cytoplasm and aquafers containing multiple salts, acids and bases.

12.6 Appendix B: Bisection Method

By now we should start feeling comfortable with writing small MATLAB scripts to solve chemistry problems that are a bit more difficult than introductory textbook examples. Throughout this book, we have been building the key skill of taking a chemistry problem and translating it into a maths problem followed by turning that into a set of commands in order to solve it in a programming environment. Along the way we have been heavily relying on MATLAB's high-level functions working in the background to execute the tasks we set. Have you wondered, for example, how MATLAB is able to approximate the concentration of H^+ for a carbonic acid solution?

Many algorithms have been developed to efficiently approximate the solutions of problems like computing the inverse of a matrix, integrating differential equations or finding the roots of an equation. MATLAB uses a variety of these algorithms, automatically choosing the one most suitable for a particular problem.

Now we are going to write a simple algorithm to approximate the concentration of H^+ for a carbonic acid solution, which essentially is a root finding problem. Our approach will be less advanced than the algorithms MATLAB uses for this purpose. However, it will follow the same general principle that all iterative root finding algorithms share.

Iterative methods start from an initial guess of what the solution to the problem could be and keep refining it until the solution is not worth refining any further because the computational cost outweighs the usefulness of subsequent improvements in the solution, or the refinement is stopped when the required precision is reached. In our scenario, this means that it is reasonable not to continue improving the approximate concentration of H^+ of a carbonic acid solution by doing further iterations beyond what we can reliably measure with a high-quality pH probe. In terms of the initial guess, we can use the simplest formula for estimating $[H^+]$ as $\sqrt{K_a[H_2CO_3]_0}$ which neglects $[OH^-]$, the second deprotonation and even the dissociation of carbonic acid, but it is quick and easy to compute.

Once we have the initial guess and the threshold for stopping the iterative approximation, we need to pick a root finding method and code it up. Here we choose the *Bisection Method* to approximate $[H^+]$ of a solution. Steps taken in each iteration are shown in Figure 12.4.

In each iteration (i), we will calculate the value of the polynomial given in Exercise 12.6 at two $[H^+]$ values, $f(^{(i)}H_1)$ and $f(^{(i)}H_2)$, and halfway in between them, $f(^{(i)}H_h)$. Then, we will need to find out which half interval, $[^{(i)}H_1, {}^{(i)}H_h]$ or $[^{(i)}H_h, {}^{(i)}H_2]$, contains the root (where the polynomial is zero). We will do this by computing $f(^{(i)}H_1) \times f(^{(i)}H_h)$ and $f(^{(i)}H_h) \times f(^{(i)}H_2)$, and check, using an *if-statement*, which product is negative. A product is negative when the quantities multiplied together have opposite signs. Here, this means that the polynomial must go through zero within the half interval for which the product is negative.

$f(^{(i)}H_1) \times f(^{(i)}H_h) < 0$	root within $[^{(i)}H_1, {}^{(i)}H_h]$ sub-domain	$^{(i+1)}H_2 = {}^{(i)}H_h$	$^{(i+1)}H_1 = {}^{(i)}H_1$
$f(^{(i)}H_h) \times f(^{(i)}H_2) < 0$	root within $[^{(i)}H_h, {}^{(i)}H_2]$ sub-domain	$^{(i+1)}H_1 = {}^{(i)}H_h$	$^{(i+1)}H_2 = {}^{(i)}H_2$

In iteration $(i+1)$, we redo the process in exactly the same way but now using the new boundaries $^{(i+1)}H_1$ and $^{(i+1)}H_2$ defining a domain half as wide as the domain in

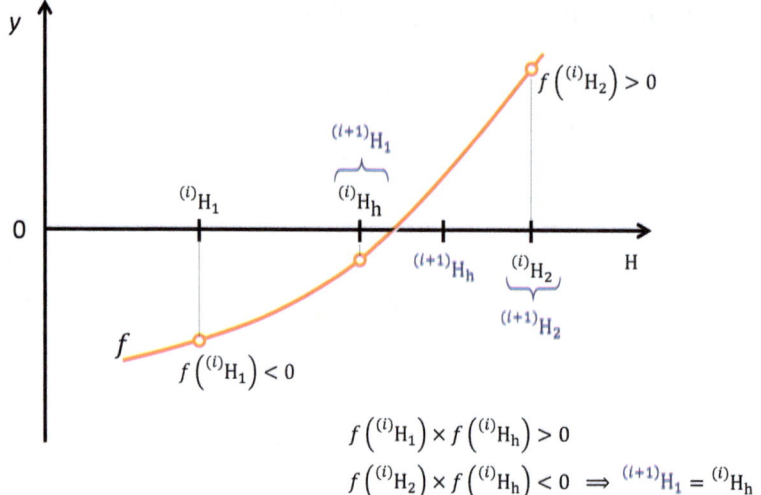

Figure 12.4 Steps in an iteration of the Bisection root-finding method. Interval $[^{(i)}H_1, ^{(i)}H_2]$ is chosen to contain the root of the function $f(H)$, the location on the horizontal axis where $f(H)$ goes through zero (the value of H where $f(H)=0$). The interval is divided into two halves, followed by deciding which half contains the root. The boundaries of the chosen half $[^{(i+1)}H_1, ^{(i+1)}H_2]$ are then assigned to become $[^{(i)}H_1, ^{(i)}H_2]$, the starting points of the next iteration involving the same steps. In each iteration the length of the domain containing the root halves, and once it goes below a pre-set threshold the code exits the iterating loop.

iteration (i). Note that (i) and $(i+1)$ have been used purely to separate two subsequent iterations in the discussion. In the code below, each iteration is computed by the same repeating block of commands inside the *while-loop* (see Section 10.1.2). The iterative process will repeat until the distance between H_1 and H_2 reaches a predefined threshold. When that happens, we take the mid-point between H_1 and H_2 as the approximate value for $[H^+]$ of a carbonic acid solution with analytical concentration 0.0001 mol dm^{-3}. Here we use 10^{-14} mol dm^{-3} for the threshold, which is lower than what pH probes can reliably measure.

The MATLAB script below approximates $[H^+]$ of the carbonic acid solution with $[H_2CO_3]_0 = 0.0001$ mol dm^{-3}

```
clear all
format longG
Ka1=4.2e-7; Ka2=5.6e-11; Kw=1e-14; H2A0=0.0001;
f=@(H) H^4 + Ka1*H^3 + (Ka1*Ka2 - Kw - H2A0*Ka1)*H^2 - (2*H2A0*
Ka1*Ka2 + Ka1*Kw)*H - Kw*Ka1*Ka2; % one-line function to evaluate
quartic polynomial at a certain [H+]
th=1e-14; % threshold: size of domain containing [H+] of carbonic acid
solution
H0=sqrt(Ka1*H2A0) % initial guess (using simplest formula to estimate
[H+])
d=H0/10; % size of domain assumed to contain root ([H+] of carbonic acid
solution)
```

```
i = 0; % current iteration number
H1 = H0 - d/2; % set low end of domain assumed to contain the root
H2 = H0 + d/2; % set high end of domain assumed to contain the root
disp([i H1 H2 d]) % output information into Command Window
while d > th
    i = i + 1; % increase iteration number
    Hh = (H1 + H2)/2; % calculate mid-point of domain containing the root
    fh = f(Hh); % compute value of quartic polynomial at Hh
    f1 = f(H1); % compute value of quartic polynomial at H1
    f2 = f(H2); % compute value of quartic polynomial at H2
    if fh*f1 < 0% if true, fh and f1 have opposite signs => root between Hh and H1
        H2 = Hh; % replace H2 with Hh
    else % if false, fh and f2 must have opposite signs => root between Hh and H2
        H1 = Hh; % replace H1 with Hh
    end
    d = abs(H2 - H1); % calculate size of domain assumed to contain root
    iter(i,:) = [i H1 H2 d]; % load iteration step details into array "iter"
end
disp(iter) % output details of iterations into Command Window
H_approx = (H1 + H2)/2% output approximate value of [H+] into Command Window
```

and returns the following output:

```
H0 =
        6.48074069840786e-06
 0   6.15670366348747e-06   6.80477773332825e-06   6.48074069840786e-07
 1   6.15670366348747e-06   6.48074069840786e-06   3.24037034920393e-07
 2   6.15670366348747e-06   6.31872218094766e-06   1.62018517460196e-07
 3   6.23771292221757e-06   6.31872218094766e-06   8.10092587300976e-08
 4   6.23771292221757e-06   6.27821755158261e-06   4.05046293650488e-08
 5   6.25796523690009e-06   6.27821755158261e-06   2.02523146825248e-08
 6   6.26809139424135e-06   6.27821755158261e-06   1.01261573412628e-08
 7   6.27315447291198e-06   6.27821755158261e-06   5.06307867063184e-09
 8   6.27315447291198e-06   6.27568601224730e-06   2.53153933531550e-09
 9   6.27442024257964e-06   6.27568601224730e-06   1.26576966765775e-09
10   6.27442024257964e-06   6.27505312741347e-06   6.32884833828451e-10
11   6.27473668499656e-06   6.27505312741347e-06   3.16442416913802e-10
12   6.27489490620501e-06   6.27505312741347e-06   1.58221208456901e-10
13   6.27497401680924e-06   6.27505312741347e-06   7.91106042284504e-11
14   6.27501357211135e-06   6.27505312741347e-06   3.95553021142252e-11
15   6.27501357211135e-06   6.27503334976241e-06   1.97776510566891e-11
```

```
16   6.27501357211135e-06   6.27502346093688e-06  9.88882552792103e-12
17   6.27501851652412e-06   6.27502346093688e-06  4.944412763537e-12
18   6.27501851652412e-06   6.2750209887305e-06   2.4722063817685e-12
19   6.27501975262731e-06   6.2750209887305e-06   1.23610319046073e-12
20   6.27502037067891e-06   6.2750209887305e-06   6.18051595653883e-13
21   6.2750206797047e-06    6.2750209887305e-06   3.09025797403425e-13
22   6.2750208342176e-06    6.2750209887305e-06   1.54512898278196e-13
23   6.2750208342176e-06    6.27502091147405e-06  7.72564487155816e-14
24   6.2750208342176e-06    6.27502087284583e-06  3.86282243577908e-14
25   6.2750208342176e-06    6.27502085353171e-06  1.93141121788954e-14
26   6.27502084387466e-06   6.27502085353171e-06  9.65705566593122e-15
H_approx =
   6.27502084870319e-06
```

where H0 = 6.48074069840786e-06 is our initial guess, followed by rows representing each subsequent iteration where the columns are: (1) iteration number, (2) H_1, (3) H_2, and (4) width of the domain containing [H^+] which represents the level of precision we have in [H^+]. It takes our rudimentary algorithm 26 iterations to refine our initial educated guess on [H^+] to the prescribed level of precision of 10^{-14} mol dm^{-3}, where our approximate [H^+] is 6.27502084870319e-06. For comparison, vpasolve() returned 6.275020851578002e-06, making the difference between the two approximate values about 3×10^{-15} mol dm^{-3} which is in line with the set threshold of 10^{-14} mol dm^{-3} and would not be measurable experimentally. If, for some reason, higher precision was required, we would have to lower th, which would increase the number of iterations and consequently the runtime.

Appendix

Predicting Titration Curves

Following on from Section 12.5, where we discussed using code to calculate the pH of a weak monoprotic acid (HA), we will look here at how to predict the titration curve when it is titrated with a strong base (MOH).

Because the base fully dissociates

$$\text{MOH} \rightarrow \text{M}^+ + \text{OH}^-$$

thus

$$[\text{MOH}]_0 = [\text{M}^+]$$

and accounting for the change in the analytic concentrations of the acid ($[\text{HA}]_0$) and base ($[\text{MOH}]_0$) in the equations describing the system, which results from the changing volume due to the base being gradually added to the flask, we will have the following simultaneous equations so solve:

$$K_w = [\text{H}^+][\text{OH}^-] \qquad \text{H}_2\text{O} \rightleftharpoons \text{H}^+ + \text{OH}^-$$

$$K_a = \frac{[\text{H}^+][\text{A}^-]}{[\text{HA}]} \qquad \text{HA} \rightleftharpoons \text{H}^+ + \text{A}^-$$

$$[\text{A}^-] + [\text{HA}] = \frac{[\text{HA}]_0 \, V_{\text{HA}}}{V_{\text{HA}} + V_{\text{MOH}}}$$

$$[\text{H}^+] + \frac{[\text{MOH}]_0 \, V_{\text{MOH}}}{V_{\text{HA}} + V_{\text{MOH}}} = [\text{OH}^-] + [\text{A}^-]$$

A First Look at Coding in Chemistry: Solving Problems Using MATLAB
By Tamas Bansagi
© Tamas Bansagi 2025
Published by the Royal Society of Chemistry, www.rsc.org

where expressions $[HA]_0 V_{HA}/(V_{HA}+V_{MOH})$ and $[MOH]_0 V_{MOH}/(V_{HA}+V_{MOH})$ represent the analytic acid and base concentrations after dilution, with V_{HA} and V_{MOH} denoting the volume of acid and base, respectively. In the charge balance equation $[MOH]_0 V_{MOH}/(V_{HA}+V_{MOH})$ represents $[M^+]$.

We make the following rearrangements and substitutions

$$[OH^-] = \frac{K_w}{[H^+]}$$

$$[HA] = \frac{[HA]_0 V_{HA}}{V_{HA}+V_{MOH}} - [A^-]$$

$$K_a = \frac{[H^+][A^-]}{\frac{[HA]_0 V_{HA}}{V_{HA}+V_{MOH}} - [A^-]} \Rightarrow [A^-] = \frac{K_a [HA]_0 V_{HA}}{([H^+]+K_a)(V_{HA}+V_{MOH})}$$

$$[H^+] = \frac{K_w}{[H^+]} + \frac{K_a [HA]_0 V_{HA}}{([H^+]+K_a)(V_{HA}+V_{MOH})} - \frac{[MOH]_0 V_{MOH}}{V_{HA}+V_{MOH}}$$

$$\left([H^+] - \frac{K_w}{[H^+]}\right)([H^+]+K_a) = \frac{K_a [HA]_0 V_{HA} - [MOH]_0 V_{MOH}([H^+]+K_a)}{(V_{HA}+V_{MOH})}$$

$$\left([H^+]^2 + K_a[H^+] - K_w - \frac{K_a K_w}{[H^+]}\right)(V_{HA}+V_{MOH}) + [MOH]_0 V_{MOH}[H^+] = K_a [HA]_0 V_{HA} - [MOH]_0 V_{MOH} K_a$$

$$\left([H^+]^3 + K_a[H^+]^2 - K_w[H^+] - K_a K_w\right)(V_{HA}+V_{MOH}) + [MOH]_0 V_{MOH}[H^+]^2$$

$$+ K_a([MOH]_0 V_{MOH} - [HA]_0 V_{HA})[H^+] = 0$$

which rearranges to the cubic equation

$$[H^+]^3 + \left(K_a + \frac{[MOH]_0 V_{MOH}}{V_{HA}+V_{MOH}}\right)[H^+]^2 + \left(\frac{K_a([MOH]_0 V_{MOH} - [HA]_0 V_{HA})}{V_{HA}+V_{MOH}} - K_w\right)[H^+] - K_w K_a = 0$$

enabling us to calculate the titration curves for a monoprotic weak acid titrated with a strong base from first principles.

Appendix

Table A.1 Acidities of selected weak acids.

Acid	Formula	pK_a
Pyruvic acid	$CH_3COCOOH$	2.50
Lactic acid	C_2H_5OCOOH	3.86
Acetic acid	CH_3COOH	4.75
Dimethylhydroxylamine	$(CH_3)_2NOH$	5.20
Imidazole	$C_3N_2H_4$	6.95
Phthalimide	$C_6H_4(CO)_2NH$	8.30
Phenol	C_6H_5OH	9.89

```
% predict titration curve for pyruvic acid
pKa = 2.5;
Kw = 1e-14;
HA0 = 0.1; % analytical concentration of acid in mol/dm^3
MOH0 = 0.1; % analytical concentration of base in mol/dm^3
VHA = 0.01; % volume of acid in flask in dm^3
VMOH = 0:1e-5:0.02; % emulating dropwise addition of 0.02 dm^3 base to
flask
Ka = 10^-(pKa);
for i = 1:length(VMOH)
    a = 1;
    b = (Ka + MOH0*VMOH(i)./(VHA + VMOH(i)));
    c = (Ka*(MOH0*VMOH(i) - VHA*HA0)./(VHA + VMOH(i)) - Kw);
    d = -Kw*Ka; % a and d could go above the loop
    r = roots([a b c d]); % find roots of cubic polynomial (We could have used
instead the methods introduced in Section 12.5.)
    H(i) = r(r > 0); % take positive root; in "r > 0" relational operator >
compares each of the three roots calculated by roots() and returns the index
of the positive root; the number stored in r corresponding to that index is
then assigned to H(i), the H+ ion concentration for the volume of base added
specified as the ith element of VMOH
end
plot(VMOH,-log10(H),'-r','Linewidth',2)
xlabel('V_{MOH}/dm^{-3}')
ylabel('pH')
legend('0.1 M pyruvic acid titrated with 0.1 M NaOH','Location',
'southeast')
ylim([1 14])
grid on
```

Using Table A.1, we obtain the titration curves seen in Figure A.1. As the pK_a increases, the pH at the stoichiometric point (the volume of base at the steepest incline) moves up. This is why the acid–base indicator used when the titration is followed visually needs to be chosen with the pK_a of the weak acid in mind, so that the pH range that the colour change of the indicator occurs overlaps the pH jump in the titration.

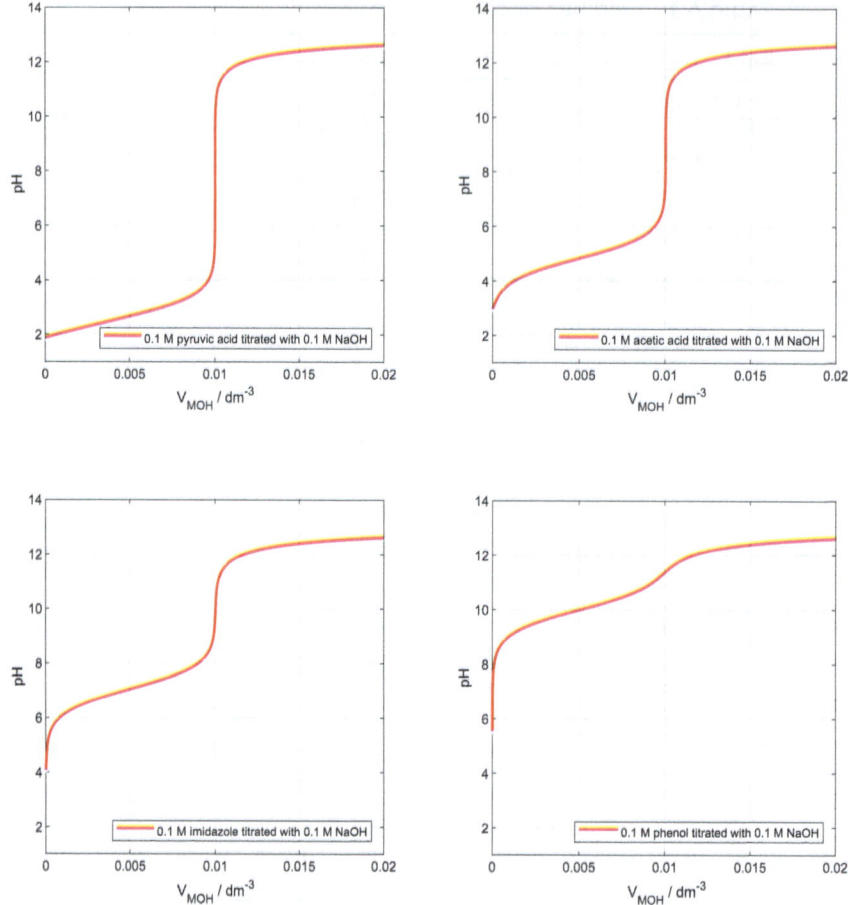

Figure A.1 Titration curves calculated for some weak acids. The pH traces can be verified *via* monitoring the progress of the titrations by a pH electrode (which is called potentiometric titration).

Exercise A.1

Using the volume of the acetic acid solution, the concentration of the standard base solution and the stoichiometric point determined, predict the titration curve for Exercise 4.2.

Answer:

With 0.010 M NaOH solution added in 1.0 cm³ amounts to titrate 50 cm³ of acetic acid solution resulting in a stoichiometric point of 26.6 cm³ and using that at the stoichiometric point

$$n_{HA} = n_{MOH}$$
$$c_{HA} V_{HA} = c_{MOH} V_{MOH}$$

yielding

$$c_{HA} = \frac{c_{MOH} V_{MOH}}{V_{HA}}$$

the analytical concentration of acetic acid is $[HA]_0 = 0.01$ M 26.6 cm^3/50 cm$^3 = 0.0053$ M. With that, we amend the code above to:

```
% predict titration curve for Exercise 4.2
V_exp=0:0.001:0.05; % volume of NaOH solution added in dm^3
pH_exp=[3.21 3.39 3.54 3.68 3.75 3.86 3.96 4.07 4.14 4.21
4.29 4.36 4.39 4.46 4.54 4.57 4.61 4.68 4.75 4.86 4.93 5.04
5.14 5.29 5.46 5.82 6.71 8.57 10.14 10.43 10.54 10.64 10.75
10.79 10.86 10.89 10.93 10.96 11.00 11.04 11.07 11.11 11.11
11.14 11.14 11.18 11.18 11.21 11.21 11.25 11.25]; % pH values
measured during titration
pKa=4.75; % pKa of acetic acid
Kw=1e-14;
HA0=0.0053; % analytical concentration of acetic acid in mol/dm^3
MOH0=0.01; % analytical concentration of NaOH in mol/dm^3
VHA=0.050; % volume of acetic acid in flask in dm^3
VMOH=0:0.0001:0.05; % emulating dropwise addition of 0.05 dm^3 base to
flask
Ka=10^-(pKa);
for i=1:length(VMOH)
    a=1;
    b=(Ka+MOH0*VMOH(i)./(VHA+VMOH(i)));
    c=(Ka*(MOH0*VMOH(i) - VHA*HA0)./(VHA+VMOH(i)) - Kw);
    d=-Kw*Ka; % a and d could go above the loop
    r=roots([a b c d]); % find roots of cubic polynomial
    H(i)=r(r>0); % take positive root; in "r>0" relational operator>
compares each of the three roots calculated by roots() and returns the
index of the positive root; the number stored in r corresponding to that
index is then assigned to H(i), the H+ ion concentration for the volume of
base added specified as the ith element of VMOH
end
plot(VMOH,-log10(H),'-r','Linewidth',2)
hold on
plot(V_exp,pH_exp,'sb')
xlabel('V_{MOH}/dm^{-3}')
ylabel('pH')
legend('0.0053 M acetic acid titrated with 0.01 M NaOH',
'Experiemntal data from Exercise 4.2','Location','southeast')
ylim([2 13])
grid on
```

which generates the output seen below.

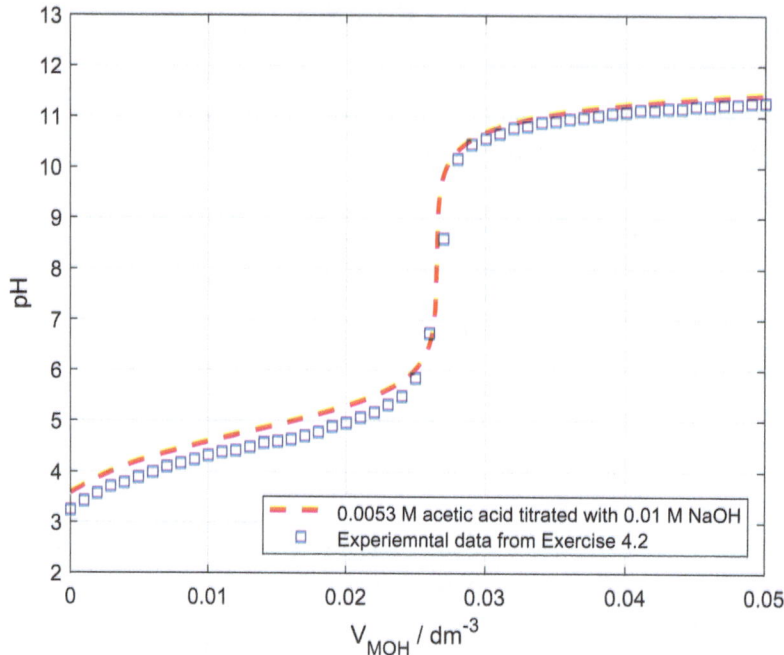

There appears to be a small but systematic difference between the measured and predicted pH traces, which indicates that the pH probe may have been not calibrated correctly, especially in the low pH range.

Kinetic Modelling in Elucidating Reaction Mechanisms

Investigating a new reaction involves elucidating the steps of the mechanism as well as establishing the rate equation for each step. We typically construct several plausible mechanisms, each composed of a set of (elementary) steps. The rate equations for the steps then form the basis of the kinetic model of the reaction. One way to choose the most likely mechanism is by comparing how well the concentrations predicted by the different kinetic models corresponding to the proposed mechanisms trace out the concentrations measured experimentally.

Here we are looking at a reaction where a compound is thought to turn into another *via* an intermediate difficult to isolate and trace its concentration. With A, B and C denoting the reactant, intermediate and product, we therefore have been only able to measure [A] and [C] throughout the reactions. We are considering two competing plausible mechanisms.

In Mechanism I, the reaction is proposed to occur through unimolecular steps

$$R1 \quad A \rightarrow B \quad r_1 = k_1[A]$$

$$R2 \quad B \rightarrow C \quad r_2 = k_2[B]$$

This would mean that both transition states involve only one species, thus the reaction most likely goes through (first order) consecutive internal rearrangements. This mechanism translates into the kinetic model

$$\frac{d[A]}{dt} = -k_1[A]$$

$$\frac{d[B]}{dt} = k_1[A] - k_2[B]$$

$$\frac{d[C]}{dt} = k_2[B]$$

In Mechanism II, the reaction is proposed to take place through bimolecular steps

$$R1 \quad A + A \rightarrow B \quad r_1 = k_1[A]^2$$

$$R2 \quad B + B \rightarrow C \quad r_2 = k_2[B]^2$$

which would mean that both transition states involve two species, hence the intermediate and product are formed in (second order) collisions. This mechanism leads to the kinetic model

$$\frac{d[A]}{dt} = -2k_1[A]^2$$

$$\frac{d[B]}{dt} = k_1[A]^2 - 2k_2[B]^2$$

$$\frac{d[C]}{dt} = k_2[B]^2$$

To generate predictions for the concentrations we need to integrate these sets of differential rate equations (for details please see Chapter 9.) This would require the initial concentrations and the rate constants. However, when investigating a new reaction, we only have the initial concentrations. Without the rate constants, we cannot generate predictions for the competing mechanisms and consequently we are unable to decide which proposed mechanism is more likely to describe correctly the set of events taking place during the reaction.

What we can do is guess the rate constants, integrate the rate equations numerically using ODE solvers, and compare the predicted and measured concentrations by calculating the sum of squared residuals. Having looked at the results, we make an improved guess for the rate constants which will yield a new set of predicted concentrations and sum of squared residuals. We will repeat this iterative process until the fit between the measured and predicted concentrations stops improving, where the sum of squared residuals (a measure of the goodness of fit) reaches its lowest possible value (*i.e.* its global minimum). This set of rate constants will ensure the best fit for the mechanism considered. The other mechanism will also need to be evaluated in the same way, and once we have the rate constants ensuring the best fit for that model we can decide which mechanism performs better. The mechanism more likely to capture the sequence of chemical events correctly will not only produce a lower sum of squared residuals but also its residuals will scatter closely around zero without showing any discernible trends. Trends in residuals indicate that some of the steps in the mechanism are not correct (*i.e.* the chemistry is different) and/or the mechanism is incomplete which may be due to some steps and/or species being left out.

The good news is that we do not have to go through with these iterations as MATLAB has a function that we can deploy to do the search for the rate constants ensuring the lowest sum of squared residuals and hence the best fit to the experimental data. We only need to supply MATLAB's `fminsearch()` function with (i) what we wish to minimise, here the sum of squared residuals,[†] and (ii) initial guesses for the parameters determining the value of what we are minimising, the two rate constants in our case. The sum of squared residuals will need to be computed in each iteration `fminsearch()` makes, which in turn requires integrating the rate equations. Therefore, we create two user-defined functions `goodness_of_fit()`, calculating the sum of squared residuals, called from `fminsearch()`, and `net_rate_of_form()`, specifying the kinetic model fitted to the experimental data, called and integrated by an ODE solver inside `goodness_of_fit()`. For solving the ODEs and finding the minimum of the goodness of fit function

```
% Proposed reaction mechanism:
% R1  A→B  r1=k1*[A]
% R2  B→C  r2=k2*[B]
%
% Kinetic model for the mechanism:
% d[A]/dt=-k1*[A]
% d[B]/dt=k1*[A] - k2*[B]
% d[C]/dt=k2*[B]
%
% Experimental data: [A],[C]
%
% Aim: find rate constants k1,k2 for the mechanism by fitting the
% kinetic model to the experimental data

tp=0:2:20; % times concentrations measured at (in min)

% experimentally measured concentrations at times given in tp
A_msd=[1.0000  0.2512  0.0534  0.0294  0.0185  0.0188  0.0138  0.0446
       0.0291  0.0209  0.0264];
```

[†] The sum of squared residuals is a function whose values represent how well a set of rate constants for a mechanism fit the experimental data (the goodness of fit). Our aim is to have MATLAB change the values of the rate constants (for each proposed mechanism) until the overall minimum of this function is found. Problems involving changing parameters until an optimal outcome is reached, *i.e.* a minimum or maximum is found, are hugely important not only in chemistry but also in industry and economics (for example, when minimizing the cost and/or environmental impact of building and operating a chemical plant). The function in an optimization task that maps different combinations of parameter values to a number is called the *objective function*. Here we are trying to minimize our objective function, the sum of squared residuals, to find the best fit.

```
C_msd=[0.0000  0.0966  0.2231  0.3581  0.4926  0.5980  0.6337  0.7285
0.7561  0.8413  0.8754];

msd_data=horzcat(A_msd',C_msd'); % merge all measured concentrations
into a single array to be passed to goodness_of_fit()
sp=[1  3]; % species monitored experimentally (measured [A],[C] will be
compared against [A],[C] predicted by the kinetic model)

% Initial concentrations for the model:
A0=1.0; % mol/dm3
B0=0.0; % mol/dm3
C0=0.0; % mol/dm3

c0=[A0 B0 C0]; % store initial concentrations in a vector

% initial guess for the rate constants k1 [1/min] and k2 [1/min]
k0=[1  1];

op_ode=odeset('RelTol',1e-6,'AbsTol',1e-6,'NonNegative',1);
% set parameters for ODE solver
op_fmin=optimset('TolFun',1e-8,'TolX',1e-8,'MaxFunEvals',1000);
% set parameters for minimising residuals

% find best estimates for rate constants k1 and k2 (at which the sum of
square errors between measured and predicted concentrations is minimal)
k_est=fminsearch(@(k)goodness_of_fit(tp,msd_data,sp,c0,k,op_ode),
k0,op_fmin)

% plot results
tp_fine=linspace(tp(1),tp(end),1000); % generate more time points for
smoother predicted concentration-time curves
[t,c]=ode15s(@(t,c)net_rate_of_form(t,c,k_est),tp_fine,c0,op_ode);
% integrate model with estimated rate constants for plotting predicted
[A],[B],[C]
plot(t,c(:,1),'-r',t,c(:,2),'-b',t,c(:,3),':k',tp,A_msd,'rs',tp,
C_msd,'ks','LineWidth',2)
xlabel('time / min'); ylabel('Concentration / mol dm^{-3}')
legend('[A]_{prd}','[B]_{prd}','[C]_{prd}','[A]_{msd}','[C]_{msd}',
'Location','east')

[t,c]=ode15s(@(t,c) net_rate_of_form(t,c,k_est),tp,c0,op_ode);
% integrate model with estimated rate constants for plotting estimated
[A],[C]
c_sp=c(:,sp);
r=msd_data-c_sp;
figure
plot(t,r(:,1),':ro',t,r(:,2),':ko','LineWidth',2)
xlabel('time / min'); ylabel('residuals / mol dm^{-3}')
legend('r_A','r_C','Location','north')
grid on
% calculate sum of square errors (SSE) between measured and predicted
concentrations (goodness of fit)
```

```
function SSE = goodness_of_fit(tp,msd_data,sp,c0,k,op)
[t,c] = ode15s(@(t,c)net_rate_of_form(t,c,k),tp,c0,op); % predict all
concentrations by integrating the kinetic model
c_sp = c(:,sp); % load selected concentration-time values (here: [A],[C])
into an array
SSE = sum(sum((msd_data-c_sp).^2)); % compute SSE from residuals
(differences between measured and predicted concentrations (here: [A],[C])
end

% kinetic model (net rates of formation of species) generated from
mechanism in preamble
function dcdt = net_rate_of_form(t,c,k)
dAdt = -k(1)*c(1);
dBdt = k(1)*c(1)-k(2)*c(2);
dCdt = k(2)*c(2);
dcdt = [dAdt; dBdt; dCdt]; % collect rates to be returned as a vector
end
```

After saving the entire script (including the two functions at the bottom), executing (by pressing "Run", ►) will generate the following output.[‡]

```
k_est =
      0.6867    0.1039
```

are the estimated values for rate constants k_1 and k_2 in min^{-1}, and Figure A.2 showing the measured concentrations of A and B overlayed the predicted concentration-time profiles for all species computed with the best estimates for k_1 and k_2. The predicted concentrations (prd) track the measured (msd) concentrations and the residuals (measured−predicted) are scattered in a small range with no clear trends (Figure A.2).

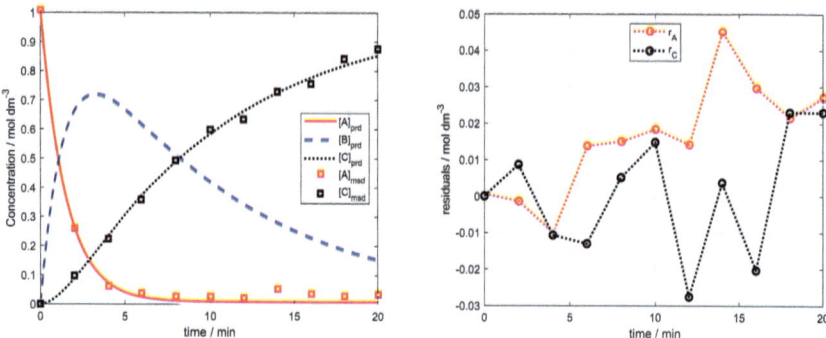

Figure A.2 (left) Measured and predicted concentrations for the A→B→C reaction assuming both steps being *first order*; (right) residuals.

[‡] On how to run this in Octave please refer to footnote on page 181.

Appendix

Assuming that both steps are second order, the above code will need to be slightly amended. First, to state the problem we are trying to solve, we change the preamble to

```
% Proposed reaction mechanism:
% R1  A+A→B  r1=k1*[A]^2
% R2  B+B→C  r2=k2*[B]^2
%
% Kinetic model for the mechanism:
% d[A]/dt = -2*k1*[A]^2
% d[B]/dt = k1*[A]^2 - 2*k2*[B]^2
% d[C]/dt = k2*[B]^2
%
% Experimental data: [A],[C]
%
% Aim: find rate constants k1,k2 for the mechanism by fitting the
% kinetic model to the experimental data
```

and accordingly we update the kinetic model in function net_rate_of_form() at the bottom of the code to

```
function dcdt = net_rate_of_form(t,c,k)
dAdt = -2*k(1)*c(1)^2;
dBdt = k(1)*c(1)^2 - 2*k(2)*c(2)^2;
dCdt = k(2)*c(2)^2;
dcdt = [dAdt; dBdt; dCdt]; % collect rates to be returned as a vector
end
```

(It might be a good idea to save the amended code under a different name to avoid confusion about which code fits which model.)

Running the code now generates the output

```
k_est =
       1.0e+07 *
       0.0000    1.1631
```

which is better displayed using the shortE number format. Typing

```
>> format shortE
>> k_est
```

results in

```
k_est =
       1.8433e+00        1.1631e+07
```

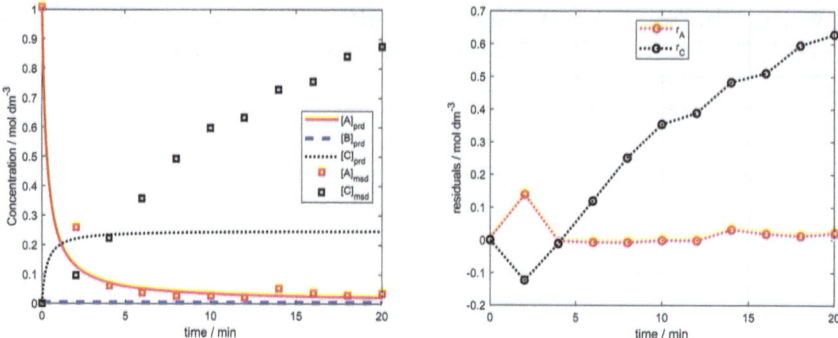

Figure A.3 (left) Measured and predicted concentrations for the A→B→C reaction assuming both steps being *second order*; (right) residuals.

Looking at these values ($k_1 = 1.8433$ dm^3 mol^{-1} min^{-1} and $k_2 = 1.1631 \times 10^7$ dm^3 mol^{-1} min^{-1}) and the graphic output shown in Figure A3, where the predicted concentrations are poorly following the measured concentrations and the residuals are not only much larger, but they also show trends indicating that the proposed model does not capture the pattern of the experimental data, we conclude that Mechanism I is much more likely. (It is possible that MATLAB failed to find the best rate constant values for Mechanism II because `k0 = [1 1]` was a poor start point. This can be evaluated by trying different initial guesses and see if that improves `k_est` and the fits in Figure A.3.)

This illustrates the power of kinetic modelling in elucidating reaction mechanisms. It must be stressed, however, that fitting kinetic models to experimental data is no replacement for developing an understanding of the chemistry involved through applying a combination of methods, such as isotopic labelling, different spectroscopic techniques, calculating the structures and energies of potential transition states, and chemical intuition developed in lectures, problems classes and by reading chemistry textbooks.

Exercise A.2

Two plausible mechanisms were proposed for the Paal–Knorr reaction based on the study by Shuliang Ye *et al.* (*Spectrochim. Acta A*, 2022, **264**, 120280). Assuming that the last step in each mechanism is much faster than the others, derive the kinetic models for the mechanisms and evaluate them against the experimentally determined concentrations given below (A: 2,5-hexanedione; B: amine; P: product)

[A]/M: 1.0000   0.7643   0.6131   0.5802   0.4847   0.4583   0.4385   0.3663   0.3936   0.3630   0.3439
[B]/M: 0.8000   0.5811   0.4371   0.3792   0.3005   0.2667   0.2450   0.2023   0.1835   0.1386   0.1496
[P]/M: 0.0000   0.1705   0.3351   0.3972   0.5195   0.5292   0.5959   0.6339   0.6502   0.6679   0.6556

measured at times:

t/s: 0 100 200 300 400 500 600 700 800 900 1000

Mechanism I

Mechanism II

Answer:

The second intermediate in each mechanism can be neglected from the kinetic model on the basis that it is removed much faster than formed. The task can be carried out by only slightly modifying the above code.

```
% Find rate constants k1 and k2 for Paal-Knorr reaction
% kinetic model for Mechanism I:
% R1   A→IM       r1=k1*[A]
% R2   IM+B→P    r2=k2*[IM]*[B]
% d[A]/dt = -k1*[A]
% d[B]/dt = -k2*[IM]*[B]
% d[IM]/dt = k1*[A] - k2*[IM]*[B]
% d[P]/dt = k2*[IM]*[B]
%
% Experimental data: [A],[C],[P]
%
% Aim: find rate constants k1,k2 for the mechanism by fitting the
% kinetic model to the experimental data
tp=0:100:1000; % times concentrations measured at (in s)
% experimentally measured concentrations at times given in tp
```

```
A_msd=[1.0000  0.7643  0.6131  0.5802  0.4847  0.4583 0.4385  0.3663
0.3936  0.3630  0.3439];
B_msd=[0.8000  0.5811  0.4371  0.3792  0.3005  0.2667  0.2450  0.2023
0.1835  0.1386  0.1496];
P_msd=[0.0000  0.1705  0.3351  0.3972  0.5195  0.5292  0.5959  0.6339
0.6502  0.6679  0.6556];

msd_data=horzcat(A_msd',B_msd',P_msd'); % merge all experimentally
determined concentrations into a single array (to be passed to )
sp=[1 2 4]; % species monitored experimentally (measured [A],[B],
[P] will be compared against [A],[B],[P] predicted by the kinetic model)

% Initial concentrations for the model:

A0=1.0; % mol/dm3
B0=0.8; % mol/dm3
IM0=0.0; % mol/dm3
P0=0.0; % mol/dm3

c0=[A0 B0 IM0 P0]; % store initial concentrations in a vector

% initial guess for the rate constants k1 and k2
k0=[0.01  0.01];

op_ode=odeset('RelTol',1e-6,'AbsTol',1e-6,'NonNegative',1);
% set parameters for ODE solver
op_fmin=optimset('TolFun',1e-8,'TolX',1e-8,'MaxFunEvals',1000);
% set parameters for minimising residuals

% find best estimates for rate constants k1 and k2 (at which the sum of
square errors between measured and predicted concentrations is minimal)
k_est=fminsearch(@(k)goodness_of_fit(tp,msd_data,sp,c0,k,op_ode),
k0,op_fmin)
% plot results
tp_fine=linspace(tp(1),tp(end),1000); % generate more time points for
smoother predicted concentration-time curves
[t,c]=ode15s(@(t,c) net_rate_of_form(t,c,k_est),tp_fine,c0,
op_ode); % integrate model with estimated rate constants for plotting
estimated [A],[B] [P]
plot(t,c(:,1),'-r',t,c(:,2),'-.g',t,c(:,3),'-b',t,c(:,4),':k',tp,
A_msd,'rs',tp,B_msd,'g^',tp,P_msd,'ks','LineWidth',2)
xlabel('time / s'); ylabel('Concentration / mol dm^{-3}')
legend('[A]_{prd}','[B]_{prd}','[IM]_{prd}','[P]_{prd}','[A]_
{msd}','[B]_{msd}','[P]_{msd}','Location','north','NumColumns',2)

[t,c]=ode15s(@(t,c) net_rate_of_form(t,c,k_est),tp,c0,op_ode);
% integrate model with estimated rate constants for plotting estimated
[A],[B],[P]
c_sp=c(:,sp);
r=msd_data-c_sp;
figure
plot(t,r(:,1),':ro',t,r(:,2),':ko',t,r(:,3),':go','LineWidth',2)
```

```
xlabel('time / s'); ylabel('residuals / mol dm^{-3}')
legend('r_A','r_B','r_P','Location','northwest')
grid on

% calculate sum of square errors (SSE) between measured and predicted
concentrations (to measure the goodness of fit)
function SSE = goodness_of_fit(tp,msd_data,sp,c0,k,op)
[t,c] = ode15s(@(t,c) net_rate_of_form(t,c,k),tp,c0,op); % predict all
concentrations by integrating the kinetic model
c_sp=c(:,sp); % load selected concentration-time values (here:[A],
[C],[P]) into an array
SSE = sum(sum((msd_data-c_sp).^2)); % compute SSE from residuals
(differences between measured and predicted concentrations
(here: [A],[B],[P])
end

% kinetic model (net rates of formation of species) generated from
mechanism in preamble
function dcdt = net_rate_of_form(t,c,k)
dAdt = -k(1)*c(1);
dBdt = -k(2)*c(3)*c(2);
dIMdt = k(1)*c(1) - k(2)*c(3)*c(2);
dPdt = k(2)*c(3)*c(2);
dcdt = [dAdt; dBdt; dIMdt; dPdt]; % collect rates to be returned as a
vector
end
```

```
k_est =
       0.0015    0.1966
```

are the estimated rate constants k_1 and k_2 in (s^{-1}) and $(dm^3\,mol^{-1}\,s^{-1})$, respectively. The predicted concentrations are poorly following the measured concentrations, the residuals are large and show trends indicating that Mechanism I does not capture the chemistry of the reaction accurately.

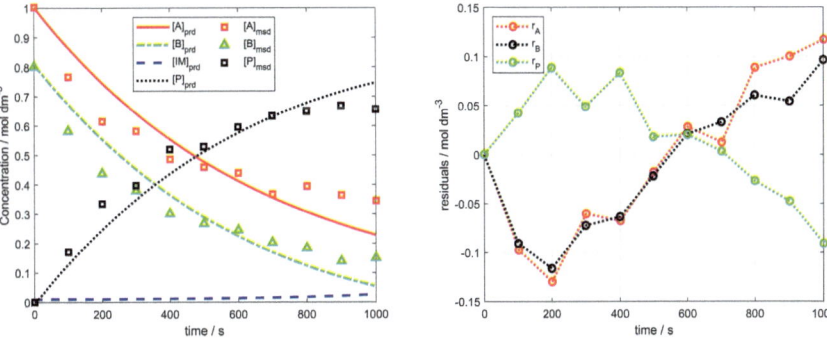

The first step of Mechanism II is 2nd order while its second step is 1st order. Accordingly, we alter the preamble as follows.

```
% Find rate constants k1 and k2 for Paal-Knorr reaction
% kinetic model for Mechanism II:
% R1   A+B→IM   r1=k1*[A]*[B]
% R2   IM→P     r2=k2*[IM]
% d[A]/dt = -k1*[A]*[B]
% d[B]/dt = -k1*[A]*[B]
% d[IM]/dt = k1*[A]*[B] - k2*[IM]
% d[P]/dt = k2*[IM]
%
% Experimental data: [A],[C],[P]
%
% Aim: find rate constants k1,k2 for the mechanism by fitting the
% kinetic model to the experimental data
```

and change the kinetic model in the `net_rate_of_form()` function at the end of the code to:

```
function dcdt = net_rate_of_form(t,c,k)
dAdt = -k(1)*c(1)*c(2);
dBdt = -k(1)*c(1)*c(2);
dIMdt = k(1)*c(1)*c(2) - k(2)*c(3);
dPdt = k(2)*c(3);
dcdt = [dAdt; dBdt; dIMdt; dPdt]; % collect rates to be returned as a vector
end
```

Running the code will result in

```
k_est =
    0.0036    0.0566
```

which are rate constants k_1 and k_2 in $(dm^3\,mol^{-1}\,s^{-1})$ and (s^{-1}), respectively. The predicted concentrations (prd) trace the measured (msd) concentrations well and the residuals (measured−predicted) fluctuate only in a narrow band without displaying any clear trends.

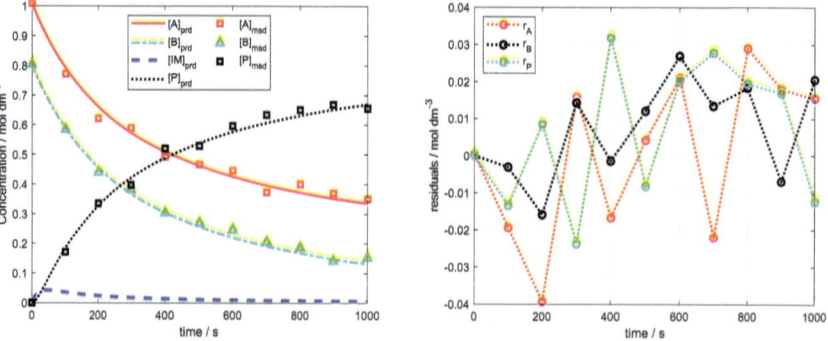

Thus, we conclude that Mechanism II is more likely to describe the chemistry of the Paal-Knorr reaction.

Reading List

The following books are great resources for further developing your coding, data analysis and maths skills.

D. Mazza and E. Canuto, *Fundamental Chemistry with Matlab*, Elsevier, 2022.

M. Maeder and Y.-M. Neuhold, *Practical Data Analysis in Chemistry*, Elsevier, 2007.

G. Beddard, *Applying Maths in the Chemical and Biomolecular Sciences: An Example-Based Approach*, Oxford University Press, 2009.

I. Hughes and T. P. A. Hase, *Measurements and their Uncertainties: A practical guide to modern error analysis*, Oxford University Press, 2010.

J. R. Taylor, *An Introduction to Error Analysis: The Study of Uncertainties in Physical Measurements*, University Science Books, 1997.

H. J. C. Berendsen, *A Student's Guide to Data and Error Analysis*, Cambridge University Press, 2011.

Subject Index

(.^) operator, 101
(') operator, 55
1D arrays, 51, 115
2D arrays, 51
2D scatter plot, 29
2D vector, 114, 117
3D diagram, 122
3σ boundaries, 90–92

abs() function, 68
absorbance plot, 94
acid–base titration problem, 25
acosd() function, 116
add comments, 17
addition, 97–98, 120–121
AFM. *See* atomic force microscopy (AFM)
aliphatic linkers, 74, 79, 80
analyte_disp_on_map() function, 27
anticommutative operation, 122
arbitrary expression, 95
arrays, 7–14
    figure formatting using, 27–28
assigning variables, 4–6
atomic force microscopy (AFM), 74, 75, 81

Beer–Lambert Law, 53, 54, 94, 102
bell-shaped distributions, 75, 77, 83, 89
bisection method, 259–262
bisection root-finding method, 259, 260
bispyrenes, 79, 80
blank script, 16
Bohr radius, 223
BondLength() function, 218

calculating volumes, 124–130
calibration curve, 94
carbon, 112
central limit theorem, 75, 82
character arrays, 11
chemical reactions, 75
chi-square, 105
colon operator (:), 7
colororder() function, 48
Command Window, 3, 5, 7, 11, 14, 16, 20, 21, 25, 59, 90, 110, 122
concentration–time profiles, 149, 171, 173, 179, 181–186, 189, 192, 210, 272
conditional statements, 197–200
confidence bounds, 56, 57
confidence intervals, 56, 57
confidence limits, 83
confint() function, 68
Coulomb's Law, 126, 130
cross product, 121–124

cubic equation, 264
curve fitting, 53
custom expressions, 64–67

data files, 12
data points, 35, 54
    with bell-shaped trend, 64
    scattered around linear trend line, 55
    spline fitted to, 63
data uncertainty, 45–47
degrees of freedom, 80, 81
disp() function, 11, 102
dissolved_oxygen_disp_on_map() function, 21, 22
division, 98
DO, 15, 17, 20, 21
dot product, 121–124
double precision, 6
DrH_st() function, 25

e-book, 4
electric fields, 130
electrons, 112
electrostatic forces, 74
element-wise division operator (./), 55
element-wise multiplication operator (.*), 24, 116
en-dash character (–), 4
env_data, 21
env_data_tiers_map() function, 209
errorbar() function, 45
Euclidean norm, 115
Euler's method, 181, 210
    one-line functions, 181–186
    self-controlling code, 210–211
exp() function, 6
expand() function, 256

feval() function, 90
fig1.XTickLabel, 40
first order polynomial, 62
fit() function, 55, 64
fitting curves
    custom expressions, 64–67
    displaying data points neglected from, 73
    lines, 54–61
    polynomials, 61–62
    quoting fitting parameters, 67–72
    splines, 62–64
flow diagram, 189
fminsearch() function, 270
force fields, 130–131
for loop command, 190–196
fourth order polynomial curve, 72
function declaration, 20, 21, 24

Subject Index

Gaussian distributions, 75
GeoCoord, 15, 17, 20, 21
glycoprotein, 85
`goodness_of_fit()` function, 270
graphical representation, 96
graphic user interface, 30, 32
Guggenheim notation, 35

"Hello, World!", 11, 12
high-level programming language, 1
`histcounts()` function, 208
histograms, 29, 49–51, 75

if-statement, 197–199
image files, 12
imaging technique, 74
`imread()` function, 13
insert dropdown menu, 30, 31
interpreted programming language, 1
ion–dipole potential energies, 126

kinetic modelling
    concentration-time curves, 171–177
    elucidating reaction mechanisms, 268–278
    equilibrium concentrations, 174
    ODE solver, 171, 172
    phenylammonium, 175
    rates of concentration changes, 170
    workspace, 172

`legend()` function, 38 linear fitting
    method of least squares, 104–111
linearised Arrhenius equation, 55
line graphs, 33
line plots, 29
lines, 54–61
line styles, 33
`linspace()` function, 9
liquid mixture, 49
`log10()` function, 6
logarithmic diagrams, 35
logarithmic plots, 35
log plots, 35–36
`logspace()` function, 9
loops
    flow diagram, 189
    for loop, 190–196
    while-loop, 196–197

magnetic fields, 123, 124, 130
magnetic forces, 74
markers, 33
mass spectrometers, 123
mathematical functions, 6–7
MathWorks website, 2

MATLAB, 1, 14, 61, 270
    addition and subtraction, 100–102, 120–121
    arrays, 7–13
    assigning variables, 4–6
    calculating volumes, 124–130
    default scatter plot generated by, 31
    desktop, 2–3
    dot product and cross product, 121–124
    drawing vectors in, 117–119
    functions, 24, 104
    mathematical functions, 6–7
    multiplication and division, 102–103
    relational operators, 199
    script, 16
`matlabFunction()` function, 216
`max()` function, 200
Maxwell–Boltzmann distribution curve, 195
Maxwell–Boltzmann distribution function, 229, 236, 247
mean deviation, 76, 78–79
`mean()` functions, 79, 81
`meshgrid()` function, 130
methane, 121
Michaelis–Menten equation, 238
molar absorption coefficient, 58
molar absorptivity, 54
molar extinction coefficient, 54, 94, 102
molecular dipoles, 225–226
multiplication, 98

nano batch, 76
`net_rate_of_form()` function, 270, 273, 278
NMR. *See* nuclear magnetic resonance (NMR)
noncommutative operation, 121
non-polar solvent, 112
normal distribution, 75–78
nuclear magnetic resonance (NMR), 64, 65
`num2str()` function, 224

ODE. *See* Ordinary Differential Equations (ODE)
one-line functions
    Euler's method, 181–186
    kinetic modelling, 169–177
    lab notebook, 167
    MATLAB, 176
    multiple expressions, 169
    nitration of aniline, 177–178
    percent yield, 166
    self-replicating peptides, 178–181
    user-defined, 165, 166
Ordinary Differential Equations (ODE), 171

Paal–Knorr reaction, 278
peak position, 68
peculiar behaviour, 4
peptide self-replication process, 178–181
PinColor, 27
PinSize, 27
`plot()` function, 32, 45, 56

plots
    combining graphs and, 36–45
    data uncertainty in, 45–47
    histograms, 49–51
    log plots, 35–36
    scatter and line, 30–35
    stacking, 51–52
    with two y-axes, 47–49
polar solvent, 112
polonium (α-Po), 8
polymerisation process, 200–209
polynomials, 61–62, 104
population mean deviation, 78–79
population standard deviation, 78–79
predicting titration curves, 263–268
`pretty()` function, 216
property name–value pairs, 32
Pythagoras' theorem, 115

quantum mechanics, 131
`quiver()` function, 117, 118
`quiver3()` function, 117, 119
quoting fitting parameters, 67–72

rate constants, 55
Rault's Law, 48
reference frame, 114
regression residual, 90
relational operators, 199
`rewrite()` function, 221
right-hand rule, 123
root-mean-square speed, 227, 228
`round()` function, 68

sample mean deviation, 79–81
`scatter()` command, 32
scatter plot in MATLAB interactively by two clicks, 30
scripts
    advantages of using, 17–19
    commands *versus*, 14–16
    as user-defined MATLAB functions, 20–27
SDOM. *See* standard deviation of the mean (SDOM)
second nano batch reaction, 76
second order polynomial, 62
self-controlling code
    conditional statements, 197–200
    Euler's method, 210–211
    loops
        flow diagram, 189
        for loop, 190–196
        while-loop, 196–197
    overview, 187–189
    polymerisation, 200–209
self-replicating peptides, 178–181
semicolons, 7, 169, 223
`separateUnits()` function, 223
simple decomposition reaction, 42
`simplify()` function, 216

single quotation marks ('), 4
sought activation energy, 56
spectrophotometer, 42
speed distribution, 227, 228
splines, 62–64
spreadsheet, 5
square brackets, 7
square root, 97
standard deviation, 76, 79–81
standard deviation of the mean (SDOM), 82
standard error, 81–83
standard normal distribution, 83
`std()` functions, 79, 81
stoichiometric coefficients, 24, 26
stoichiometric point, 59
Student's t-distribution, 83–84
`subplot()` function, 38
subplots, 39
    XTick properties of, 46
subtraction, 97–98, 120–121
`sum()` function, 101
symbolic maths
    approximation, 252
    bisection method, 259–262
    Command Window, 213, 214
    definition of, 212
    differentiation, 229–232
    exact solutions, 249–252
    integration, 233–237
    lab notebook, 213
    pre-defined units, 224
    propagation of uncertainties, 242–246
    simplifications, 252–258
    solving algebraic equations, 237–242
    unit algebra, 218–225
    Workspace, 215
Symbolic Math Toolbox, 212, 221, 249, 250
`symunit2str()` function, 224
syntax error, 4

tertiary haloalkane (R_3CX), 7
third order polynomial, 62
titration curves, 263–268
Toolstrip, 2, 30
top pan balance, 103
trigonometric calculations, 116
trigonometric functions, 6, 126

uncertainty, 68, 94
    addition and subtraction, 97–98
    comparing values, 86–89
    constants with negligible uncertainties, 98–100
    identifying outliers, 89–93
    in MATLAB
        addition and subtraction, 100–102
        functions, 104
        multiplication and division, 102–103

multiplication and division, 98
normal distribution, 75–78
population mean and standard
   deviation, 78–79
sample mean and standard deviation,
   79–81
size of error bars, 85–86
small sample sizes and Student's
   t-distribution, 83–84
standard error, 81–83
uniform arrays, 10
unitInfo() function, 222
universal gas constant, 88
user-defined functions, 21, 23, 25, 100

variables, 14
vector algebra, 113–117
vectors, 112
   addition of, 116
   algebra, 113–117
   fields, 130–131
   in MATLAB
      addition and subtraction, 120–121
      calculating volumes, 124–130
      dot product and cross product,
         121–124
      drawing vectors, 117–119
vertcat() function, 208
vpasolve() function, 247, 252, 262

while-loop command, 196–197, 260
wmmarker() function, 15
World Imagery, 22
wrapped sections, 17
wrapping scripts, 18

X-ray crystallography, 114

z-distribution, 83, 86
zeros() function, 10